Introduction to Marine Geology and Geomorphology

Introduction to Marine Geology and Geomorphology

Cuchlaine A. M. King

Professor of Physical Geography, University of Nottingham

Edward Arnold

First published 1975 by
Edward Arnold (Publishers) Ltd
25 Hill Street, London W1X 8LL

This book is a fully revised edition of the first part of
Oceanography for Geographers (*Introduction to Oceanography*),
first published 1962 by Edward Arnold (Publishers) Ltd

Reprinted 1964, 1965, 1968, 1969 (twice)

Cased Edition ISBN 0 7131 5731 3
Paper Edition ISBN 0 7131 5732 1

Printed in Great Britain by
Butler & Tanner Ltd, Frome and London

Contents

Preface

During the ten years since the first edition of *Oceanography for Geographers* was published there has been an explosive development in nearly all aspects of the subject. The revolution in our ideas of global tectonics, which was just starting a decade ago, has now become firmly established and is the basis on which submarine geology and geomorphology now rests. The work of geophysical oceanographers has played a vital role in the development of the new ideas concerning the mobility of continents and ocean basins and in the development of theories of plate tectonics. So much has been written concerning this aspect of oceanography in the last few years that it has been necessary in preparing a new edition of *Oceanography for Geographers* to divide the work into two books. The first of these concentrates on the geomorphological aspect of oceanography. The morphological features of the ocean basins are considered in the light of the new ideas of global tectonics, which form the basis of the second chapter. The first one introduces the subject by means of a historical survey of marine geomorphology and a mention of some of the more recent methods of investigation. The second and third chapters explore in a little more detail the nature and origin of submarine geomorphological features, dealing respectively with the marginal zones and the open ocean features. The fifth chapter deals with the sediments of the oceans, which provide valuable information concerning many problems, relevant both to the development of the oceans and their basins and to other geological and geographical matters, such as climatic change and glacial chronology. The final chapter is concerned with the water of the ocean, its source is considered briefly, as well as more recent changes in sea level, which are also of interest in other aspects of geomorphology.

The final chapter of the first book forms a link with the second one, which covers, at a rather more general level, aspects of physical and biological oceanography. Some problems of current importance, such as pollution and conservation of the oceans, are considered. These problems cannot be solved successfully until more is understood about the circulation and movement of ocean waters, and the intricacies of the ecological systems that maintain the balance of life in the oceans.

Acknowledgements

Citations of the copyright holder of the plates are given in the plate captions. The publishers and author owe a particular debt of gratitude to David Stoddart, A. S. Laughton, The Institute of Oceanographic Science (Wormley), the Scripps Institution of Oceanography and to James T. Morgan for their help in providing illustrations.

The publishers and author also gratefully acknowledge permission received to reprint or to modify copyright material as follows:

The American Association for the Advancement of Sciences and the authors for figures 2.3,[1] 3.15, 5.8, 5.9, 5.10 and 6.5; the American Geophysical Union for figures 2.10, 4.8, 4.9 and 4.10; the American Association of Petroleum Geologists and the author for figure 3.2; the editor for figure 6.6 taken from *Arctic and Alpine Research*; Cambridge University Press and the author for figure 4.15; the Colston Society, University of Bristol, for figure 4.3; the editor for figure 4.4 taken from *Deep Sea Research*; the editor for figure 3.5 taken from *Die Erde*; the Elsevier Scientific Publishing Company for figures 3.13, 3.17, 4.7 and 6.2; the editor for figure 4.11 taken from *Experientia*; the Geological Society of America for figures 2.6, 4.5, 5.6, 5.7 and 5.11; the Geological Society of London for figure 3.10; the Institute of Geological Sciences for figure 3.1; the editor for figure 3.16 taken from *Geological Magazine*; Pergamon Press Ltd and Microforms International Marketing Corporation figures 2.4, 6.3 and 6.4; the editor for figures 3.3 and 3.9 taken from *Petermanns Geographische Mitteilungen*; Prentice-Hall Inc. for figures 5.3 and 5.4; the Royal Geographical Society for figure 3.11; the Royal Society for figures 2.2, 3.14, 4.12 and 4.14; Scientific American and W. H. Freeman and Co. for figures 2.8 and 2.11 (copyright © 1969 *Scientific American*, all rights reserved); John Wiley Inc. for figures 4.1, 5.1, 5.2, 6.1; and the editor for figure 3.4 taken from *Zeitschrift für Geomorphologie*.

[1]The figure numbers refer to those in this book. Full references to author and source of publication may be found by consulting the caption acknowledgement and the references at the end of the book.

1 Introduction to marine geomorphology

Only 200 years ago, in 1773, the first sounding beyond the edge of the continental shelf was made successfully. In the succeeding 100 years there was relatively little advance in the knowledge of the geomorphology of the ocean basins, while in the last 100 years the pace of growth of knowledge of marine geomorphology has been accelerating at an exponential rate. Since the first edition of this book was published in 1962, the growth of knowledge has been so rapid that ideas that were first suggested in the early 1960s have now completely revolutionized our knowledge of the character of the ocean floor and the part it plays in determining the character of the whole earth's surface. In this decade marine geology and geomorphology have come to occupy a central place in the developing knowledge of global tectonic processes. A new unity has been established between the processes operating in the ocean and on land. This integration of the oceanic and land sectors of the earth's surface has been made possible by the development of new techniques for exploring these processes and the forms they produce under the sea surface. Studies of marine geomorphology have been shown to have relevance in many fields of enquiry. These interactions apply to all aspects of oceanography, so that it is worth looking at some of these links, both between the aspects of oceanography, and between oceanography and geography. Current developments have grown out of earlier work and it is useful to consider the development of techniques of study of marine geomorphology leading to the sophisticated modern devices that are revealing more and more about the land beneath the sea. This immense area of hidden ground can only effectively be explored by the combined efforts of many scientists, backed by governmental support. Thus the growth of international cooperation has been an important development in the successful expansion of submarine research and knowledge. The discovery of more and more exploitable riches beneath the sea has added practical value to this work, but at the same time has raised problems of a legal nature that still have to be solved to the optimum benefit of all people.

1 The scope of oceanography

The scope of oceanography has been expanding with the growth of detailed knowledge of the oceans from all points of view. The subject may be broadly divided into marine geology and geomorphology, physical oceanography, and biological oceanography. The first aspect will be considered in this volume, while the second and third aspects will be dealt with in the second volume. There are, however, several important ways in which these different aspects interact with each other, and all of them are relevant to the broader field of geographical enquiry at many points, some of which will be mentioned briefly in this section and developed further at appropriate points.

Marine geomorphology attempts to understand the nature of submarine relief in terms of its character and the processes that have given rise to it. The basic points concerning the relief of the ocean basins are given at the beginning of chapter 3. It is true that the oceans differ fundamentally in structure from the continents, but nevertheless there is a unity of process that makes the whole earth one unit. This unity is explored in the new global tectonics, which give an exciting, fresh insight into the earth's structure as a whole, and have linked the land and sea in a complex of mosaic blocks that have been shown to be much more mobile than was considered possible at one time.

The new global tectonic theory grew out of a study of the oceanic ridge system and owes a great deal of its development to the work of oceanographers, as much of the evidence for the theory lies beneath the sea. This development and the results of recent studies are examined in the next chapter. The active processes that are operating within the earth's mantle have been explored mainly by geophysicists, working in the fields of magnetism, seismology, and vulcanology. These processes exert a strong effect on the morphology of the ocean floor, and it is in this field that the greatest geomorphological interest lies. The form of the ocean floor cannot be understood without some reference to these major crustal processes that give rise to it; thus it is essential to examine briefly these processes and the evidence by which they have been deciphered.

The global process of sea floor spreading has given a completely new picture of the age of the ocean basins, and this fundamental reappraisal of the part the ocean basins have played in the development of the earth's relief has repercussions on many branches of geomorphology concerned both with the submarine and the subaerial surface. The mobility of the crustal plates, with their continental blocks riding upon them, has meant that land surfaces have changed latitude over geological time by considerable amounts. The changing form of the ocean basins, and the development of rifts within the continental blocks have also imposed major changes upon all aspects of the geography of the continents as well as the oceans. The continental changes in latitude and position, for example, have meant that climatic changes have been fundamental in most areas over geological time. These changes in climate have played an important part in the development of landforms on the earth's surface and have been of significant proportions, particularly where they have led to the development of major ice-sheets and continental glaciation.

One of the currently favoured theories of the origin of the glacial period of the Tertiary and Pleistocene period is that based on the changing positions of land and sea with respect to the positions of the poles. The fact that there is now a high continent over the south pole and an almost land-locked ocean over the north pole is of fundamental importance in explaining not only the present climate of the earth, but also many aspects of its oceanography, such as the pattern of oceanic circulation, both on the surface and in the deeper water. It is also this pattern of land and sea distribution that has been held responsible for the development at intervals of major ice-sheets in the northern hemisphere. This particular pattern of land and sea has not been present during much of geological time.

A fundamentally important conclusion reached as a result of the development of the new global tectonics is that the ocean basins are comparatively recent in date. Nowhere have ocean floors older than the beginning of the Mesozoic been discovered. Thus practically all of the ocean basins as they exist at present are younger than 250 million years old, which is a small proportion of all geological time (a period of about 5000 million years). Thus the present oldest ocean floor is only about one-twentieth of the age of the earth. The oldest ocean floor is found in the northwest Pacific and parts of the Arctic ocean basins, while all the Atlantic and Indian oceans have formed since the Permian or Triassic by rifting.

The rifting process can be related to happenings on the land. Thus it has been suggested that the reversal of drainage that geomorphologists have inferred from the landforms of the Appalachian Mountains was related to the development of the rift between America and Africa. The pattern of world-wide erosion surfaces have also been interpreted in terms of the splitting and drifting apart of the main continental masses. Studies currently being undertaken in the Afar triangle on the north coast of Ethiopia reveal an early stage in the development of a rift ocean. The newly developing rift in this area is at present on land so that it provides an ideal laboratory to examine the details of the process of ocean spreading, a process that has given rise to the Red Sea and the Gulf of Aden. It is interesting to note that experts have difficulty in distinguishing between cores obtained from the northern Red Sea, which are modern and derived from an actively developing ocean, and those of lower Cretaceous age from the western Atlantic Ocean. Both show the same type of sediment, characteristic of development in a closed but expanding ocean basin. Further details of this evidence are considered in chapter 5, which deals with ocean sediment.

The study of ocean sediment provides many contacts between the different branches of oceanography and between oceanography and geomorphology. One connection has already been mentioned—where the type of sediment currently forming can be used to reconstruct former conditions in sediments of much greater age. The character of ocean sediment also reflects the general ocean morphology. The two major types of deep water sediment, the pelagic particle by particle sedimentation and turbidity current sedimentation, are characteristic of different oceanic environments. The former is more common in the Pacific, with its fringing pattern of deep sea trenches which intercept land-derived sediment; while the latter is common in the Atlantic with its trailing edge margins and well-built continental shelves, which give land sediment access to the deepest basins.

There is also an interaction between the form of an ocean basin, the character of the water it contains, and the sediment that collects in it, through the fauna and flora that can inhabit it under the conditions prevailing. Thus organic sediment in particular can tell us a lot about various aspects of oceanography and their mutual interaction. It links the fields of marine geology and geomorphology with those of physical and biological oceanography. Physical oceanography is concerned with the water character and its movement. These in turn affect the surface life, reflected in the organic sediment that collects on the floor of the ocean. This relationship is illustrated in the observation that equatorial cores from the Pacific Ocean show higher values of calcium carbonate (derived from living organisms) during periods of colder climate, while those from the equatorial Atlantic have higher values of calcium carbonate during the periods of warmer climate. This variation can be explained in terms of the circulation of the oceans. That in the Pacific tends to be more vigorous during colder periods when deep water formation is more active. In the Atlantic the organisms respond to variations in temperature by greater growth rates during warmer periods, because circulation is more rapid in the Atlantic than the Pacific at all times and this factor does not impose a limit on productivity. Thus sediments reflect the conditions existing at the surface to a certain extent, as well as the morphology of the bottom.

Chemical oceanography is also connected with sediment character. This branch of the subject is concerned with the salts and other elements that make up the total salinity of sea water. A study of the salt and other elements in the sea can provide evidence both for the formation of the ocean water (a matter considered further in chapter 6) and for the character of the bottom sediment. Oceanic life can only grow where certain essential nutrient materials are present in the water, and their presence depends upon the circulation of the ocean water, and is thus a matter of concern to physical oceanographers, as well as biological oceanographers. The ultimate deposition and preservation of the organic matter in deep sea sediment depend also upon the chemical characteristics of the water at depth (other variables such as temperature and pressure are themselves a function of depth). Thus the character of the sediment discloses both surface and subsurface chemical conditions in the water, and both of these are related to the nature of the circulation.

The circulation of the oceans exerts a wide influence and is affected by many outside influences as well. It depends mainly on the wind system on the surface, so that it is related to atmospheric circulation, but the pattern of land and sea is also very important in the details of oceanic current patterns. At depth the circulation depends also upon the distribution of density in the water, which in turn reflects surface conditions of temperature and precipitation affecting the surface salinity. Deep water circulation is also influenced by the nature of the oceanic basins in which it takes place. The circulation in turn affects the climate, so there is a feedback process in this interrelationship. It is well known that the oceanic climate differs fundamentally from the continental one, and that eastern and western coasts of the continents have very different climates because of the oceanic influence.

Oceanic circulation also plays a very important part in determining the distribution of fertility in the ocean, and this aspect has very far-reaching consequences in a great number of fields. Perhaps its greatest impact is in the field of biological oceanography, but through this impact it also influences the nature of the sediment, and therefore the geomorphological aspect. The distribution of organic sediment in the deep sea can be explained to a large extent in terms of oceanic circulation and its variation, as the example discussed above indicates. This information was derived from a study of deep sea cores, the analysis of which provides evidence of wide interest and application.

The study of deep sea cores furnishes some of the most valuable information that is available concerning the changes of climate. The evidence provided by deep sea cores may be compared to a wall that has been repeatedly painted, while the evidence provided by land deposits is more like a blackboard that has been repeatedly wiped clean. The completeness of the record in suitable deep sea cores is a very valuable asset. Analysis of these cores has given rise to a fairly well-established sequence of climatic changes that have been accurately dated for the last 70,000 years at least, while magnetic reversals allow the time scale to be extended considerably further back in time in conjunction with other dating techniques. The results of some of these studies (described in chapter 5) provide a valuable link between marine and land chronological studies of the Pleistocene climate and its changes. The evidence is often provided by a study of the micropalaeontology of the sediments and so it is related to biological oceanography. The organisms that form the bulk of deep water sediments are the small members of the plankton, and these depend on the external conditions of temperature as well as on the character of the water in which they live.

Biological oceanography is a very large field of growing importance. The oceans are being looked to increasingly to provide the protein essential to the adequate nourishment of the growing world population. A study of biological oceanography is thus relevant to human geography because it links directly with the fishing industry and other forms of biological exploitation of the resources of the oceans. The organic life of the sea is extremely diverse and complex; each ecological niche has developed some suitable adaptation in the creatures that inhabit it, so that they can live together in a stable ecological system. The ecosystem approach is being extensively developed in biogeography; this approach is particularly relevant to marine biology. It is also one which geographers are able to appreciate from their experience of complex interrelationships between many different variables that make up the spatial patterns on the earth, which are the concern of geographers.

The oceanographical ecosystems depend fundamentally on the plants of the sea, as only these are able to synthesize organic matter from the nutrients in the water by the aid of sunlight in the process of photosynthesis. Marine productivity thus depends on the nature of the water in the upper layers where alone light can penetrate. The circulation of the water, bringing the nutrients to the surface, is thus of primary importance in considering the geographical distribution of fertility in the oceans. Many of the food chains in the ocean are long and complex, and at each stage in the chain there is a great loss of original input energy and substance, so that efforts to obtain nourishment from lower in the chain are worthy of consideration. The oceans are, however, large and mainly uncontrollable and efforts to regulate biological exploitation from both the point of view of controlling the fauna and flora themselves and of controlling man's use of them have so far not met with much success. Developments in this field will be of increasing importance in the future.

Other matters of great current concern in the field of biological oceanography relate to the danger of pollution, which damages the marine environment. One of the major sources of pollution is connected with the use of the oceans as a means of transport and the problem of potential oil pollution from tankers, such as the *Torrey Canyon* disaster. Oil pollution also has another cause that is directly related to the exploitation of the geological resources of the ocean. This type of pollution is exemplified by an accident that took place off Santa Barbara, California, when an offshore oil well discharged oil prematurely, with serious consequences in the vicinity.

The exploitation of the mineral resources of the sea, however, is and has been instrumental in the development of many useful techniques and a great deal of oceanographic research in marine geology. The exploitation of the offshore oil reserves is a good example of commercial exploration and scientific research helping each other towards a much better knowledge of the geomorphology and geology of the continental margins. There are two major types of mineral resources that have initiated much speculation and survey work of a geophysical and geomorphological nature. One of these is offshore oil, located around many of the continental margins. This resource is intimately linked with processes of sedimentation, mainly in connection with land-derived sediments and associated organic remains. Oil is thus likely to be found in those environments to which turbidity currents have access, and where large volumes of sediment have accumulated during the fairly recent geological past (since the Palaeozoic period, and mainly during the Tertiary period).

By contrast, the other type of mineral resource that has given rise to much speculation is the large quantity of manganese nodules that are so widely distributed in parts of the ocean. These deposits occur mainly in areas of slow pelagic sedimentation, where plenty of time is available for the necessary chemical reactions to take place before the nodules are buried beneath a mantle of sediment. They are, therefore, found mainly in the Pacific and in those areas of the Atlantic which form rises and plateaux, to which turbidity currents do not have access. A knowledge of the processes operating in the ocean helps to locate the most favourable areas for exploration with a view to exploitation, while the methods of exploration in turn provide valuable information concerning the processes. There is thus a feedback relationship in these aspects of ocean use and study.

2 History of marine geomorphology

The first stage in the development of the knowledge of marine geomorphology was to ascertain the shape of the ocean basins. Their areal extent had first to be established and then their three-dimensional form needed to be observed. When their form had been established, it was then necessary to discover the nature of the material on their beds and to delineate and classify their morphological forms. Such a classification can usefully be made first in terms of descriptive morphology, without implying any specific origin. A more valuable classification can, however, be made in genetic terms. Such a genetic classification can only be provided when the processes that give rise to the observed forms have been established. Once the processes that are operating have been examined and explained, then it is possible to put forward a unifying theory to account for the forms in terms of basic, world-wide controls. These in turn provide a means of predicting future changes in marine morphology.

The achievement of the first step in this process of understanding marine geomorphology could only be fulfilled after accurate means of fixing positions at sea had been made possible by the development of accurate chronometers. For a very long time sailors have been able to measure their latitude fairly exactly by means of observations of the sun's altitude at noon. Even the early Norse voyagers, who reached the shores of North America, could find their latitude reasonably accurately. For many years the delimitation of the extent of the oceans was a secondary concern of those who sailed across the seas. They mostly set out for trade or conquest, rather than for oceanographical exploration for its own sake. Gradually, however, the map of the world took on a more realistic shape.

One of the first voyages of scientific oceanography was that of James Cook (1728–1779). During his voyages accurate longitude measurements could be made for the first time, because he was able to take a chronometer with him to observe the difference of time between the transit of the sun at Greenwich and at the point of observation. His first expedition took place between 1768 and 1771 in the *Endeavour*, its main purpose being to observe the transit of Venus. His second voyage took place between 1772 and 1775 when he set out to explore the possibility of a southern continent. The southern ocean was the only large blank still remaining on the world map by this time and Terra Incognita Australis still loomed large on the contemporary charts. One of the first to make direct contact with this land-mass was the Dutchman, Abel Tasman. During 1642, over a century before Cook sailed, Tasman had circumnavigated Australia and Tasmania, now named after him. He was mostly out of sight of land so that the true nature of Australia was still doubtful when Cook sailed. Cook penetrated much further south; at his first attempt he reached 67°s where he was among ice floes, and his maximum south latitude was 71° 10′s, where he again encountered ice. He charted the coast of New Zealand and discovered and mapped many Pacific islands, as well as the east coast of Australia. He established that the southern continent was quite small if it existed at all, as he did not sight the mainland of Antarctica. He also carried out oceanographical observations, and established that the southern ocean was dominated by the great west wind drift in high latitudes.

After the time of Cook the outline of the ocean was adequately known and the next stage was to explore the ocean in depth. Cook had carried out some work at depth, such as temperature measurements, but he had not made accurate soundings in the deep ocean over most of his route. Following the voyages of Cook in the southern hemisphere, the main activity of oceanographic mapping and surface exploration was concentrated in the polar regions, which by this time were the only areas still not fully known. The northern polar sea routes had early been explored in the search for the northwest and northeast passages, which it was hoped would give a quick and easy route to the wealth of the East Indies. Hudson, between 1607 and 1611, extended the earlier voyages of Davis and Frobisher in the northeast Canadian archipelago. The northwest passage was not penetrated for very many years, but during the search for the fate of the unfortunate expedition of Sir John Franklin in the middle of the nineteenth century, William E. Parry finally succeeded in doing so. The difficulty of this northern sea route has more modern connotations, witnessed in the difficulties experienced by the *Manhattan* tanker, even though she was accompanied by an ice breaker, the *John E. MacDonald*. Her passage was a feasibility study on the possibility of taking oil from the rich north Alaskan oil field by sea. The risks of pollution in the fragile Arctic environment (which is very susceptible to damage of this type since rehabilitation would be very slow and difficult in such a climate), do not encourage the use of this method of transporting oil. This voyage is an interesting example of the changes that take place through time in the value of different parts of the world ocean from the point of view of transport routes and exploitable assets.

The second stage of exploration of the oceans required the development of methods of sounding so that the submarine relief could be established. The first sounding beyond the edge of the continental shelf was made by Captain Phipps, later Lord Mulgrave, when he was in command of HMS *Racehorse* on a voyage towards the north pole in 1773, contemporaneous with the voyages of James Cook. The sounding recorded a depth of 683 fathoms (1250 m). It was made with a rope to which a 150 lb (68 kg) lead weight was attached, and Phipps noted that the lead sank nearly 10 feet (3 m) into very fine, blue, soft clay. Little further sounding was carried out for the next 50 years partly because of difficulties with the sounding lines. The currents and winds carried ships sideways so that recorded depths were often much exaggerated. During some sounding attempts, up to 6000 fathoms (11,000 m) of line were let out before the bottom was reached. One of the first deep soundings was made by Sir James Clark Ross in 1840 during his voyage to the Antarctic. He sounded a depth of 2425 fathoms (4440 m) at a position 27°26′s and 17°29′w. He carefully noted the rate at which the line was paid out over the drum and noted the change of speed when he presumed the weight had reached the bottom. He had 4 miles of rope on a large drum and the whole operation took 4 hours to accomplish. It was at about this time that wire line started to replace rope for sounding. The process was still, however, extremely slow and laborious.

Even the earlier oceanographers had appreciated the problems of sounding with a rope. They were concerned to experiment with a lineless technique and several ingenious devices were suggested, one dating from 1650. The essence of the method was to attach a buoyant object to the sounding lead in such a way that it was released when the weight hit the bottom. The time it took to return to the surface was recorded, which would give the interval between the dropping of the weight and the return of the light object to the surface. The obvious problem, however, was to sight the object the moment it reached the surface again. Much of the early sounding was related to observations of temperature and obtaining samples of water from depth, rather than being a direct attempt to record the morphology of the ocean floor. None of the early sounding devices were very successful, compared with the lead and line method. Indeed Ross's sounding of 2425 fathoms (4430 m) was re-sounded by the USSS *Discoverer*, and a depth of 2100 fathoms (3840 m) was recorded. The main difficulties in the early lead line sounding was the recognition of when the lead touched the bottom. One sounding Ross recorded was a depth exceeding 4000 fathoms (7330 m), the full length of his sounding line, because he failed to notice the first slight decrease in the rate of descent of the weight, the true depth at the site being about 2200 fathoms (4030 m). During the period 1840 to 1860 a number of oceanographers made a series of random soundings, but frequently the same problem led to erroneous results or else to breakage of the rope or wire. The next step was to use a much smaller weight of only a few pounds on a much finer silk or twine line, but exaggerated results were obtained (reaching 15,000 fathoms (27,400 m) in one instance), and many depths of 7000 (12,800 m) and 8000 fathoms (14,600 m) were recorded.

These problems led to the search for new methods, often based on the increase of pressure with depth. The change of temperature with pressure was also used in attempts to assess depth more reliably. The first hint of echo-sounding came from experiments made in Lake Geneva in 1826, when sound waves were propagated under water by ringing a bell and listening for the sound 9 miles away, the waves travelling horizontally to the recording device. In 1833 H. F. Talbot suggested that sound from an exploding shell should be propagated to the sea floor and back to the surface. The method, however, could not become feasible until hydrophones had been invented and electronic amplifying equipment was available. In fact, it was not until the years after the First World War (1920s and 1930s) that echo-sounding became a practical proposition. Before the days of echo-sounding the knowledge of ocean floor relief was limited to relatively few isolated soundings, often of doubtful accuracy.

One technological advance that stimulated the exploration of underwater relief was the invention of the telephone, which led to the laying of cables across the ocean floor. An accurate knowledge of the sea floor relief was needed in this operation. The first trans-Atlantic cable was laid in 1858 after a systematic survey of the route by the ship HMS *Cyclops*. The first attempt at a bathymetric map of the Atlantic Ocean was that published by M. F. Maury (1806–1873) in his book *The physical geography of the sea* (1855), which can lay claim to be the first textbook on oceanography. This bathymetric map gives the first hint of the presence of the mid-Atlantic Ridge, at least in the northern half of the ocean, which was better known than the southern part. By 1895 the total number of soundings over 2000 m in length was only 7000, and only 550 of these were in water deeper than 5500 m. Maury called the first indication of the mid-Atlantic Ridge the Telegraphic Plateau, in connection with schemes for laying the trans-Atlantic cables.

One of Maury's assistants, Midshipman Brooke, first overcame the problem of relieving the strain of the sounding line and of bringing a sample up from the bottom. He developed an apparatus that consisted of a cannon-ball, with a hole bored in it, into which a tube was inserted. Hooks attached to the apparatus allowed the cannon-ball to be separated from the tube so that the tube was drawn back to the surface filled with sediment from the bottom, where it and the cannon-ball sank in. This apparatus collected the first specimens from the floor of the Atlantic and Pacific Oceans in 1855.

In the 1850s and 1860s soundings and samples became much more numerous, while the introduction of steam winches facilitated the use of sounding lines. Much of the oceanic research in the nineteenth century up to about the 1860s was spasmodic, and not clearly planned or organized towards recognizable goals. During the later years of the nineteenth century much of the emphasis in oceanography was on the biological side of the subject, partly arising from the stimulating ideas of the origin of species by natural selection proposed by Charles Darwin, after his visit to various Pacific islands during the voyage of the *Beagle*. Darwin published the *Origin of species* in 1859.

In the nineteenth century it was a common assumption that life could not exist in the deep ocean, but this view became increasingly questionable as samples in which foraminiferal remains were common were obtained from deeper and deeper water. Interest in the origin of these fossils led to attempts to solve this major query. It led to the expedition by Charles Wyville Thomson (1830–1882) in the *Lightning* under the auspices of the Royal Society in 1868 to dredge for samples of life in the depths. The work was carried out north west of Britain and interesting observations of variation in bottom temperatures were obtained, as well as dredged samples from depths greater than 500 fathoms (915 m) in which there was considerable evidence of life. This direct evidence settled the problem of the possibility of life at depths of great magnitude, but it raised many other matters that led directly to one of the most important oceanographical expeditions so far undertaken. Many new types of creatures had been found for the first time in the dredges, as well as the foraminifera, such as *Globigerina*, which Wyville Thomson thought lived on the bottom. He appreciated the similarity between these microscopic creatures and the fauna that make up much of the Chalk of the Cretaceous outcrop in eastern England. The discovery of different temperatures at depth in close proximity led to the development of ideas of deep water circulation. Cold water at depth in low latitude must mean movement of water from high to low latitude below the surface.

Wyville Thomson, however, was more concerned to dredge to great depths to extend the growing evidence of bottom deposits and benthic life. These problems led to the setting up of the Committee on Marine Researches by the Royal Society in 1869. The committee recommended further marine exploration and the Navy was approached to provide a ship; committees were set up to examine the best possible types of thermometers for deep water observations. The *Porcupine* was used for the observations carried out in 1869, and samples were obtained from a depth of 2435 fathoms (4450 m) in latitude 47°38′N, 12°8′W. These samples contained mud and marine animals, so that the azoic theory was effectively killed. The expedition was successful in showing that different water masses existed at depth, each with a distinct fauna. As a result W. B. Carpenter put forward a theory of ocean circulation based on temperature variations. In the following year, the *Porcupine* worked in the Mediterranean, studying the water movements through the Straits of Gibraltar. Wyville Thomson published the results of these series of expeditions in his book *Depths of the sea* (1873). The remark that 'we are still living in the Cretaceous epoch' led to some argument, since some people considered this an attack on the views of the stratigraphical succession and the theory of evolution. Wyville Thomson, however, assured people that all he intended to say was that the sediment accumulating at the bottom of the Atlantic at that time resembled the Cretaceous rocks on land, so that he suggested that the deposition had gone on uninterrupted in the interval in the ocean. Differences between the chalk fossils and the animals dredged from the sea floor were recognized, and modifications were established.

In 1871 W. B. Carpenter, who had carried out much of the work on the earlier expeditions, had ideas for a more ambitious venture, no less than a circumnavigation of the world to undertake oceanographical observations in all the major oceans. The Royal Society and the Government gave the project a favourable reception, and a committee was set up to prepare for the expedition to sail in 1872. The ship chosen for the project was HMS *Challenger*, under the command of Captain Nares. Guns were removed to make room for laboratories and scientists. The scientists were led by Wyville Thomson, and included John Murray (1841–1914). Hemp lines were used for nets, trawls, and dredges, but steel piano wire was taken for soundings. However, the drum collapsed and this new method was not workable.

The *Challenger* sailed on 21 December 1872, on a voyage that lasted for nearly 3½ years (figure 1.1). This long expedition laid the foundations on which modern oceanography is essentially based. The routine work of the ship included measurement of temperature at different depths, the collection of water samples from various depths, the sounding of depth, collection of living fauna from the water and bottom, as well as the collection of bottom samples. As the expedition moved south in the Atlantic it gradually found that the familiar *Globigerina* ooze gave way to a darker deposit in which the foraminiferal remains became fewer and fewer. On 26 February 1873, the depth was recorded at 3150 fathoms (5760 m) and Wyville Thomson reported a deposit of pure Red Clay with no organic matter. He interpreted this as a deposit derived from the rivers of South America, rather than due to a depth greater than that at which the *Globigerina* could live. The *Globigerina* ooze was again encountered on the Dolphin Rise in shallower water. They dredged up a new substance, a nodule consisting of almost pure manganese peroxide. This is the first record of nodules that were later to be found to occur very extensively, particularly in the Pacific. They obtained live worms from depths in excess of 3000 fathoms (5500 m), and made a sounding of 3875 fathoms (7080 m) in the Puerto Rico Trench, their deepest so far. They also found that the deposits of Red Clay were much more extensive than previously thought and they could not be explained as a purely local phenomenon, although much of the clay was dark brown or grey rather than red. After leaving Bermuda they sailed north to explore the Gulf Stream and recorded currents at various depths in order to find out if lower currents flowed in the reverse direction to the surface ones.

During the voyage Murray became convinced that the foraminifera that made up the

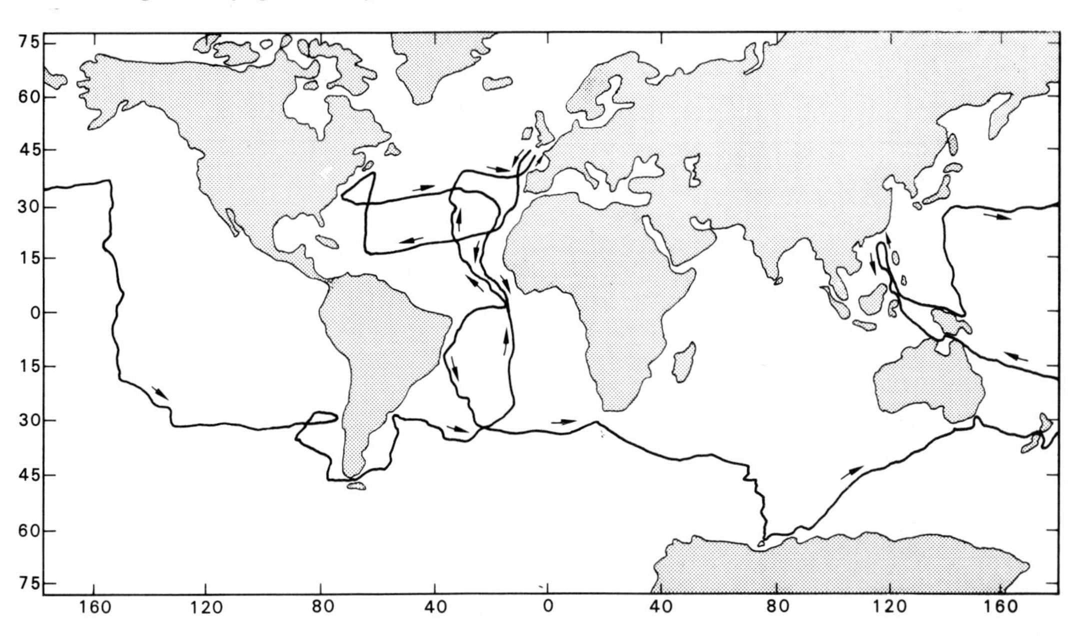

Figure 1.1 Map of the route of the ship HMS *Challenger* 1872–6

Globigerina ooze lived on the surface, and that their distribution on the surface was reflected in the distribution of the ooze on the sea floor. Murray was also able to show that as they approached the ice around the Antarctic the *Globigerina* ooze gave way to a deposit consisting mainly of Diatoms, where these organisms were dominant also on the surface. The scientists began to appreciate the greater solution that the *Globigerina* tests underwent as they were recovered from increasing depths, so that this factor was used to explain the relative distribution of calcareous ooze and Red Clay, the latter being restricted to the deeper water, in which solution had been more complete. They made a mistake, however, in considering that the Red Clay was the insoluble remains of the calcareous tests.

By 1874 the *Challenger* had reached Australia and in 1875 she was working in the Pacific, together with several other ships. Research showed that the Pacific was different from the other oceans in being deeper, larger, and having a much more extensive area of Red Clay. The Red Clay also differed in character from that in the Atlantic, having larger particles and other characteristics that pointed to a volcanic origin (according to Murray) as many volcanoes were also found. Much more extensive areas of manganese nodules were also discovered, and it was found that they had grown around a nucleus, such as shark's teeth. The variation of thickness of the coating led Murray to the conclusion that sediment must accumulate very slowly in the Pacific in comparison with oozes in the shallower water. The origin of the nodules was not explained, although the idea of chemical exchanges with sea water was suggested. The *Challenger* sounded depths of 4475 fathoms (8200 m), where radiolarian ooze was found as their siliceous skeletons did not dissolve in the very deep water. Early in 1876 the *Challenger* sailed through the Straits of Magellan back into the Atlantic on the way home.

When the *Challenger* re-entered the Atlantic, she visited Tristan da Cunha and Ascension Islands and discovered the zone of relatively shallow water between the islands. The concept of an underwater mountain range gradually emerged and is clearly shown on the map reproduced in the preliminary report of the expedition published in 1876. The southern part of the ridge was called the Challenger Ridge and the northern part the Dolphin Ridge, with the Connecting Ridge shown trending west-northwest to south-southeast between them. Subsidiary ridges were also surmised linking the main ridge to the continents, because of the distribution of bottom temperatures. The Walvis Ridge was shown to prevent the cold Antarctic water from reaching the eastern Atlantic Basin in the southern hemisphere.

Considering that the weight of the expedition had been mainly on the biological side, it is remarkable how much valuable information on marine geology and relief had been collected. The expedition gave a reasonably clear picture of the distribution of deep sea sediments throughout the world ocean, while the major relief feature of the mid-oceanic ridges were recognized at least in the Atlantic Ocean. On the biological side, the expedition had amply justified itself and shown that life existed at all depths, and was remarkably uniform in all oceans at depth. The measure of success of the voyage is seen in the immense collections that were made, which took many years to process and publish. This scientific expedition paved the way for modern oceanography and it has not been equalled for volume of work achieved and new concepts advanced. Wyville Thomson died before the results were finally published and John Murray was appointed to succeed him in supervising publication. Murray himself worked on the marine deposits with the help of A. F. Renard. Fuller investigation confirmed the early conclusions that Murray had reached during the voyage. The report appeared in 1891.

3 Recent techniques in marine geology and geomorphology

The epic voyage of HMS *Challenger* laid the foundations of modern oceanography. One contribution in the field of marine geology and geomorphology was the better knowledge of the relief of the ocean floor, although the advances in this aspect were limited by the equipment available. The greatest advance probably was in the pattern and distribution of bottom sediment types that were revealed. After the cruise of the *Challenger* scientific oceanographic cruises became more numerous and between the two world wars the following cruises were undertaken:

Table 1.1 Scientific oceanographic cruises

Ship	Year	Nationality	Area investigated
Carnegie	1909–1929	USA	World cruises
Pourquoi-Pas?	1912–1936	French	Atlantic–Arctic
Dana	1921–1932	Danish	World cruise
Michael Sars	1924	Norwegian	Arctic
Discovery	1925–1926	British	Arctic
Meteor	1925–1927	German	South Atlantic
William Scoresby	1926–1932	British	Antarctic–Atlantic
Armauer Hansen	1928	Norwegian	North Atlantic
Godthaab	1928	Danish	Baffin Bay
Marion	1928–1935	USA	Arctic–North Atlantic
Willebord-Snellius	1929–1930	Dutch	Moluccas
Discovery II	1930–1935	British	Antarctic
Norvegia	1929–1931	Norwegian	Antarctic
Atlantis	1931–1932	USA	Atlantic
Catalyst	1932–1936	USA	Pacific
Hannibal	1933–1936	USA	Antilles–Pacific
Mahabis	1933–1934	Egyptian	Indian Ocean
E. W. Scripps	1937–1940	USA	Gulf of California

The list shows that at the beginning of the post-Second World War period the relatively neglected parts of the ocean were the Pacific and Indian Oceans. Much of the work carried out by these expeditions was more related to physical and biological oceanography, rather than geological oceanography or geomorphology, although the introduction of echo-sounding in the 1920s did ensure that the ocean floor relief was becoming known in increasing detail and with increasing accuracy, as profiling replaced single depth determinations.

The exploration of the deep sea was advanced by the US Fish Commission steamer *Albatross* in the eastern Pacific in the first two decades of the twentieth century by means of bottom samples. The German ship *Meteor* was one of the first to be equipped with echo-sounding apparatus, and she did much survey work in the south Atlantic after the First World War, while the United States Coast and Geodetic Survey revealed for the first time (in the 1920s and 1930s) the existence of submarine canyons on the continental shelf off the eastern seaboard of North America. A major advance in the knowledge of deep sea relief came with the establishment of oceanographical institutes, such as the Lamont Geological Observatory, the Scripps Institution, California, and the Woods Hole Institute in 1930. The latter's ship, the *Atlantis,* collected much data in the 1930s and 1940s. Observations included sediment sampling by means of the Piggot-gun, which provided cores much longer than those previously available. The pace of exploration rapidly accelerated after the Second World War with the introduction of new and more sophisticated equipment. Soviet scientists and ships entered the field, using their large ships *Vitiaz* and *Lomonosov,* especially in the Black Sea and the Arctic Ocean and around Antarctica. Photography of the sea bottom started in 1940, originally with military use in fairly shallow water, but was extended to deeper water in the 1950s. This phase was followed by the descent of men to the ocean depths. Thus modern developments can be subdivided into those that took place under the ocean and those from the ocean surface; still more recently remote sensing from above the oceans has been exploited.

3.1 Underwater observations

Underwater observations can be made indirectly by photography or television, or directly by men, either diving freely or in diving vehicles. The capabilities of underwater photography are well illustrated in the sample photographs published by Heezen and Hollister in their book *The face of the deep* (1971).

One of the problems of undersea photography is the opacity of the water at shallow depth and the darkness at greater depth. Five metres is the limit of clear visibility in deeper water and 3 m in shallower water, thus undersea pictures cannot show extensive views. Bottom photographs are normally taken within 5 m of the bottom. This is not such a disadvantage when it is realized that most objects on the sea floor are small, being less than 2 cm in diameter. Until recently 35 mm film was used, with attendant resolution limitations. Cameras are attached to cables, with suitable lighting, and protected from the great pressure of the deep sea. Once the camera cable touches the bottom, a series of photographs can be taken. The range of deep sea photographs is world-wide and most underwater features have been recorded, all environments examined, and many interesting items revealed. Many of these are illustrated by Heezen and Hollister (1971). They include such features as ripple marks that have been located at all depths. More recently underwater television has been developed. Television tubes that amplify light by a factor of 30,000 have made artificial lighting unnecessary in some circumstances, so that back-scatter is eliminated.

Diving has also made rapid advances in the post-war period. In the relatively shallow water of the coastal areas and continental shelf, free diving equipment is now available that allows unencumbered movement, known as SCUBA (self-contained underwater breathing apparatus), but at depth divers must be protected from extreme pressure by diving saucers or bathyscaphs. The freedom and independence given to divers by the SCUBA equipment has greatly increased the use and value of this method of underwater observation. Divers are freed from necessary contact with a ship via a life-line, which is essential with all conventional 'hard-hat' diving. The use of the SCUBA equipment began in 1949. The aqualung was introduced from France, where it was invented by J-Y. Cousteau, and became generally available in 1951. It is of particular value for geological and geomorphological investigations in shallow water.

One of the problems of investigating the movement of sediment on the sea floor is that most pieces of equipment inserted to measure the movement, themselves disturb it. When free divers can operate, however, they can make direct measurement with the minimum of interference. Some very valuable results have been obtained by means of thin pegs pushed into the sea bed in depths up to over 30 m by divers. They can return to the pegs after a lapse of time and measure the amount of material deposited or eroded at the site. Observations of this type have been carried out off the coast of California and they have shown that there is little change in sea bed level in depths in excess of about 12–15 m. The sea floor is not static, however; ripple marks are common. Disturbance of the bottom by the activities of benthic creatures also can be examined, an important aspect of some sedimentary structures. The size and pattern of ripples can readily be recorded by direct observation. Interesting observations on the movement of sediment at the heads of some submarine canyons have shown that turbidity currents do in fact exist and can move considerable volumes of sediment through the canyons. In fairly shallow water SCUBA divers have been able to add much valuable information to the geological maps in areas where solid rocks outcrop offshore. Rock samples can be obtained directly and dips and strikes measured *in situ* down to depths of about 45 m. Special instruments have been adapted for underwater geological mapping, and efficient exposure suits make work even in the Arctic seas possible. There is, however, a limit to the time a diver can work at depth. A stretch of 15 minutes is the maximum at the limiting depth of 45 m. There are dangers in diving, such as nitrogen narcosis and oxygen poisoning, making decompression a necessity when rising from great depths (Woods and Lythgoe, *Underwater science*, 1971).

A certain amount of diving has been carried out around the coasts of Britain, but observation in this area is seriously hampered by the very poor visibility. Kidson, Steers, and Fleming (1962) record that even in depths of only 12 m really good visibility off north Norfolk was only 1·5 to 4·5 m, and visibility was sometimes as low as 60 cm, or less with onshore winds. Such very poor visibility necessitates working with a life-line, which restricts the scope of the work in turbid seas very considerably compared to that in much clearer seas, in California for example, where observations have been made to much greater depths and more extensively. The presence of strong tidal currents around Britain is another hazard in free diving work in this area. Off California the visibility renders underwater observations difficult beyond a distance of 30 m, and it is often less. Despite its difficulties and limitations, SCUBA equipment has given shallow water observations another dimension and great impetus.

In deeper water protection from pressure is essential and deep diving craft have been developed to provide this protection. A large number of different submersible craft have been developed to probe directly to the ocean floor. The most famous of these craft is the bathyscaph (meaning deep ship) called the *Trieste* II, which was constructed by A. Picard. This craft has descended to a depth of 10,950 m, the greatest depth known, in the Challenger Deep of the Mariana Trench off Guam, in the southwest Pacific, in 1960. Essentially the *Trieste* is a metal-covered balloon filled with petrol instead of air. The observation chamber, which holds two people, is entered through a shaft and is a very strong steel sphere. The craft is ballasted with shot to increase its density enough to enable it to sink to the bottom. The ballast can then be jettisoned to enable the craft to return to the surface. The bathyscaph is 20·4 m long, 4·5 m beam, and its pressure hull is 18·2 cm thick, made of forged steel. It can move at 2·4 knots over a range of 14 nautical miles. There are automatic grabs to collect samples, cameras to take photographs, a television camera, a depth meter, and sonar equipment. Portholes allow vision of the surroundings of the craft, which are illuminated by searchlights.

This bathyscaph has been followed by the *Archimede*, which belongs to the French Navy, and has descended to a depth of 9450 m in the Kurile Trench. A new craft, the *Trieste* III, is being designed and other bathyscaphes are also under construction or ready to work. One is called the *Aluminaut*, as it is made of forged aluminium. It is designed to go to 4500 m, but insurance problems have meant that its maximum depth has been 1800 m. The craft is 15·5 m long. It has a manipulator arm that can be used to explore the bottom in its vicinity.

One of the problems of these elaborate deep diving vehicles is the great running expenses (about $1000 per hour diving time). The *Aluminaut* has a complement of 3 or 4 scientists and a speed of 3·8 knots over a range of 40 nautical miles. It can spend up to 72 hours below water, which is more than twice as long as the *Trieste* II and three times the length of underwater stay of the *Archimede*.

The tendency to devise craft that can stay below water longer is being further developed by the design of semi-permanent laboratories beneath the sea. Off the coast of California oceanographers have spent up to 15 days continuously beneath the sea at a depth of 62 m in *Sealab* II, which belongs to the US Navy. *Sealab* III has also now been developed. Still longer spells underwater have been spent in the Mediterranean by J-Y. Cousteau. He spent 3 weeks at a depth of 100 m, working at even greater depths. The advantage of living for longer periods at depth is the time saved by eliminating the requirements of decompression that are necessary when returning to the surface. Divers can go out of the underwater laboratories for periods of observations and then return to them again; thus better use can be made of time devoted to underwater observations. Underwater structures may become more important in the near future as exploitation of the ocean increases, especially for oil and gas on the continental shelf, and perhaps later on the continental rise.

The search for oil offshore has led to a great development of techniques for exploring submarine structure and for drilling. These techniques have also benefited pure research. Samples of the sea bed can provide much valuable information of interest to many different branches of oceanography. At first the character of the bottom was determined by greasing the hollow base of the sounding lead so that some of the bottom sediment stuck to it and the depth and type of bottom were obtained together. This method was not very reliable, however, as some types of sediment would not adhere readily to the greased lead. The next stage was the development of various types of grab to obtain samples from the bottom. Some of these operate on the same principle as the clam closes its two valves, such as the Pettersson grab that can obtain fairly large samples from the bottom up to depths of between 180 to 360 m. Another type has 4 petals that close around the sample when the grab reaches the bottom. One problem with this type of sampler is that the jaws may get bent so that they do not close tightly and some of the sample is lost. Where there is large material to procure the jaws may be unable to close and finer material can then escape on the way to the surface.

Coring devices that enable a long core of sediment to be obtained are of considerably greater value for many purposes. Studies of long cores have already shown great value in elucidating many problems, for example in connection with climatic change. One of the essentials of coring, whether the sample be short or long, is to obtain it with as little disturbance as possible. The Russians have developed a vibro-corer for obtaining fairly short cores in shallow water to study sedimentation by waves. This type of instrument can obtain good cores in any type of bottom material, including gravel, sand, and finer sediment.

The simplest form of coring device is the gravity corer, which is allowed to fall freely to the bottom where its weight drives it into the sediment, allowing a sample to be withdrawn when the tube is raised to the surface. The Phleger corer can be used to obtain short cores from a small boat. These cores are about 2·5 cm in diameter. The standard corer for taking long cores in deep water is the Kullenberg piston corer. This was developed soon after the Second World War and has become a standard piece of equipment for obtaining deep sea cores. Some cores up to 20 m in length have been acquired with this instrument. One problem with coring is that it is unwise to lower the coring tube too fast or it may become entangled with the wire cable supporting it. If it is lowered too slowly, however, it will not sink into the bottom very far. In the piston sampler this problem has been overcome to a certain extent by suspending a weight below the bottom of the coring barrel which will hit the bottom first. This releases a lever allowing the core barrel to fall freely for some distance so that it penetrates into the sediment. The piston inside of the core tube remains fixed to the wire, and the core tube falls past the piston. It is difficult to obtain an undisturbed core when the core barrel is not completely filled as the instrument sinks into the sediment. In order to achieve as deep a penetration as possible, the corer must be very heavy, and weights up to 1000 kg are used at times. One of the difficulties of using apparatus of such weight is that it puts a great strain on the winch of the ship and extra strong, and therefore heavier, wire must be used. The coring tubes are also very long compared with their width, so they are very liable to bend or break. Another problem is the compaction of the material in the core, which is particularly great in the gravity type corer. This instrument can, however, obtain undistorted samples from the top of the sediment. The upper part of the sediment is liable to get lost in the piston corer, so that it is usual to obtain a short gravity core alongside the piston core, the former being attached to the weight which hits the bottom first. The upper layers can then be seen in one core and allowance can be made for the compaction of the gravity core. Kullenberg has further developed his coring device so that the trip only operates after the core tube has sunk a short distance into the soft sediment. The problem of preventing the core leaking out on the way to the surface is greater when there is a large amount of sand in the sediment. It is in fact very difficult to obtain cores in hard packed sand with either the gravity or piston corer. This problem is overcome to some extent with the vibro-corer. There is still room for considerable improvement in the technique of obtaining long cores from the deep sea and some techniques that have been used may have disturbed the samples so that the results require great care in interpretation. However, improvements are possible and recent work has greatly extended sediment cores.

Problems of coring are small compared to those of drilling much longer cores through both sediment and hard rock, a process that is necessary in exploitation of oil resources on the continental shelf and further offshore. One of the earliest attempts at deep drilling for scientific purposes was the Mohole project, designed to obtain a sample of the earth's mantle material from beneath the Mohorovicic discontinuity. The Moho is much nearer the surface under the deep oceans than in the continents, so that the best site for the Mohole would be in the ocean. The problems of drilling in deep water through a depth of about 6 km of basalt below the sediment layer are formidable. Total drilling depth would be about 5000–10,000 m in water about 4000–5000 m deep. One problem of deep drilling is to keep the ship steady so that the drill pipes remain vertical through this great depth of water. Some experimental work has already been carried out by drilling in a depth of 1000 m of water in the San Diego trough. Other deep boreholes have been put down in 4000 m of water east of Guadalupe. The ship was maintained on site by a series of tautwire buoys anchored by large submerged floats. The buoys were maintained in position around the drilling site and the position of the ship was fixed by radar ranging between the fixed buoys. The ship had outboard motors that were used to keep its station. The vessel used for the preliminary test was the *Cuss* I, a ship of barge type, 100 m long. Its relatively large size meant that it provided a stable platform as long as the waves were not too long. The derrick from which the drilling was carried out was nearly 30 m high. The first experimental bore penetrated to 300 m into a large fan at the mouth of the San Diego trough. The second borehole passed through 170 m of upper Tertiary strata and then penetrated through 15 m of basalt, making a total distance of 185 m.

These experiments provide information needed to develop improved techniques that will eventually be necessary to drill to much greater depths. The coring was done by lowering core barrels on a cable. Four different types of core barrels were used experimentally. The most successful of these in recovering cores was the rotary core barrel. It drilled 9 m into the sediment at each run. One of the results obtained from this experimental drilling through sediment was the direct measurement of sound velocity through this type of material. The rate was found to be 1·6 km/sec, which is lower than earlier estimates of 2·2 km/sec. This result would mean that previous interpretation of sediment thicknesses of the sedimentary cover on the ocean floor, derived from geophysical surveys, would have been too great. The actual sediment thickness is, therefore, probably somewhat less than currently thought.

The early experiments have shown that a drilling ship can maintain its position successfully without anchors, and that both soft sediment and hard basalt can be drilled through at considerable depth below the ship. Penetration of the basalt was, however, very slow, so that improvement in drilling would be essential before the great thickness of basalt to the Moho could be penetrated.

One of the most recent and valuable additions to the deep drilling capacity for ocean sediment observation is the work of the *Glomar Challenger*. This ship is operated by the Scripps Institution of Oceanography under contract with the National Science Foundation of the USA. The deep sea drilling project is part of the National Ocean Sediment Coring Program and is planned by the Joint Oceanographic Institutions Deep Earth Sampling (JOIDES) group. The ship is able to drill through 760 m of sediment in water depths of 6000 m. Developments to keep the ship on station include acoustical position-sensing devices and automatic control of the propulsion units, giving more reliable dynamic positioning. The ship has already completed most of its proposed drilling program in all the major oceans and very valuable results have been obtained, some of which are mentioned later. The discovery of the salt structure in the Sigsbee deep, with hydrocarbons, has changed views on the availability of oil and gas from the deep oceans. The first legs of the *Glomar Challenger* cruise were situated in the Atlantic Ocean and she has since drilled in the Pacific, obtaining cores that penetrate right through the sedimentary blanket into the basement in both oceans. The results of this work are extremely valuable and the whole venture has been outstandingly successful.

3.2 Surface observations

In exploiting oil reserves offshore, technological developments in drilling platforms as opposed to free ships have been important. Some of these have legs fixed to the bottom, others are semi-submersible, such as *Blue Water* 3, which has drilled for oil off Trinidad.

Plate 1 The *Glomar Challenger* (*above opposite*) is specially designed for deep sea drilling to obtain sediment cores; under the direction of the Scripps Institution of Oceanography the deep-sea drilling project has covered every ocean. (*Scripps Institution of Oceanography.*) The US Navy's *Sealab* III (*below*) is shown here during testing at Long Beach, California. (*Keystone Press Agency.*)

The platform is towed into position and floats at the corners are filled with water so that part of the structure lies below the water level to give it greater stability. The platform is about 60 m square and is designed for work in the open ocean.

Many other observations of interest to marine geologists and geomorphologists are made from surface research ships. Much of this work is of a geophysical nature and the main geophysical techniques, which include seismological studies, gravity measurements, heat observations and magnetic surveys, are discussed in more detail in the next chapter, in which the results of such surveys are outlined.

Another method of obtaining greater stability for surface observations is the FLIP ship, which is partly ship and partly buoy. The craft is 108 m long, weighing 600 tons. It can move as a ship in the horizontal position, and when it reaches a station for observing, it can be turned through 90 degrees to stand vertically in the water, where it can be anchored and can serve as a fixed buoy. Because of its great length of over 100 m draft below the surface, it is very stable even with high waves. There are two watertight containers that allow scientists to descend to a depth of 45 m below the surface.

Geomorphologists are particularly concerned with details of the morphology of the underwater relief, as well as the larger features. Advances in surveying these features from surface survey ships have made it possible to obtain detailed pictures of the shallow subsurface structure. Echo-sounding equipment that is set at an angle, rather than pointing vertically downwards, can reveal the details of the submarine relief and shallow structure clearly. A. H. Stride of the National Institute of Oceanography and his collaborators have obtained a great deal of valuable detailed information concerning the surface form and shallow structures of the offshore zone around Britain, using oblique asdic and profiling techniques. The equipment on RRS *Discovery* II provides a valuable means for rapid survey of the shallow waters of the continental shelf. It has a frequency of 36 kc/sec and pulse length of 1 m/sec. The acoustic beam is fan shaped, $1\frac{1}{2}$ degrees wide horizontally and 11 degrees vertically. It points normal to the length of the ship and is tilted down a few degrees below the horizontal, according to the depth of water and the maximum slant range of 733 m. Side lobes of lesser range point down more nearly vertical; one, being actually vertical, gives a profile of the sea floor. The results can be interpreted to give a detailed picture of the sea floor and its structure, so that rock outcrops, faults, sand, waves, and other features can be recognized.

An important aspect of many oceanographical observations is the ability to fix the position of the ship accurately. Sonar was one of the early developments in this field, similar in principle to the echo-sounder in that waves are used to assess distance. This technique has now been considerably improved. Frequencies of the waves emitted have been increased, narrowing the beam width, so that distances and depths can be measured more accurately. Sonars employing the Doppler effect, which is the change in pitch of sound due to the relative motion between the source and the observer, have been developed. These allow the direct measurement of the ship relative to the bottom, an ability that is essential for detailed determination of gravity and other observations that require exact fixes.

Plate 2 FLIP (*opposite*) in action (see text above). (*Scripps Institution of Oceanography.*)

3.3 Remote sensing methods

In connection with accurate navigation and position fixing at sea, further advances have come about through use of satellites, by means of which it should be possible to navigate as accurately in the open ocean as it is now possible to navigate in coastal waters. In 1967 the US Navy's satellite navigational system was extended from its use by Polaris submarines to commercial shipping and oceanographical exploration work. The system consists of a number of satellites orbiting the earth in a 1160 km high polar orbit, each orbit taking 90 minutes; three satellites were operational in 1968. The principle on which the system works is that if two positions of a satellite in space are known, and the difference in range of an object on earth from each is measured, the object will be on a hyperbola whose foci are the two known points. The measurements can be repeated at another point and the desired position is where the two hyperbola intersect. In fact three hyperbola are measured to obtain a check on the position. A computer is necessary to calculate all the complex equations that enable the ship's position to be accurately located. In order to operate the system the ship requires an antenna, a receiver for the correct frequencies, a stable oscillator, a data processor, and a digital computer and print-out mechanism. Despite the complex calculations necessary, the operation on the ship is simple when the equipment is functioning correctly since all the measurements are made automatically. The operator only has to insert the local information, consisting of the date and GMT time to the nearest 15 minutes, the estimate of latitude and longitude to within 3 degrees, the receiving antenna height, and the ship's course and speed. The accuracy with which the result is printed out by the computer is about 0·1 nautical mile, which is adequate for most oceanographical work.

Aircraft and satellites can also be used for many other types of oceanographical remote sensing operations. In the field of marine geology and geomorphology satellites also have their uses, and with the likely growth of satellite facilities, details of the type of information that they could obtain have been collected by the Woods Hole Oceanographic Institution (G. C. Ewing, editor 1965). Magnetometers can be towed behind aircraft thus greatly increasing the range and speed with which such observations can be collected.

At the other end of the scale, a method of surveying beach profiles with the aid of a helicopter has been devised. A sledge is carried offshore of the breaker zone by the helicopter. The sledge is attached to the shore by a line and a surveying pole is fixed to it. The helicopter then drops the sledge into the water and it is dragged towards the shore from the land while observations are made on the pole and the profile of the beach recorded. The process is then repeated on another line of profile. This technique allows observations to be made in the active surf zone in storm conditions (not possible by other means), when conditions are changing most rapidly under extreme processes.

Satellites can best carry out large-scale, often-repeated observations. Synoptic studies can also be carried out well by this means because satellites have wide coverage over a short time span. Satellites are probably best suited to studies in shallow water above about 20 m, while deeper observations can be made more efficiently from other types of platforms. In order to make the most of the data available, it is necessary to familiarize workers with the interpretation of different types of satellite material, such as infra-red, radar from rockets, U-2 planes, TIROS satellite, and similar techniques.

Before effective use can be made of satellite material it is essential that all the material is accurately located and calibrated. Calibration can be carried out in areas where the variable that it is desired to calibrate is accurately known. This applies to such variables as beach material, currents, and tides, for example. The sensors that are now available could provide evidence on such variables as beach width, beach slope, characteristics of waves, angles of wave approach, dimensions of the surf zone, nearshore water circulation, and features in tidal marshes. Well-known areas could be used for calibration, such as Scripps beach, Cape Cod, and Virginia beach, Virginia, which are all sites of intensive ground research. Radar and infra-red techniques could probably be used to distinguish between different types of beach sand, such as quartz, calcite, rock, and organic mats. Heavy minerals of economic importance could possibly be distinguished. Coastal currents and zones of upwelling could be located, and coastal zonation could be carried out in suitable localities, such as coral reefs.

One example of synoptic use could be the effect of a storm. The advantage of a long stretch of coast as well as an area inland and seaward of the coast being observable at the same time is evident. The storm could be identified, its progress and that of the waves to which it gives rise could be followed. Precipitation associated with it might cause slumping inland, thus increasing the river load, which in turn could increase the material at the coastline, and also initiate submarine slumping by turbidity current activity, all of which would be apparent.

One of the problems is the resolving power of the sensors for this type of observation, but it is a field in which continual advances are being made, so that this aspect should improve steadily. Beach width should ideally be measurable to 1·5 m, but 7·5 m would be useful, especially if variability of width over a season could be assessed. Measurement of beach elevation should preferably be to 30 cm, although a value of 1·5 m would be effective. An ability to record offshore relief, and details such as sand waves would be valuable, if the results were accurate to about 30 cm. Dispersal of fresh water in salt can be determined by infra-red photography. The determination of the wave spectrum is another field in which aircraft or satellite observation has been carried out successfully, both in the open ocean and close to the shore. An estimate of the rate of sediment transport along the coast would be very valuable, as well as an indication of the type of material being transported. The ability to see a large area at the same time is one of the main advantages of satellite observations; such a view would allow regional comparisons of the type that are not at present possible. Orbiting satellites could provide an answer to the problem of the tempo of coastal change in different areas and under different conditions. Such information would provide a more reliable estimate of the optimum time intervals over which different observations could be made. The problem of spatial scale of observations could also be examined by means of satellite data, since in many areas in order to understand coastal changes in one area, it is necessary to extend the observations to neighbouring areas. Such data would provide information on the optimum sampling plans for many different types of observation.

Satellites could also provide data on the differences of conditions in the three main oceanic zones—the shore zone, the shelf zone, and the basin zone. The extent to which these zones influence various oceanic phenomena, such as currents and wave patterns, could be estimated. The relative effectiveness of tidal and other types of currents in the nearshore zone could be assessed. The extent of coral reefs and their character could be assessed by infra-red sensing, while the pattern and movement of submarine bars could also be examined by sensing devices. Observations from even greater distances from the earth can show large-scale structures, and these give useful insight into the pattern of the structures from the point of view of the new global tectonics, such as lineations.

4 Developments in oceanographical investigation

As oceanographical exploration and exploitation become increasingly necessary and expensive, it will become more important to have investigations supported on an international cooperative basis, both amongst the scientific community of the world and amongst the governments of those countries able to participate. Together with the opportunity for cooperation, the freedom of the high seas implies the possibility of conflict in the exploitation of oceanic resources. The legal and social implications, as well as controls over biological exploitation, will be considered in the second book, but mention will be made of the possible future path of development in the field of marine geology and geomorphology in connection with the exploitation of the inorganic resources of the oceans.

Multi-national ocean research first took place on a large scale during the International Geophysical Year of 1957, for which a programme was developed by the International Geological and Geophysical Union. From this work the Special Committee on Oceanic Research was established. This institution organized one of the first major oceanographical international expeditions for research work, the International Indian Ocean Expedition, in which 40 ships of 23 countries took part in 180 cruises. The expedition lasted from 1959 to 1965 and accomplished a great deal of valuable basic scientific observation in one of the oceans previously least well known. The work covered all fields of oceanography, and many of the results should be of direct benefit to the heavily populated and underdeveloped countries in this area. The work of the Special Committee on Oceanic Research in the Indian Ocean was taken over by the Intergovernmental Oceanographical Commission in 1960, established by UNESCO.

On 8 March 1968, the President of the United States of America proposed the launching of 'an historic and unprecedented adventure—an International Decade of Ocean Exploration for the 1970's.' Since this suggestion was made, a steering committee has been set up to consider the best way to implement the proposal. A group of specialists met at the Woods Hole Oceanographic Institute in September 1968 to discuss possible programmes of research in all fields of oceanography. This is to be a cooperative venture, both within the United States and internationally. The emphasis of the studies will be on utilization of the ocean and its resources, which can only be optimized as knowledge of the oceans grows. Only those proposals directly related to marine geology and geomorphology will be mentioned, but the programme includes all aspects of oceanography.

The Decade research should aim to investigate on a broad basis the nature of the ocean basins, rather than attempt a delineation of immediately exploitable resources. This type of broad survey is likely to produce evidence of value in resource exploitation, such as the discovery of suitable structures for oil and gas in the Gulf of Mexico and offshore in the Atlantic, for example. Oil will be one of the most important resources required from the offshore zone in the future. The discovery that conditions in the deep marine basins off southern California are very similar to those in the oil-bearing basins of Los Angeles and Ventura (except that the offshore basins are not yet filled with sediments) is important. Observations in the deep basins provide very valuable evidence for the study of the formation of oil, and the part played by turbidity currents and organic processes.

Sea floor mapping is another task that should feature in the Decade studies. At present adequate sea floor maps are confined to the shallow marginal areas, but detailed maps will be increasingly required for deeper water. At first the most essential areas should be mapped and effort should be devoted to improving mapping methods to speed up and make the process more accurate. The inorganic resources that can be, and in many cases are already being, obtained from the oceans include the following:

Table 1.2 Inorganic ocean resources

	Annual value of world production 1967–8 in million $	
Resource	Total (US)	Offshore (US)
Petroleum	26,000	3900
Sulphur	340	37
Sand and gravel	900	150
Heavy heavy mineral		
Gold	1900	—
Tin	460	24
Platinum	150	—
Light heavy minerals		
Ilmenite	54	—
Rutile	16	—
Zircon	10	—
Monazite	1·8	—
Gems		
Diamonds	290	4
Precious coral	2	2
Subsurface consolidated deposits		
Coal	18,500	35
Iron ore	4300	17

Source: Committee on Oceanography, *An oceanic quest* (1969), p. 30.

Manganese nodules and phosphorite deposits also constitute resources of great potential value. The former are widespread, particularly in the deep ocean floor of the Pacific, where the slow deposition of sediment allows time for them to grow by accretion of the minerals from the surrounding sea water. Their potential value lies in their nickel and copper content mainly, although cobalt, chromium, molybdenum, and manganese could eventually be significant. Phosphorite occurs mainly on the edge of the continental shelf and the upper slope, and its major value lies in its qualities as a fertilizer. The discovery of hot-brines in the Red Sea is also of potential interest for their lead, zinc, and copper. In order to recover these and other minerals from the oceans, widespread surveys are needed, including seismic, gravimetric, and magnetic data, and also bottom samples and chemical analyses of ocean water.

One area where detailed surveys would be profitable is the eastern side of the Atlantic Ocean, which was the subject of a recent symposium organized by the Scientific Committee on Oceanic Research, a committee of the International Council of Scientific Unions. The symposium (Institute of Geological Sciences, Delany (ed.) 1970) aimed to report on present knowledge of the area and to identify gaps that should be filled by coordinated marine research programmes of national or international character. The symposium concluded that work was needed on the continental slope and rise to supplement that already done on the continental shelf. The African part of the continental margin is less well known than the European part so that effort should be concentrated in this region, which is a good example of a continental margin that has been built out by sedimentation over a long period of time since its separation from the western Atlantic margin when rifting started. The latitudinal range makes such studies of particular interest. The topics that could usefully be studied include foraminifera from Novaya Zemlya to Capetown, sediment texture, and seismic profiles across the continental margin.

The deep drilling programme that has been so successful in the deep ocean basins could be extended to cover the smaller ocean basins, such as the Gulf of Mexico, the Mediterranean, Aleutian, and Black Seas, and then to the continental margins. Fundamental changes in the concepts of the origin and character of the ocean basins that have come about through the development of the new global tectonics, based on ocean floor spreading, give rise to many important lines of enquiry, and require the collection of many fundamental data. Geophysical work will be necessary to solve some of the problems, such as that involved in a study of the way in which new mantle is emplaced. Other regions where further data are required are the convergence zones of trenches and young mountains, which are relatively less well known, but potentially more important. These areas represent resource potentials, as well as scientific problems. The problem of undeformed sediment in the trench floors, and the balance between new crust emerging at the ridges and being lost in the collision zones requires further investigation. The clearest convergence zone should be selected for study, and the Peru–Chile Trench could provide suitable conditions. The relief is spectacular and the aridity of the Andes means that the sedimentation rate is low; the equatorial situation is important for certain chemical tests. Oil seeps provide a possible resource element as well. Complete geophysical profiles should be run across the continental marginal zone. The deep sea type of structure could be examined in the Tonga–Kermadec Trench, the Solomon–New Guinea convergence zone, or the Aleutian Arc. There is a proposal for USA and USSR collaboration in the latter area.

These projects will entail the development of ever more efficient and sophisticated instruments, including the perfecting of some of the methods already mentioned, such as satellite navigation, platforms of various types, both on the surface and beneath it, or on the bottom.

Further reading

COMMITTEE ON OCEANOGRAPHY. 1969: *An ocean quest.* Washington: National Academy of Science—National Research Council. (Recommendations for the International Decade of Ocean Exploration.)

SCIENTIFIC AMERICAN. 1969: *The Ocean.* San Francisco: Freeman. (Reprints of a special number devoted to oceanography.)

DEACON, M. 1971: *Scientists and the sea, 1690–1900. A Study of marine science.* London: Academic Press. (A scholarly account of the development of the science of oceanography, including an account of the *Challenger* expedition.)

DELANY, F. M. (editor). 1970: *The geology of the east Atlantic continental margin. 1: General and economic papers.* Institute of Geological Sciences Report 70/13. ICSU/SCOR Symposium held at Cambridge, 1970. HMSO. (Contains some general papers and some on more specific topics.)

FAIRBRIDGE, R. W. 1968: *Encyclopedia of oceanography.* New York: Reinhold. (A valuable work of reference, covering nearly the whole field of oceanography.)

MENARD, H. W. 1969: *Anatomy of an expedition.* New York: McGraw-Hill. (A personal account of a deep-sea geological expedition in the Pacific Ocean.)

MOORE, J. R. 1971: Introductions by J. R. Moore to a series of *Scientific American* reprints. *Oceanography.* San Francisco: Freeman. (Section III deals with the floor of the ocean.)

RICHARDS, A. F. 1967: *Marine geotechnique.* Urbana: University of Illinois Press.

SHEPARD, F. P. 1973: *The earth beneath the sea.* (A reasonably general account of submarine geology.)

TUREKIAN, K. K. 1968: *Oceans.* Foundations of earth science series. Englewood Cliffs: Prentice-Hall. (A short general account of oceanography, concentrating on geochemistry and sediments.)

2 The structure of the ocean basins

The impact of the new ideas of global tectonics on the knowledge of the structure of the ocean basins has been of immense importance, and has revolutionized the study of marine geology and geomorphology. The ideas are only about a decade old, but already the amount of evidence assembled from a wide range of fields has given striking support to the theory of plate tectonics and ocean floor spreading. The basic outline of the theory is explored in this chapter and the more detailed evidence for it, in terms of ocean floor morphology, is considered in the two subsequent chapters, which deal with the oceanic margins and the deep ocean respectively.

The oceans form one of the fundamental divisions of the earth's surface, their volume being considerably greater than that of the land above sea level. If all the inequalities of the earth were levelled off there would be enough water in the oceans to cover the whole surface of the earth to a depth of about 2600 m. Most of the land lies fairly close to sea level, the mean height of the continents being 700 m above sea level, while the oceans, on the other hand, have their greatest area between 3660 and 5500 m below sea level. In fact 41 per cent of the whole surface of the earth lies between the last two depths. In terms of the diameter of the globe (which is nearly 12,800 km), however, the relief features of the continents and oceans are relatively insignificant. The highest mountains and greatest deeps are only about 20 km apart vertically, 1/320 of the radius. There are, nevertheless, very important differences between the structure and relief of the ocean basins and the continental areas.

The new evidence accumulated during the last decade has completely changed ideas on the nature and formation of the ocean basins. Recently acquired data show that the oceans are dynamic, and that they continually change as a result of the operation of processes on a world-wide scale. In this chapter the nature of the oceanic crust is examined by means of various types of geophysical surveys, which will be briefly reviewed first. The results of the studies are then summarized in terms of general inferences and with more specific examples.

Much of the evidence for the new global tectonics has come from the ocean basins. It has been derived from a variety of geological and geophysical observations, which, together with a growing knowledge of the relief of the ocean floor, have enabled a unified synthesis to be developed. The most important geomorphological discovery has been the location of the world-wide oceanic ridge system. This system lies central in the Atlantic and Indian Oceans, but not in the Pacific or the Arctic Ocean. There are three points where the ridge system runs into the land. One place is where the Arctic Ocean Ridge runs into the north Siberian land-mass, probably reaching the Lake Baikal depression eventually. This Arctic Ridge then links with the mid-Atlantic Ridge, which covers the whole length of the Atlantic Ocean and continues into the Indian Ocean, bisecting the distance between South Africa and the Antarctic continent as it turns east in the south Atlantic. The ridge diverges into two parts in the Indian Ocean, one branch continuing southeastwards across the southern Indian Ocean, while the other runs northwards and then northwest towards Arabia before turning into the Gulf of Aden and the Red Sea, thus forming the second point of contact between the ridge system and the land. The other branch continues eastwards, again bisecting the distance between Antarctica and Australia, before crossing the south Pacific Ocean and turning northwards to run finally into the land at the third point of contact in the Gulf of California. The ridge system is not continuous all the way, but is frequently broken by lateral faults, normal to its elongation that transpose it sideways. These important ruptures are known by the new name of transform faults. Other major dislocations have been found, particularly in the northeast Pacific Ocean, where a series of so-called Great Faults run from east to west.

The significance of the ridge system is that it is the locus along which new ocean crust is forming. It is the line of sea floor spreading. The evidence from which this important concept arose was related mainly to the geophysical and geological properties of the ridge and its surroundings; the relief is also important, tying in with the other evidence. The ridges are usually characterized by a central rift-valley, on either side of which are irregular stepped plateaux of gradually decreasing amplitude and elevation.

The geophysical evidence that first drew attention to the possible process of sea floor spreading was the pattern of magnetic reversal anomalies that were found to be symmetrical on either side of the central rift-valley. The same pattern was identified in many parts of the ridge system, a particularly regular pattern occurring on either side of the Reykanes Ridge to the southwest of Iceland, which lies on the centre of the mid-Atlantic Ridge. The magnetic reversal pattern has enabled the rate of spreading to be determined with a fair degree of accuracy. The discovery of reversals in the earth's magnetic field has also aided the dating of deep sea cores, in which a similar pattern is recognized to that which occurs laterally on either side of the ocean ridges. Other geophysical techniques have all added evidence in support of the global tectonic system. Heat flow studies have indicated that the central parts of the ocean ridges are regions of abnormally high heat flow, supporting the view that these are the lines along which hot material is reaching the surface from greater depths. The seismic surveys, which have played such an important part in increasing our knowledge of the structure of the earth's crust and mantle, have also added considerable weight and valuable evidence to the new ideas. They have shown that the oceanic

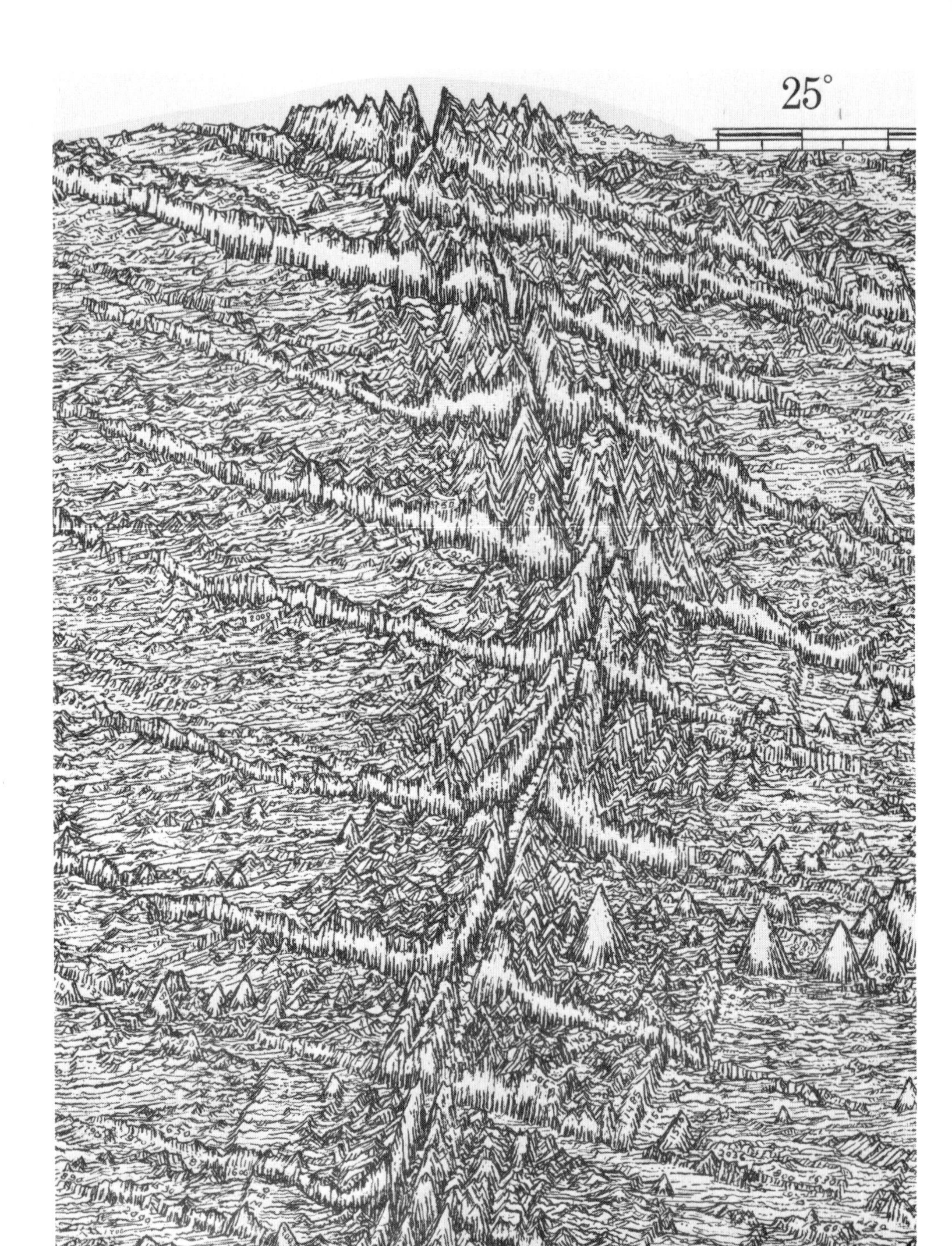
25°

crust and the continental crust is fundamentally different, and also that within the oceanic sector there are variations as the ridges are approached. The positive relief of the ridge features can be explained by the nature of the material of which they are formed. The relatively less dense rising matter allows the ridges to form positive features, but as the crust moves outwards from the ridges, it becomes denser and, therefore, begins to sink, with the result that the depth of water above it increases. The deep ocean basins that occupy the greatest area of the surface of the earth eventually result from this process. Thus geophysical, geological, and geomorphological evidence can all add to the growing knowledge of the underlying forces that create the pattern of land and sea, and to the details of the major relief features within these two major components of the earth's surface.

It is clear that if new oceanic crust is being created along the ocean ridges in their central rift-belts and if the total area of the earth's surface is not increasing, then oceanic crust must be consumed at a rate similar to its production. The evidence that this is indeed true is also found in the oceans. The consumption of oceanic crust takes place in easily recognized areas of the earth. These elongated belts can be recognized by their distinctive relief and geomorphological character, by their distinctive geological character, and also by the manifestations of a geophysical nature.

Geomorphologically these long zones are recognized by the presence of narrow, deep trenches and their accompanying island arcs or mountain chains. These features form a nearly complete ring around the Pacific Ocean, with the exception of the southern margin; they protrude into the Atlantic in two places, the Scotia arc in the south pushes between South America and Graham Land, while the West Indian arc pushes through the former gap between North and South America at Panama; and they occur in a complex from in the eastern Indian Ocean. The deep trenches provide the greatest depths recorded, attaining values in excess of 10,000 m.

Geologically the volcanic rocks of the island arcs and the structure and character of the mountain chains, such as the Andes in South America, provide additional evidence of the process of crust absorption. From the geophysical point of view the pattern of earthquakes is closely related to the region of crustal consumption. On either side of the Pacific Ocean is an unique arrangement of earthquake foci. The depths of the earthquake foci increase in an orderly fashion on a plane that slopes downwards under the continental margin away from the deep ocean, where the earthquakes are relatively shallow. The depths of the earthquake foci descend to about 700 km, indicating that in these zones the earth's crust must have strength to this depth, but only in these zones is this true. It is here that the cold crust sinks down to be re-absorbed into the mantle at greater depth, and its relative coolness allows the exceptional depth at which earthquake activity takes place.

Gravity observations were perhaps the first indication that these long narrow belts of instability in the earth's crust were abnormal. As long ago as the 1930s geophysical observations revealed long narrow zones of large negative gravity anomalies along the lines of the deep sea trenches, indicating that in these zones the crust is out of isostatic adjustment.

Plate 3 Detail from a physiographic diagram (*opposite*) illustrates the characteristics of the central part of the Atlantic Ocean. The central rift valley and rough relief that flanks it can be seen. Further from the spreading line along the central rift the relief becomes increasingly blanketed with sediment, and hence progressively smoother. The transcurrent faults that form at right angles to the ridge can be seen to form conspicuous scarps. (*Copyright © 1968 Bruce C. Heezen and Marie Tharp; published by the Geological Society of America.*)

These, therefore, are the zones along which the crust is sinking, causing deep water as well as structural instability.

In between the zones of activity marked out by the ridges and the trenches, where crust is created and consumed respectively, the earth's surface is much more stable. These large stable zones are the plates, formed of both continental and oceanic sectors. Although they are stable, they are not immobile. It is the movement of these plates across the earth's surface that is a major element of the new global tectonics. This movement indicates that the earth's crust is mobile and that the surface pattern of land and sea is continually changing. Only two decades ago, in the 1950s, the ideas of continental drift, as originally put forward by Wegener, were not acceptable to the majority of the scientists at the time. Now, however, the compelling evidence assembled from so many fields of enquiry has completely altered the situation. The grand simplicity of the basic pattern that has emerged as a result of the global tectonic scheme is such that few would question its basic correctness. There are, however, many details to be filled in and some awkward questions still remain. Nevertheless, it does provide a coherent and eminently acceptable view of the basic nature of the processes by which land and sea patterns change with time. It is a dynamic picture that points to the youth and mobility of the ocean basins. The water of the ocean is probably older than the basins in which it lies. In fact one of the most fundamental implications of the global tectonics is that most of the ocean basins are no older than Mesozoic, and much of the ocean floor is Tertiary or younger. Indeed an appreciable portion of the ocean crust is less than 10 million years old, as at least some of the evidence points to an exceptionally rapid period of ocean crust formation during this period.

This chapter presents some of the evidence, derived mainly from geophysical and related studies, that has given weight to the new ideas of the global tectonic system. Even though the pattern of plate movement and the evidence for crustal formation and absorption are strong, there is still much to be learned about the mechanism whereby these movements are achieved. The strength of the case for the global tectonics rests partly on the way in which a wide variety of evidence from many fields supports it.

1 The character of the major structural units of the earth

The thickness of the crust, which is much thinner under the oceans, is the most fundamental structural difference between oceans and continents. It is directly related to the depth of the Mohorovicic discontinuity (Moho), which separates it from the mantle beneath. The Moho can be identified by seismic surveys, as it is associated with a sudden acceleration in the speed of seismic waves, which increase from 6·7 km/sec to 8·1 km/sec in passing below it. The Moho lies at a depth of 10–12 km below the sea surface, or about 6 km below the ocean floor, in the oceanic sectors of the crust. On the other hand, in the continental areas it is found at a depth varying between 30 and 50 km below the surface. Above the Moho the crust in the continental areas is much more complex than it is in the true oceans. There are layers of different density and often complex structure in the continental areas. The crustal rocks are thick and of fairly low density. They are associated with relatively slow seismic waves, as they contain much granitic type of rock, which is entirely lacking in the true ocean areas.

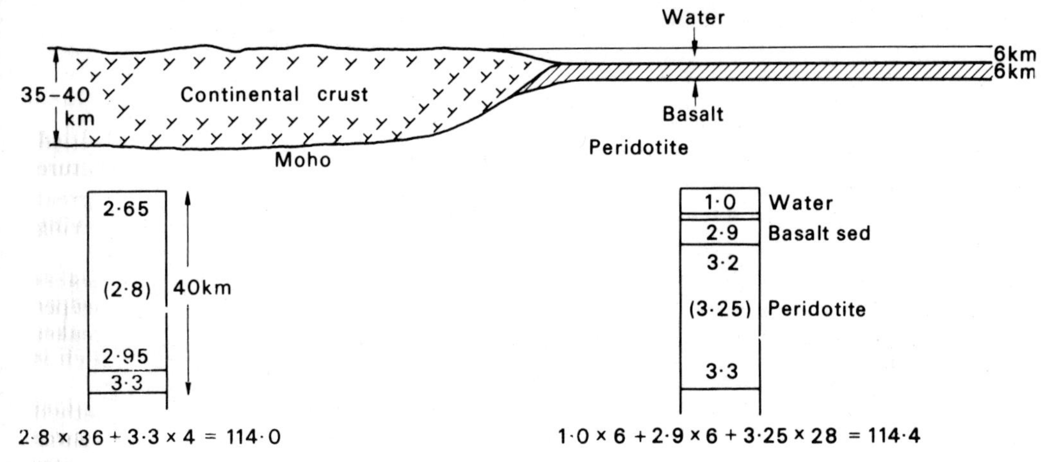

Figure 2.1 Diagram to show the general characteristics of the oceanic and continental crustal types and their approximate isostatic balance. (*After Hess, 1934.*) 54

The generalized picture of the true oceanic structure, as suggested by Hess (1954) (figure 2.1) shows that under about 6 km of water there is a layer of about 6 km of basaltic type rock. The Moho separates this crust from the ultrabasic rocks of the mantle beneath.

1.1 Methods of investigation

A number of different techniques are used to study the structural character of the ocean basins. Seismic surveys investigate interior structure; gravity surveys also yield useful information. The newer methods of magnetic and heat flow studies provide valuable additional knowledge.

1.1a Seismic studies

Information concerning the structure of the deep oceans can be obtained by either reflection or refraction methods. In order to deduce the structure from a seismic reflection shot, it is necessary to know the speed of travel of the primary, P, waves (which are like sound waves) through the layers of rock. The refraction method of seismic survey depends on the refraction of waves from layers of differing density—the denser the layer the greater the speed with which waves pass through it.

The twentieth-century *Challenger* expedition used the refraction technique during the 1950–1952 cruise. Four buoys, spaced 0·5 miles (0·8 km) apart were used. The distance between the explosion and the receiver buoys can be obtained accurately from the time taken for the sound waves to travel from the explosion, detonated near the ship, to the receiver buoys directly through the water. These buoys, to which hydrophones are fixed, are linked by radio to the ship. This method provided fairly detailed information about the upper layers beneath the ocean floor.

In order to distinguish between waves refracted by different layers, it is necessary to place the receiving buoys some distance from the sound source. The waves travelling through the deepest layer then reach the receiving apparatus first, because, although they have travelled further, they have also moved faster. It is possible to estimate the number of layers of material present and to assign to each a speed of travel of the P waves emitted from the explosion. The method is also used to obtain information on earth structure from natural earthquakes. Observations made with artificial explosions have the great advantage, however, that the exact position and time of the source relative to the receiving apparatus is known.

Refraction studies can also be made with two separate ships, one to send out the waves and the other to receive them. This method provides more information of the deeper structure, but less detail of the upper layers. The ships can be further apart because weaker waves can be received by apparatus on a ship than by that on a floating buoy, which is more affected by the surface movement (Gaskell, 1960, pp. 54–73).

A development of the reflection method of seismic work is the continuous profile method in which a hydrophone is towed behind the ship. It receives signals from a sound source that is also towed behind the ship. Low-frequency sound penetrates better in deeper water than the high-frequency sources normally used in shallow water. The continuous profiler has supplied much valuable information concerning shallow structures, such as the sedimentary layers on the continental shelf, but it is also producing valuable results from deeper water. These include surveys across abyssal plains and over abyssal hills. Nearly all the results show multiple layering, which throws some doubt on the common generalizations of oceanic structure. The problem still remains of identifying the materials delimited in this way, and this can only be done with certainty if cores can be obtained. The new reflection instruments can penetrate to several thousand feet in fine detail. This method may supersede the reflection and refraction methods.

The use of seismic surface waves requires different techniques and provides different information. Long natural earthquake waves and nuclear explosion waves can reach hundreds of kilometres in depth and thousands of kilometres in length. Surface wave studies have established that the thickness of the crustal layers is relatively uniform over the earth. One very significant feature confirmed by this method is the fact that the mantle rises to shallow depths in all the major oceans. The average thickness of sediment has been found to be 0·5–1 km. The top of the low-velocity zone within the mantle is shown to be shallower under the oceans. The data are obtained from seismograms on which the long surface waves, which have lengths of 10–2000 km, are recorded.

The results of surface wave analysis suggests that the structure of the three major oceans is similar. Some anomalies occur in the Easter Island Rise and Polynesia. The technicalities of seismic and geophysical work need not be considered, but the results of the seismic surveys have been shown to corroborate the data obtained by other geophysical means in supporting the new global tectonics (reviewed in the next section).

1.1b Gravity surveys

Gravity anomalies can be measured by gravimeters, which give the difference between the observed and computed values of gravity. The anomaly can be stated in three ways, the Bouguer anomaly, the free air anomaly, or the isostatic anomaly. These are given as follows:

$$\text{Bouguer anomaly} = g_{obs} - g_{\phi} + FAC - BC + TC$$
$$\text{Free-air anomaly} = g_{obs} - g_{\phi} + FAC$$
$$\text{Isostatic anomaly} = \text{Bouguer anomaly—computed anomaly of root,}$$

g_{obs} is the observed value of gravity at the point under consideration, g_ϕ is the theoretical gravity on the spheroid at latitude ϕ, FAC is the free air correction, allowing for the variation of gravity with height, BC is the Bouguer correction (which is the attraction of the rock between sea level and the point under consideration, on the assumption that it is uniform), TC is the correction for deviations of the relief from a flat plateau (Bott, 1971). The anomaly is expressed in milligals, where one milligal is 1/1,000,000 part of gravity, which is 987·048 cm/sec/sec at the equator, or 987,048 milligals. An acceleration of gravity of 0·1 cm/sec/sec equals 100 milligals. The extra mass equivalent to about 1 km of surface rock is about $2{\cdot}5 \times 10^5$ grams/cm^2, and this increases gravity by 105 milligals.

Over most of the earth the gravity anomalies are fairly small, showing that the isostatic balance is nearly attained. This is achieved by an excess of density where the mass is reduced, as under the deep oceans. Under high mountains the density is less to compensate for the extra mass.

Vening-Meinesz (1934) pioneered the study of gravity anomalies at sea. He made his observations, using an accurate pendulum, from a submerged submarine to avoid the disturbing effect of surface wave action. Some of his most interesting results were obtained from the marginal parts of the western Pacific, where narrow zones of considerable isostatic anomaly show that the earth's crust is not in adjustment. These long narrow zones of negative anomaly are closely associated with the deep sea troughs. Within the truly oceanic segments of the crust the gravity anomalies are generally positive.

An important development in the field of gravity surveying was the development of a gravimeter that could be used on surface ships. As long as the ship exceeds 1000 tons and the sea state is no more than moderate, continuous recordings of gravity can be made. Variations in the acceleration of gravity are measured at sea level and correction due to water depth can be made if the gravity survey is combined with depth recording. Any remaining variations can be interpreted in terms of differences in density of the rocks beneath the sea bed. The solution is usually not unique and assumptions must be made. These can often be based on seismic data so that the two techniques are complimentary. Modern methods are quick and so provide a useful preliminary indication of unusual structural conditions. Measurements must be made to an accuracy of 5 mgals if possible. The range of values experienced on a ship vary between 10,000 and 100,000 mgals, but the range is less than 5000 at a depth of 76m. Spring-type gravimeters have been developed for sea surface use, and were first used at sea in 1957. Gravimeters can also now be used in the air; in stable conditions aircraft provide a steadier platform than a ship and the operation is more rapid.

Methods of reduction can use either the free-air or Bouguer anomaly. The latter is positive over the deep ocean, owing to the addition of mass. On the whole the Bouguer method is to be preferred for small-scale problems and the free-air method for larger ones. Major conclusions drawn from early work at sea included the fact that the oceans are close to isostatic equilibrium. An asymmetry between the northern and southern hemispheres was shown. The geoid is depressed below the ellipsoid between 50°N and 15°S, but is slightly above between 15°S and 40°S. Details relevant to specific oceanic features are considered in the next chapter.

1.1c Heat flow studies

Heat flow studies give valuable evidence concerning the processes at work beneath the ocean floor. The surprising result of these measurements is that the heat flow under the oceans is almost the same as that under the continents. It would be expected that the heat flow under the continents would be greater, owing to the concentration of radioactive rocks in the continental crust. Heat reaching the floor of the oceans appears to come from beneath the crust, and not from radioactive disintegration.

Heat flow is measured at sea by means of several thermistors attached to a probe that penetrates into the soft sediments to a depth of up to 20 m. Allowance must be made for the thermal conductivity of the sediment. One problem in the comparison of heat flow between oceans and continents is the uncertainty of heat values generated by radioactive decay in different rocks. It is generally agreed that granitic rocks produce about six times as much heat as basic rocks and more than a hundred times as much as ultrabasic rocks. The pattern of heat flow is often highly localized in the oceans, particularly on the ridges. Bott (1971) gives the following values:

Table 2.1 Heat flow observations

Continents	Number of values	Mean m-cal/cm^2/sec	Standard deviation
Pre-Cambrian shields	26	0·92	0·17
Post pre-Cambrian non-orogenic areas	23	1·54	0·38
Palaeozoic orogenic areas	21	1·23	0·40
Mesozoic–Tertiary orogenic areas	19	1·92	0·49
Tertiary volcanic areas	11	2·16	0·46
Ocean basins	273	1·28	0·53
Ocean ridges	338	1·82	1·56
Ocean trenches	21	0·99	0·61
Other oceanic areas	281	1·71	1·05

The normal temperature gradient is about 50°C/km, so for two points 4 m apart the temperatures must be recorded to values better than 0·2°C.

1.1d Magnetic studies

The earth's magnetic field is continually changing. At any one time the dip and orientation of magnetic particles in suitable modern rocks are orientated in sympathy with the magnetic field of the time. Therefore, by studying the orientation and dip of magnetized minerals in rocks of different geological periods, it is possible to arrive at a picture of the magnetic field in past periods. Assuming that the geographical pole, or axis of the earth's rotation, has remained set at an angle of about 11° to the dipole magnetic field, it is possible to obtain information concerning the movement of the pole relative to the crust in earlier geological time. If observations can be obtained from several continents, the relative movement of land-masses and the resulting change in the shape of the ocean basins, can be assessed. These data are of considerable importance in considering the permanence of the oceans, but they give little information on subcrustal structure.

The magnetic field of the earth is related to the flow of electric currents in the core of the earth, and it has been suggested that the cause of the changes in the magnetic field are probably due to convectional movements in the fluid core. Without fluid motions in the core, it appears that the magnetic field would be dissipated in a few thousand years. A further suggestion is that these movements within the earth's core are caused by the rotation of the earth and its movements relative to those of the sun and moon. This would provide a continual source of energy to maintain the magnetic field of the earth.

Cox and Doell (1960) consider that during much of the Tertiary period the earth's magnetic field was much as it is now. The observations for the Mesozoic and early Tertiary could be explained either by relatively rapidly changing magnetic field or by large-scale continental drift. The former explanation may be more plausible. During the Permian and Carboniferous periods the magnetic field was very steady and very different from the present. The pre-Cambrian field was consistent for all continents, but was different from the present.

Recent results of various types of magnetic surveys have provided the most compelling evidence in favour of the new tectonics. Many of the studies are based on palaeomagnetic

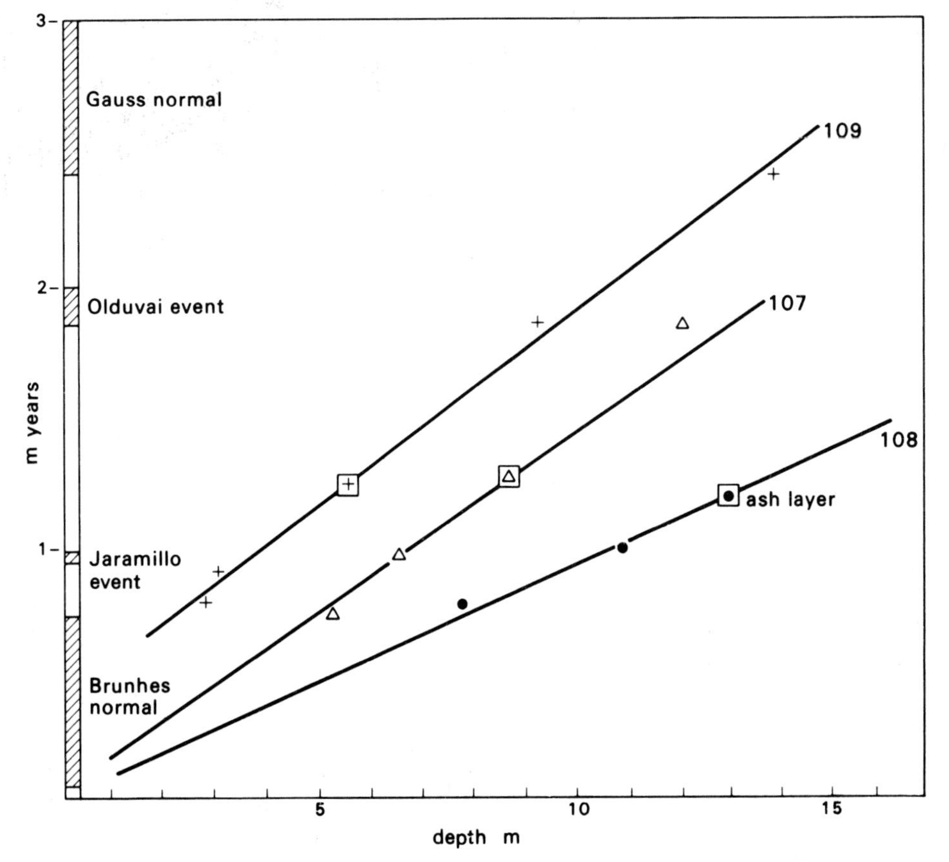

Figure 2.2 The relationship between magnetic reversals and depth of samples from deep sea cores. (*After Bullard, 1968.*)

observations. The pattern of magnetic reversals is particularly important. In order to define the magnetic field, the value of the field strength, and the direction of the field, the true vertical and the direction of true north must be established. Magnetic measurements are now made with gyro-stabilized platforms and the field is measured electronically. Over most of the earth the field is within 20 per cent of that which would be produced by a dipole at the centre of the earth. There is no significant difference between continents and oceans, owing to the origin of the field in the core of the earth. Centres of rapid change, however, tend to avoid the Pacific Ocean. There is considerable variation in the magnetic properties of rocks in the ocean. An analysis of the variations may reveal information concerning temperature relations at depth as rocks become non-magnetic when heated above the Curie point. This varies from rock to rock, but is generally less than 575°C, a temperature which is attained between 20 and 50 km depth.

The reversals of the magnetic field were revealed over 60 years ago by B. Brunhes in 1906, but were ignored for 40 years. In the last decade, however, magnetic reversals have been widely mapped. They provide a valuable dating system as well as strong support for the new global tectonics. Reversals were first found in deep sea sediments in 1964 by Harrison and Funnell in the equatorial Pacific in Radiolarian ooze. The pattern of reversals can be established in cores, and they provide a measure of sedimentation rates as shown in figure 2.2. The reversal is accomplished by a change to zero and then a build-up in the opposite sense over a period of about 1000 years (Bullard, 1968). Bullard's table of the normal and reversed fields is given below (table 2.2). The frequency of reversals in-

Table 2.2 Magnetic reversals

	Normal	Reversed	N/R	Probability
Pre-Cambrian	49	23	2·1	0·0015
Cambrian to Devonian	31	37	0·84	0·27
Carboniferous	24	34	0·70	0·12
Permian	2	50	0·02	3×10^{-3}
Trias to Cretaceous	71	17	4·20	3×10^{-9}
Tertiary	37	42	0·88	0·33
	214	203	1·1	0·31

creased about 25 million years ago. It is revealed in the pattern of reversals found in the areas adjacent to the central oceanic ridge crests. The area 100 to 400 km beyond the continental edge in the Atlantic shows no change in polarity and this could reflect the stable conditions in the Permian as shown in table 2.2. The pattern for the last 3 million years is well established. It shows two normal periods, the Gauss from 3·32 to 2·40 million years, the Matuyama reversed from 2·40 to 0·7 million years, and the Brunhes normal up to the present. Earlier the Gilbert reversed extends to 4·9 (?) million years. There are minor reversals, called events, within the main phases. These include the normal Olduvai and Jaramillo at about 2 and 1 million years ago respectively. Details according to Cox (1969) are shown in figure 2.3. Vine (1966) has shown that most of the remanent magnetism is in layer 2. The dating is based on Potassium–Argon dates of volcanic material.

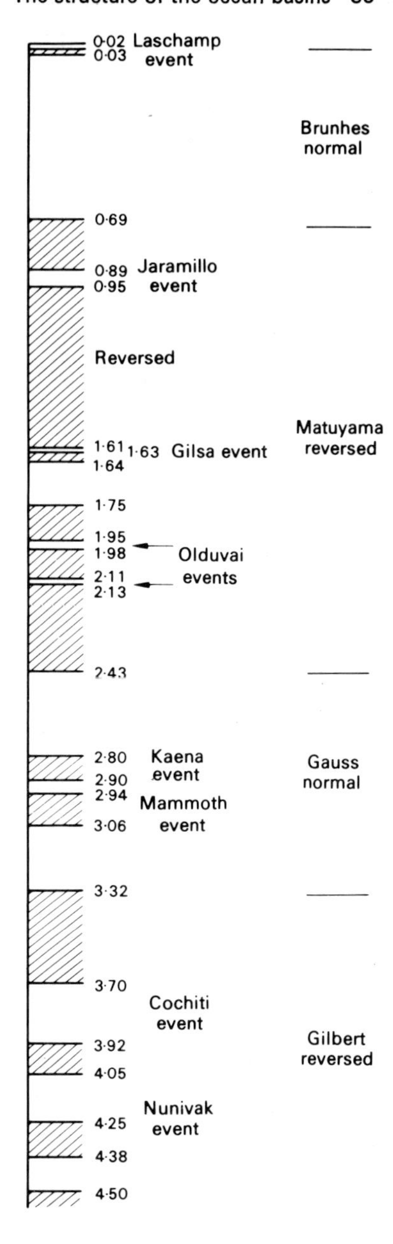

Figure 2.3 Geomagnetic reversals. The reversed periods are shaded and dates are given in millions of years. (*After Cox, 1969.*)

2.2 General structural pattern in the oceans

The crustal rocks have been explored by the methods already mentioned. Most of the basic information on the crust is derived from the first time of arrival of refracted waves. Christensen (1970) has studied the elastic properties of rocks in the laboratory in order to analyse the geophysical results to determine the nature of the crustal material. Four layers have been established. Layer 1 is made of unconsolidated sediments or semi-consolidated sediments. The average thicknesses of the layers are given in table 2.3. Layer 2 is controversial,

Table 2.3 Thickness of crustal layers

	Thickness km	Velocity km/sec	Thickness km	Velocity km/sec
Layer 1	0·45 [a]	2·0		
Layer 2	1·75	5·0	1·71 ± 0·75 [b]	5·07 ± 0·63
Layer 3	4·70	6·71	4·86 ± 1·42	6·69 ± 0·26
Layer 4		8·09		8·13 ± 0·24

[a] (Hill, 1957) [b] (Raitt, 1956)

the recorded velocities could occur in most sedimentary rocks, volcanics, and granites. The range of porosity and variability could produce the range of values recorded. Layer 2 could be consolidated sediments or it could be the upper part of layer 3. As layer 2 often thickens near volcanic islands it may well be igneous rock. Magnetic studies support its basaltic composition. The occurrence of layer 2 is not universal. It is present everywhere in the Pacific, but is only found at $\frac{1}{3}$ of the stations in the Atlantic. This may be due to its masking in the Atlantic by thicker sediments. Different techniques may also account for its apparent absence. Methods most likely to record it have shown it to be often equally present in both oceans. The wide range of velocities recorded in layer 2 support its variability in different areas.

Layer 3 is, by contrast, relatively uniform in terms of wave velocity and thickness, and is probably homogeneous. Layer 3 makes up 2/3 of the volume of the oceanic crust. H. H. Hess (1954) has suggested that layer 3 is serpentinized peridotite. It may originate at the oceanic ridge and spread laterally to form the oceanic crust. The Moho is at the level of the 500°C isotherm along the ridge crest. Away from the ridge the discontinuity could be fossilized at a lower temperature of 150–200°C. The inverse thickness of layers 2 and 3 in the Atlantic suggest that they may be the same material. Dredge hauls indicate the occurrence of serpentinite, but basaltic material also occurs. Tholeiites are the most common type of oceanic basalt. Heat flow relationships are more compatible with the material of layer 2 being basaltic than serpentinite. The most likely composition of layer 3 is amphibolite, which gives the correct velocities. This rock and basalt are the most likely oceanic crustal material in layers 3 and 2 respectively.

Layer 2 probably originates from sea floor eruptions. The amphibolite forms in the temperature range 550–750°C, so both materials must form in zones of high temperature gradient. The magma solidifies above the zone of fusion and is carried away from the ridge crest (where it forms) to make layer 3. Water vapour is needed to produce amphibolite. It could also be called hornblende gabbro, a material that has been obtained from the Carlsberg Ridge. This material could be stable under the continental crust, if it were moved beneath the continents. Along the active margins of the Pacific, melting could occur to form calc-alkaline rocks and eclogite. Such material could include andesite.

Layer 4 makes up the upper part of the mantle below the Moho. Bott (1971) concludes that the Moho is almost certainly not a phase change boundary. He considers it most likely to be a boundary between peridotite below, and its hydratious product, serpentinite, above. He does not consider it likely that it is a basalt/eclogite boundary. Layer 4 below the

Moho is thicker by a factor of 10 than the core and crust combined (J. I. Ewing in Hill (editor), 1963). Its outer surface is at a mean depth of 12 km under the ocean surface and 35–50 km under the continents. It extends about half-way to the centre of the earth. Olivine is an important constituent of the mantle. Waves travel through the upper mantle at a mean speed of 8·1–8·2 km/sec. Some velocities recorded are lower than the modal values, probably indicating a contamination with the overlying material, or where fractional phase change has occurred. The lower velocities tend to be concentrated in the Atlantic, and Raitt (1956) has recorded no velocities below 8·0 km/sec in the Pacific. The low Atlantic velocities were obtained near the Bermuda Rise, and differences in the deep ocean are probably not significant.

The results of geophysical surveys show that the deep ocean structure consists of 4 layers, the sediments in various stages of consolidation being the uppermost. Layer 2 consists of variable material, but seems more likely to consist of igneous material over the greater part of the ocean. It thus more often represents the upper part of layer 3 than the lower part of layer 1. Layer 3 is the most uniform of the crustal layers and is basic or ultrabasic igneous material. It is separated from the mantle beneath by the Moho discontinuity. Layer 4 constitutes the ultrabasic mantle, which extends down to the core at 2900 m depth.

1.2a Atlantic Ocean

Results of gravity and seismic work off the eastern United States (Worzel and Shurbet, 1955) show that the continental boundary ends at the 1830 m depth contour. Towards the ocean the continental crust thins abruptly to merge into the thin oceanic crust in a horizontal distance of about 200 km. M. Ewing, and his co-workers (1950) have made many seismic refraction studies on the ocean margin, particularly in the western Atlantic. Their results indicate a wedge of unconsolidated sediments, thinning offshore and overlying a lens of consolidated sediments, beneath which the continental basement thins out rapidly towards the ocean. The sediments and sedimentary rocks approach 9 km in thickness. The structure is shown in figure 2.4. The sediments fill the marginal trench and are geosynclinal in nature. The feature becomes more mature towards the north. Off Puerto Rico the margin appears young in character. In the Blake Plateau region it is in late youth or early maturity, while off northeast USA the continental margin is in late maturity. It is senile off Newfoundland. These deposits may at some future geological period form a new mountain range.

In the deep Atlantic Ocean true oceanic structure is found, and various sources produce consistent results. The 1950–52 *Challenger* results in the west Atlantic show three layers of material. In the region around Bermuda the sediments were between 500 and 700 m thick. Layer 2 was between 1·7 and 2·6 km thick. The second layer may well be volcanic in type, as the island of Bermuda (which itself contains andesitic material, and is not truly oceanic) is volcanic. The third layer extends down to the Moho at depths between 11·2 and 18·2 km, rather deeper than in the Pacific. There is no evidence of continental type granitic rocks. The western Atlantic, with its genuine oceanic type of structure, is separated from the eastern Atlantic by the mid-Atlantic Ridge, which is a zone of great seismic activity.

The *Challenger* seismic work in the eastern Atlantic gave results in general similar to those

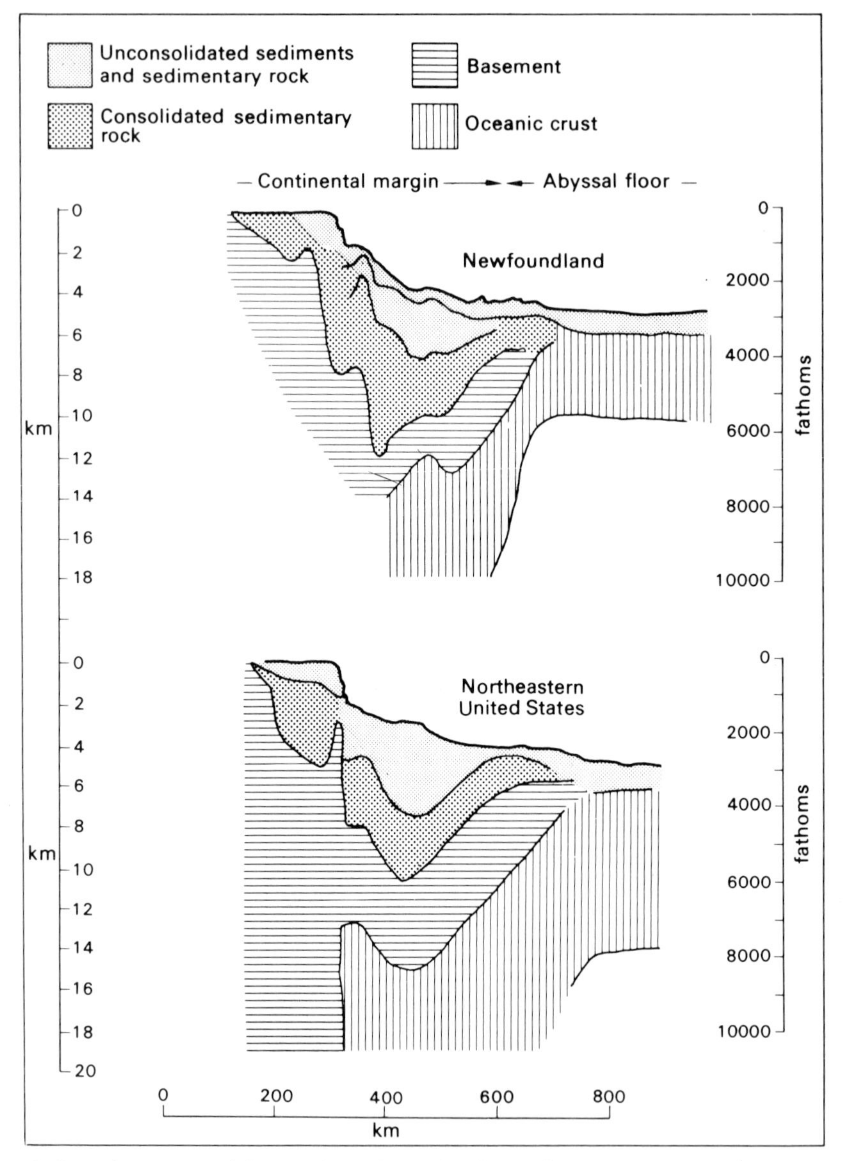

Figure 2.4 Crustal structure of the continental margin off Newfoundland and the northeastern United States. (*After Heezen, Tharp, and Ewing, 1967.*)

of the other deep ocean basins. They showed a layer of unconsolidated sediments overlying a second layer and a basement layer. The most northern stations were situated in rather shallower water on the plateau extending from Greenland to Scotland, south of Iceland. The sediment layers here ranged around 700 m in thickness, being rather thinner than values to the southeast. Beneath this a volcanic layer was identified, which outcrops at the surface in Iceland and Rockall. The second layer was rather thicker than in the western

Atlantic. In the third layer the travel velocity of waves was greater than that in continental rocks. It was assumed, therefore, that the North Atlantic Plateau is not a foundered land-mass.

Observations further south, some of which were near the mid-Atlantic Ridge, suggest no appreciable thickness of granitic type rock. The depth at which the Moho was found in three places was 13·2, 11·4, and 8·8 km, all values more typical of the oceans than the continents. The larger values were located nearer to the continental slope. The results suggest that there is no fundamental difference between the structure of the deep oceans in the western and eastern Atlantic, and that this is similar to the Pacific structure.

Data for the north Atlantic are not so plentiful as those in lower latitudes. Nafe and Drake (1969) summarized the results of geophysical observations in this area. The whole of the north Atlantic is dominated by the mid-Atlantic Ridge. The seismic refraction studies show greater homogeneity within the ridge province than between it and those obtained in the deep ocean areas. In only one instance was a velocity as high as 8 km/sec recorded, values of 7·4–7·7 km/sec being normal for the upper mantle. The results suggest that the ridge is made of material that is less dense than the normal mantle material. The mantle boundary rises towards the mid-Atlantic Ridge from its normal level of 10–13 km below the sea surface, reaching about 7 km near the Azores. Velocities between 7·2 and 7·7 km/sec occur under both the ridge and the continental margins, and near the large sea mounts and the Bermuda Rise. Northwards the Atlantic narrows and the sediment layer becomes thin or absent. A 2–4 km layer with velocities of 5–5·8 km/sec overlies the 7·2–7·7 km/sec layer. Ridge-like velocities have been recorded in the Labrador Sea. Layer 2 has been shown to be double in the Reykanes Ridge area near Iceland.

This area represents one of the best examples of the magnetic anomaly pattern associated with the major ridges. The linear trends lie parallel to the crest of the Reykjanes Ridge, which is part of the mid-Atlantic Ridge system. The pattern of anomalies correlates well with small-scale relief features. Magnetism is mainly associated with a layer about 500 m thick of high magnetization at the surface. A somewhat similar pattern has been found in the Labrador Sea, but it lacks the symmetry about a central ridge of the mid-Atlantic pattern. There is a distinct magnetic boundary in the north Atlantic seaward of the continental slope. The field here is smooth and its possible cause has already been mentioned in the lack of reversals of the Permian period. The interruption of this zone is probably due to transcurrent faulting.

Gravity surveys over the mid-Atlantic Ridge have helped to elucidate the possible structure. Three alternative interpretations of the gravity data all indicate a wedge of less dense material, with a density of between 3·05 to 3·20, consisting of altered mantle material, beneath the ridge. This less dense material also fits in with the seismic results. The data could be explained by a phase change and one model suggests low-density material rising along the axial zone of the ridge.

A summary of the heat flow observations for the Atlantic Ocean give a mean of 1·13 $\pm$ 0·24 m cal/cm^2/sec for 74 values in the basins, and 1·48 $\pm$ 1·48 for 87 ridge values. Thirty-two values within 100 km of the ridge crest gave a mean of 3·00 $\pm$ 2·00. The world average for comparison is 1·5 $\pm$ 1-m cal/cm^2/sec. The heat flow is rather lower over the northern part of the Atlantic (1·15 m-cal/cm^2/sec) than over the equatorial Atlantic (1·50 m-cal/cm^2/sec). The high values of heat flow are confined to a narrow axial zone along the ridge crest in contrast to the East Pacific Rise, where the zone of high heat flow is much wider. The high heat flow zone is even narrower on the Reykanes Ridge southwest of Iceland, where high values only occur within 30 km of the axis. Values on the flanks were lower than normal oceanic values. No observations could be made on the crest owing to the absence of sediments.

Earthquakes are concentrated along the ridge axis, mainly occurring in the transform fault offset zones. Evidence on the whole supports the new geophysical theories, but there are some unresolved problems, such as the presence of material with a wave velocity of 7·2–7·7 km/sec under the ridge and its absence beyond it. There are also areas on the ridge where heat flow is lower than the mean.

Krause (1965) has given details concerning the east-west fracture zones in the north Atlantic near the Azores. The fracture zone east of the Azores is complex and has some of the properties of a median rift and some of a fracture zone. It is seismically active. The whole system stretches from Gibraltar to the northeast USA and is composed of fracture zones, volcanic chains, block-faulting, and rift formation. Displacements also occur at the continental margins. Large-scale tectonics are involved and the features have probably been developing and active from mid-Mesozoic times to the present. The east Azores fracture zone is now active, but the west one has probably been quiescent since Cretaceous or early Tertiary times.

Further north the mid-oceanic ridge has been the dominant feature in the formation of the Norwegian and Greenland sea (G. L. Johnson and Heezen, 1967). It is cut by west-northwest to east-southeast fracture zones, which originated early in the history of the basin. Between Iceland (and the area north of it) and Greenland the ridge forms a sediment dam and the western flank has been buried under the sediment of the Scorseby abyssal plain, which is 1640 m deep. The Jan Mayen Ridge lies along the eastern border of the main ridge. Strong magnetic anomalies are associated with the main ridge. The rugged crestal zone of the Mohns Ridge and the mid-oceanic ridge north of Iceland were apparently created in the last 70 million years. The whole sea must be about $1–2 \times 10^8$ years old, by extrapolation. Sediment is virtually absent in the rugged crest zone. The Mohns Ridge lies nearly equidistant between the parallel portions of the Greenland and Norwegian continental slopes, while the section between the Jan Mayen fracture zone and Iceland is very asymmetrical in form and position, and this property seems old. While the Norwegian Sea grew to a maximum of 800 km, the Atlantic expanded to 4800 km. The mid-oceanic ridge increases in width from 20 to 50 miles (32–80 km) from Iceland to the Jan Mayen fracture zone. The pattern is opposite to the pie-shaped growth of the Icelandic Plateau, where the greater growth was to the south. This pattern suggests that the relative movements of Greenland and Europe have radically changed during the latest geological periods. The point of pivoting of Greenland, which probably lay well north of Greenland earlier, may have moved south in very recent geological time, as shown in figure 2.5.

1.2b Indian Ocean

Most of the observations made by the *Challenger* survey in the Indian Ocean, with one exception, were made in the eastern half of the ocean. The only observation near the western side was made in the Seychelles. These islands contain granite, which supports

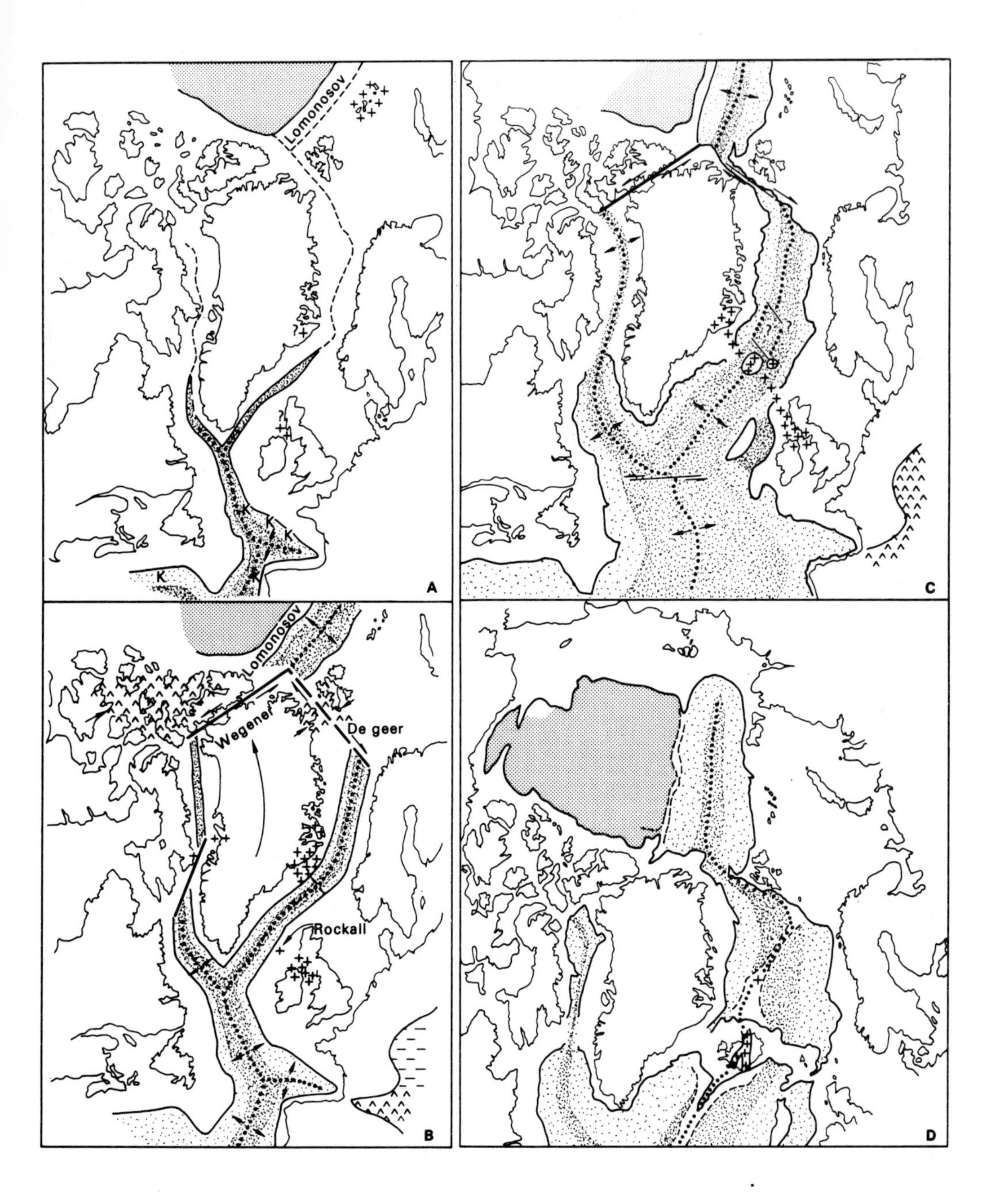

Figure 2.5 Stages in the growth of the Norwegian Sea **A:** Minimum opening of the north Atlantic in late Cretaceous time. K = Upper Cretaceous ocean sediment. Volcanoes marked by crosses. **B:** Early Tertiary compression in Spitsbergen due to northward movement of Greenland and general opening of the north Atlantic and Arctic Oceans. Active transcurrent faulting and extensive volcanic activity is taking place. **C:** The mid-Tertiary pattern, showing the position of proto-Iceland. **D:** The later Tertiary and Quaternary movements and the pattern of the mid-oceanic ridges. (*After Harland, 1967.*)

the view that continental structures may be present in the western Indian Ocean. Seismic studies in the eastern Indian Ocean, just south of the equator, showed that the thickness of sediment varied between 400 m and 1600 m. Beneath the sediment at two stations a thick layer was identified, nearly 3 km thick in one place. Waves travelled through this layer faster than in the equivalent layer of the Pacific. It probably consisted of basic volcanic material or lithified sediments. The great thickness may have been due to the proximity of a sea mount. The deeper layer, at a point nearer the equator and further west, appeared to be a rock which might be granite under considerable pressure, as waves passed through it only slightly quicker than through the Seychelles granite. Elsewhere layer 3 was similar to the same layer in the Atlantic and Pacific.

The Indian Ocean is of particular interest from the structural point of view as it is one of the three places in the world where mid-oceanic ridges meet a continental land-mass. The Carlsberg Ridge runs up through the Gulf of Aden into the Red Sea in the northwest Indian Ocean. The Carlsberg Ridge is very similar to the mid-Atlantic Ridge in its relief and magnetic pattern. In the part surveyed in detail by Laughton and Matthews (1965), a displacement of the median valley by a right-lateral tear fault with a 16 km displacement was found. Hydrothermally altered and brecciated basalts and dynamically metamorphosed gabbros were dredged from the fault scarp. The structure changes abruptly 320 km southeast of Socotra. Northwest of this there is no median valley and no earthquakes. The mountains are larger and non-magnetic. Where the Carlsberg Ridge changes structure it is crossed by the 2400 km long Owen Fracture zone, which extends from the continental shelf off Karachi to the middle of the Somali Basin. An asymmetrical ridge along this line is flanked by a sediment-filled trough. It is similar to the fracture zones in the Pacific. Matthews has examined the Owen Fracture zone and concluded that its underlying structure is a system of parallel transcurrent faults, affecting only the ocean floor. The mid-oceanic ridge suffers a net right lateral displacement of 170 km. The whole Owen system of displacement probably occurred in the late Tertiary, as fresh basalts of Tertiary age occur. The Murray Ridge is also a strike-slip zone continuous with the Owen Fracture zone. The Carlsberg Ridge can be followed by its rough relief, magnetic signature, and earthquake belt down the axis of the Gulf of Aden. It then runs, as a single valley, into the Gulf of Tadjura. Here it is offset by several left-lateral displacements, which show in the continental margin.

The evidence suggests that this is the site of a newly forming ocean basin, which has been developing since the Miocene. The feature is associated with the African rift-valleys. The heat flow values are higher in the Gulf of Aden than elsewhere in the Indian Ocean. The mean value is 3·89 m-cal/cm^2/sec, with maximum values of 6·15 and 5·98. Seismic refraction studies showed a layer of 7·7 km/sec material below the crustal layer 3 of 6·4–6·5 km/sec material, which was 6 km thick, with its base at 10 km below sea level. Layer 2 above it has a velocity of 3·94–4·60 km/sec, and layer 1, which was 1·5–0·5 km thick, had a velocity of 1·85 km/sec. There are transform faults across the axis of separation. If separation took place at a uniform rate since the early Miocene, 20 million years ago, the rate of spreading has been 2 cm/year. The total displacement between Hadramaut and Africa is 400 km. Support for spreading is also found in the Carlsberg Ridge at 5°21′N. Magnetic survey revealed parallel strips of magnetic anomalies. The Seychelles Bank showed continental structure, with a crustal thickness of 30 ± 3 km of continental rocks extending to the edge of the bank. The Moho slopes upwards at 19° along the northern edge of the bank. Between the Seychelles Bank and Lamu in Kenya the crust become very thin and layer 2 is missing. Layer 3 is only 6–9 km below the surface, being only 3 km thick. On the edge of the bank the structure is typically oceanic, with layers as follows: water 4·38 km (1·54 km/sec); layer 1—0·30 km (1·78 km/sec); layer 2—1·72 km (4·86 km/sec); layer 3—4·70 km (6·86 km/sec); and layer 4 had a velocity of 8·14 km/sec.

On the whole the heat flow on the Indian mid-oceanic ridge is the same as that on the south Atlantic Ridge, with a mean value of 1·35 m-cal/cm²/sec. There are low values on the flanks of the ridge. The values are rather lower between the Seychelles and Africa, with a mean of 1·17 m-cal/cm²/sec. The Indian Ocean thus has a well-developed ridge system that diverges in the centre.

During the recent International Indian Ocean Expedition the ocean has been studied intensively. An unusual feature is the scattered linear micro-continents, which are mostly meridional in position. There are also aseismic ridges in the ocean, of which the straight Ninetyeast Ridge is the most notable. It runs from 15°N to 35°S in a nearly straight north–south line. Another ridge of this type is the Laccadive–Maldive Ridge, which is studded with islands. Madagascar and its associated ridge extends over 2400 km. The Kerguelen Plateau, which is broad and flat-topped, and the plateau near the Crozet Islands lie in the southern part of the ocean. These large block features must have crustal thicknesses of continental proportions, a structure which has been confirmed for some of them by seismic measurements. These aseismic zones may be either small relics of former continents, or nuclei of growing continents. The Seychelles and Madagascar seem to be of great age. They are probably fragments of continental crust that have become distributed in a meridional pattern. They are similar to the Walfisch Ridge in the Atlantic, which lies parallel to them.

The mid-oceanic ridge of the Indian Ocean is forked as it winds through the aseismic plateaux. Movement may have taken place parallel to the long Ninetyeast Ridge and the Owen Fracture zone, both of which indicate strike-slip movement. In the equatorial Indian Ocean the mid-oceanic ridge is parallel to the aseismic ridges, but their trends diverge to north and south.

The relief and structure of the Indian Ocean are not compatible with a simple spreading movement as the micro-continents should have been carried away to the edge of the ocean. There is doubt concerning the part of the mid-ocean ridge that links the southern mid-Atlantic Ridge to the mid-Indian Ocean Ridge. Where the three ridges join appears to be the meeting point of three distinct crustal plates. The structural pattern of the Indian Ocean is considerably more complex than that of the Atlantic Ocean, and parts of it are less well known.

1.2c The Pacific Ocean

The Pacific Ocean is unique amongst the major structural features of the earth. It covers about half the globe and is the largest single unit of truly oceanic structure. The boundary of the deep ocean type structure is demarcated by the Andesite Line (shown on figure 4.4); inside this line the volcanic material is entirely basic in character. The Andesite Line, so called after the volcanic rock of this name, runs parallel and close to the west coast of South and North America. It then swings westwards, following the ocean side of the arcuate string of the Aleutian Islands west of Alaska, and turns abruptly southwest along the ocean side of the peninsula of Kamchatka and the string of the Kurile Islands to skirt the main islands of Japan as far as Tokyo. From there it turns south, running on the ocean side of the Mariana Islands to reach a position north of New Guinea, where it turns east-southeast towards the Fiji Islands, which lie on the continent side, although the Ellice Islands are in the true ocean. It then turns south between the Tonga and Cook Islands, the latter being inside it, to leave New Zealand outside. Its position in the south Pacific is not clearly determined. The many islands which are found within the zone are all basaltic volcanoes, of which the Hawaiian Islands are the largest. Some of the volcanic islands are crowned with or buried beneath coral, to form atolls.

Zones of deep focus earthquakes are associated with the edge of the Pacific Ocean. They indicate that this is a fundamental structural division of great significance. Normal earthquakes originate within 35 km of the surface; intermediate ones occur at depths between 35 and 250 km, while deep focus ones originate at depths between 250 and 700 km. The earth's crust and mantle must, therefore, have strength at least to this depth. The deepest earthquake detected occurred at 720 km (Bott, 1971).

Deep focus earthquakes at present are limited to two zones, both of which border the Pacific, one in eastern Asia (see figure 4.6) and other in South America. They increase in depth along a plane sloping down under the continents. It has been suggested (Umbgrove, 1947; Stille, 1955) that these steeply dipping zones of earthquake activity are associated with deep-seated shear zones at the boundary of the true oceanic and the continental structures. A zone of fold mountains on the eastern side of the Pacific is associated with this structural border. Island arcs and deep sea trenches border much of the western Pacific.

The unity of the Pacific basin has been mentioned by Cotton (1958). He compares the character of the Pacific rim at opposite sides of the ocean in New Zealand and California, 11,200 km apart. Similar large transcurrent or tear faults occur in both areas. The San Andreas Rift in California has caused many earthquakes, and along this zone of dislocation the movement has been in general to the north of the ocean side of the fault relative to the continental side. Both in California and in New Zealand there has been much recent faulting and upheaval of the land bordering the ocean margin. In New Zealand there is also a major fault system, the Great Alpine Fault, which is partly transcurrent and partly vertical in its movement. The vertical movement has been very considerable even during the last million years, while a lateral movement of about 500 km, acting over a much longer period, has been suggested. The ocean side has moved south relative to the western side.

The rim of the Pacific, therefore, shows a unity of structure, which can be associated with the general character of its unified structure. The direction of the major transcurrent movements give some support to the tentative idea that the body of the Pacific may be rotating anti-clockwise in relation to the lands around it.

Within the true oceanic Pacific, inside the Andesite Line, the seismic surveys show a fairly uniform type of structure beneath the deeper parts of the ocean. Nine stations all showed a basement layer at depths varying between 0·7 and 2·7 km below the sea floor. At some stations the thickness of sediment appeared to be very small, amounting to only 100 m, which is less than that found in the Atlantic. Volcanic outpourings were found on some sections, while in other are as there was evidence for the presence of consolidated sediment. At two stations outside the Andesite Line near the Mariana Trench, the Moho was found at the unusually shallow depth of 4 km below the ocean floor. Two stations in the west Pacific outside the Andesite Line showed an intermediate type of structure, in which a layer of less dense material was located. These stations were on banks, and this material (which underlies a relatively great thickness of semi-consolidated sediment) may consist of sedimentary rocks. It may in fact represent foundered continental type structure. The study of surface Rayleigh waves, however, does not indicate that there are any continental rocks in the Philippine Sea, west of the Andesite Line. Stations near the American coast also show indications of the presence of continental rocks at depth, although the sea in the area is not so deep.

One difference between the Pacific, and the Atlantic and Indian Oceans is the asymmetry of the position of the major Pacific seismic ridge. The East Pacific Rise lies far to the southeast of the median line and continues northwards to run into the land through the Gulf of California. Another seismic ridge diverges from it to run southeast towards south Chile. Linear strings of islands, consisting of volcanoes, are associated with the ridges. The volcanoes get progressively older away from the ridge. They include the Hawaiian group, the Revilla Gigedo Islands, the Marquesas, the Tuamotu–Gambier group, the Society Islands, Austral Islands on the west of the ridge, and the Galapagos and Easter Island group on the east of the ridge. Ten out of 11 groups increase in age away from the ridge, and they are strung out at an angle almost normal to the ridge alignment. The only exception is the Samoa group, which increases in age from the central zone to both east and west. They lie on the northern side of the great strike-slip fault zone that runs through New Zealand.

Marginal faults are common in parts of the Pacific, for example along the southern part of South America and virtually along the whole length of North America. Present evidence does not support a general anti-clockwise circulation of the central Pacific, as suggested by Cotton. Movement out from the East Pacific Rise is more likely.

The northwest Pacific is probably older than the southeastern part. The northwest Pacific and the North American part of the Arctic Ocean are probably the only parts of the ocean older than Cretaceous or possible Jurassic. No ridge separates the old and new parts of the Pacific, as in the Arctic Ocean. This may be due to the split in the Pacific, from which it widened, crossing the original ocean. In the Arctic Ocean the split separated the continental shelf from its parent continent.

Hayes and Pitman (1970) have prepared a map (figure 2.6) illustrating the age of the basement rocks and, therefore, the maximum age of the overlying sediment of the northern Pacific Ocean. The pattern suggests that the tectonic development of this area was dominated by a Y-shaped system of migrating ridge axes until the Pleistocene, when the ridge further east started to open up. The ridges and associated trenches were affected by periods of alternating activity and inactivity. A new system of magnetic lineations, trending northwest and bounded partly by the Shatsky Rise, the Emperor Seamounts, and the Hawaiian Arch, may have formed by sea floor spreading before the late Cretaceous.

Menard (1971) has discussed the late Cenozoic history of the Pacific and Indian Oceans on the basis of global tectonics. He estimates the area of new oceanic crust created in the last 10 million years to be $2{\cdot}6 \times 10^7$ km^2, or 9 per cent of the ocean area. This has been a period of intense crustal formation, which if continued would renew the whole ocean crust in 110 million years, or about half the estimated time during which the existing crust

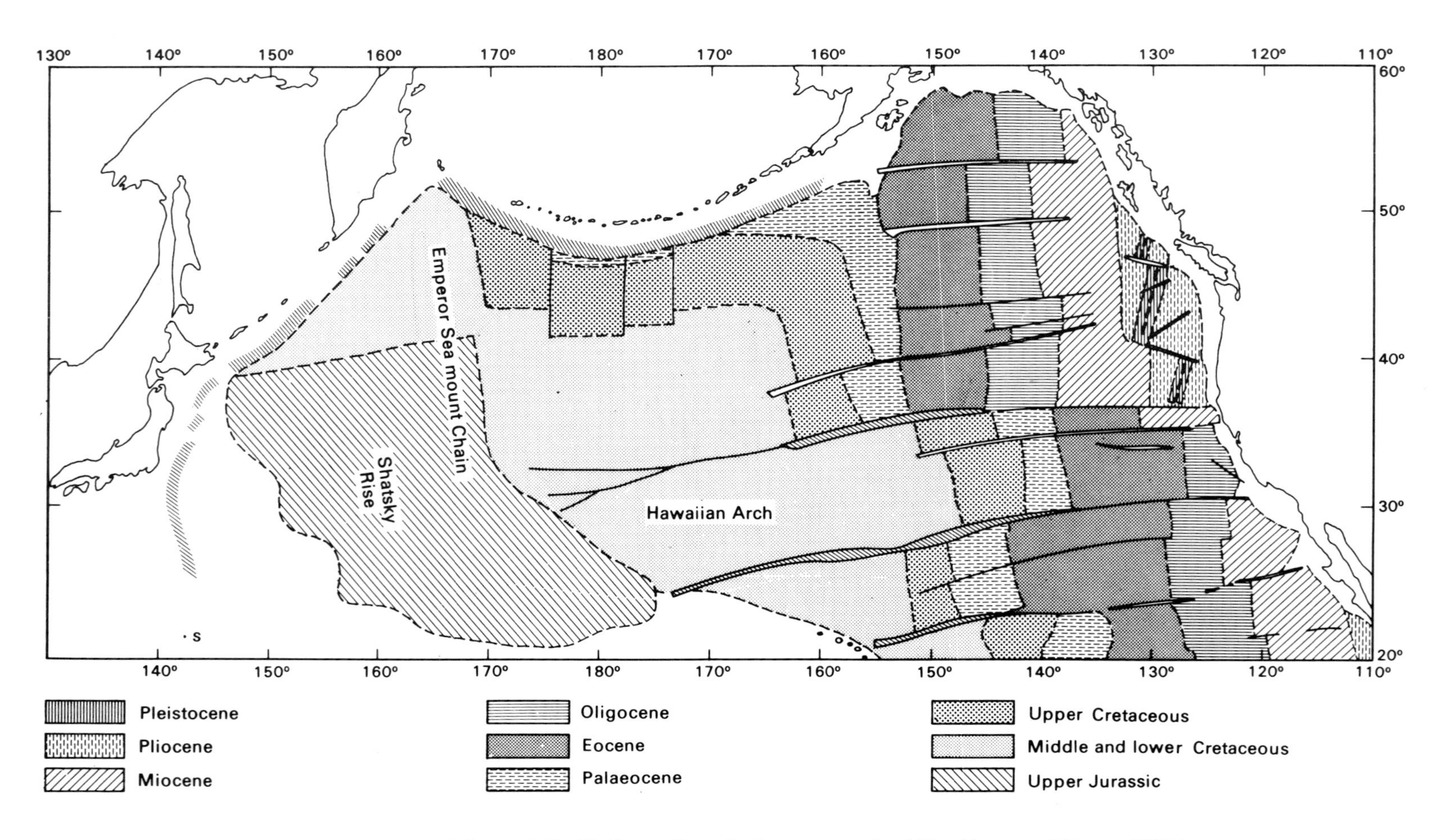

Figure 2.6 The age of the north Pacific Ocean floor, the basement rocks. (*After Hayes and Pitman, 1970.*)

formed. The Pacific is the principal locus for new crust formation, with very fast spreading taking place in the southeast and between Fiji and the New Hebrides Islands. The oceanic crust is elevated in this area and heat flow is high. At the spreading centres there is about 1 km of volcanic rock, 2–5 km of oceanic crust, and a few kilometres of rigid, cooling lithosphere. The oceanic crust eventually becomes about 5 km thick and the lithosphere between 50 and 100 km. The oceanic layer has remarkably uniform velocities of $6{\cdot}81 \pm 0{\cdot}16$ km/sec.

The area of crust destroyed can be estimated from the dip and area of the Benioff zone. This area is $2{\cdot}9 \times 10^7$ km^2, only slightly larger than the area of new crust. The relative areas suggest that the island arcs of Indonesia and New Britain and the mountains of New Guinea are migrating northeast, and part of the extension of the Indian ocean is balanced by the convergence in the north Pacific. A complex history of changes in direction of plate migration is indicated in the edge phenomena of the Pacific and Indian Oceans.

The composition of destroyed material differs in type according to latitude, being carbonate-rich near the equator and more siliceous in higher latitudes. The pattern of crustal movement is such that the area in high latitudes has increased. During the late Cenozoic, vulcanism also increased, while the volume and depth of the ocean basins could have increased significantly during late Tertiary time, when the continents arose. All these factors could have played an important part in initiating glaciation.

The East Pacific Rise differs from the other mid-oceanic ridges in the Atlantic and Indian oceans. It appears to be younger, has a normal oceanic structure, and no median rift-valley, although the heat flow is high on both sides of the crest. The seismic velocity below the Moho is 7·5 km/sec. It may be that the ridge is still normal sea floor that has yet to split. The crust has started to rise, but has not yet gone beyond this stage. There are still sediments on the ridge, and Hess (1965) suggests that it may be less than 1 million years old. This interpretation does not, however, agree with some of the findings of the pattern of magnetic reversals, which are well developed on either side of the ridge in some sections. Spreading may have started from the two zones of high heat flow on either side of the ridge.

The Darwin Rise in the western Pacific, on the other hand, is a much older feature. Hess suggests that this ridge once extended northwestwards to the present coast of Asia. Southeast of the East Pacific Rise it may still be present as the South Chile Rise. The loss of the central portion of the rise may be due to westward spreading of the Pacific floor. Atolls formed by subsidence provide evidence that the Darwin Rise has been subsiding. On the whole the subsidence increases towards the northwest, as the material moves down the crest of the ridge towards the ocean margin.

One of the other features of interest in the Pacific is the widespread distribution of volcanic islands. The Hawaiian Islands are formed of lava derived from a depth of at least 50 km. The source of lava seems to be relatively fixed to account for the plume of islands stretching away in age from the constant point of eruption. The source may lie in a stable core of a convection cell, across the top of which the more mobile upper crust moves in the upper part of the cell.

In the northeast part of the Pacific occur the largest series of transcurrent faults in the ocean floor. These fracture zones run east–west and form a series of at least 8 major faults. Their lateral movement can be fairly accurately established because of the pattern of displacement of the magnetic anomalies that run at right angles across them. These great fractures lie on the western side of the marginal spreading zone along the western side of North America. They link with the San Andreas and north–south transcurrent rifts. The faults were first discovered by bathymetric data, and show by the offset of the deep water contours, such as the 2500 fathom (4580 m) line. This contour is offset along the Mendocino Fault by the same amount as the magnetic anomalies. The movement is 1185 km. The Murray Fracture zone does not extend east of 125°W. It is covered by undeformed sediment, which must post-date the formation of the fracture. Evidence suggests a date of pre-late Miocene or possibly pre-late Cretaceous. This fracture ceased to be active before the San Andreas rift, which is still active. The two systems do not seem to be closely related, although the fracture may exert some control on the structure of the Transverse Ranges and the San Andreas system. The Murray Fracture could be either a transform or a transcurrent fault. If the fault is transform it must precede the San Andreas and Transverse Ranges. There are a number of features that would be difficult to explain on this hypothesis (von Huene, 1969).

The combined displacement of the Mendocino and Pioneer faults is left lateral of 1400 km magnitude. It can be traced over a distance of 2200 km from 140 to 165°W. The latitudinal extent of the pattern is from 28°N to 41°N, covering 1450 km. Some of the faults can be traced onto the land. The Mendocino Fault, for example, can be traced into Nevada to 115°W. The movement probably took place deep beneath the continental crust inland, as magnetic anomalies do not form a recognizable pattern. The pattern of faulting suggests that the upper crust must be moving above a laterally mobile upper mantle, which is active to a considerable depth. The consistency and length of the magnetic anomaly strips in the Pacific suggest that they may have been generated by a different process from those on either side of the mid-Atlantic Ridge.

Orowan (1965) suggests that the east-west faults that displace the mid-Atlantic Ridge in equatorial latitudes could be due to local acceleration in the creep rate. Both in this area and in the northeast Pacific the ridges run north–south and the fractures east–west. The fractures should lie along the maximum shear-stress direction. Where upwelling increases, it tends to wedge apart adjacent crustal strips and shear-fractures form. These will be east–west where the extension is taking place in an east–west direction also.

1.2d Arctic Ocean

The Arctic Ocean was tectonically active in the early Tertiary, with the mid-Atlantic Ridge spreading up between Scandinavia and Greenland, while further movement took place on the west side of Greenland. A major transcurrent fault developed between northwest Greenland and Ellesmere Island. The aseismic Lomonosov Ridge crosses the entire ocean. The structures developed further during the mid-Tertiary period and remained active into the later Tertiary and Quaternary, although transcurrent faulting was not active in this phase. Volcanic activity in Iceland and Beerenberg on Jan Mayen continues along the axis of the ridge, which passes between Svalbard and northeast Greenland. The Arctic can be divided into the Palaeoarctic and the Neoarctic by the Lomonosov Ridge. The latter part of the Arctic Ocean has probably had an evolution broadly similar to that of the Atlantic Ocean. The Palaeoarctic basin existed at least through phanerozoic time, although its shape must have been changed by the Caledonian compression and the

Svalbardian transcurrent movements. Folding of the Franklinian geosyncline could be mid-Devonian. This geosyncline borders the Palaeoarctic ocean basin. The northward movement of Greenland and Ellesmere Island changed the shape of the geosyncline to give it an arcuate form. Thus the oldest extant parts of the ocean basins occur in the western Arctic and the northwestern Pacific oceans (Harland, 1969). The age of the western Arctic Ocean is shown by the Franklinian geosyncline, which contains material from Pennsylvanian to Cretaceous. This material was probably derived from the Appalachians before the Atlantic Ocean formed, so that rivers draining from these mountains reached the Arctic Ocean. This part of the Arctic is separated from the eastern part by the Lomonosov Ridge. Evidence for the youth of the other oceans will be explored in the brief review of the new tectonics considered in terms of process. The older ideas of permanence are, however, still worthy of some comment.

2 The pattern of land and sea

There are various possibilities to be considered. The continents and oceans could have remained more or less in their present positions throughout geological time. Alternatively, the continents could have started as two units, originally under the oceans, and subsequently appearing above the sea to form the land-masses. This is the basis of Hills (1947) theory of the formation of the continents by convection. Another idea is the view of Wegener (1922) and others that the continents originated as one large unit which subsequently split up and drifted apart to form the present pattern. Various patterns of polar wandering have also been suggested to account for different palaeogeographical facts. The latest views consider that both oceanic and continental crust is mobile.

2.1 Theories of permanence

The distribution of marine sedimentary rocks on the land show that much of the area which is now dry land has at one time been beneath at least a shallow sea. The present shorelines are very far from being permanent. On the other hand, there is much evidence that the deep oceans have never been dry land, nor the continents true deep oceans. Since the oldest rocks now visible on the continents were formed, none of them have escaped severe disturbance by earth movements. Large areas, now forming the Shields, have however been relatively stable since the beginning of the Palaeozoic period, about 600 million years ago.

The original state of the earth, as pointed out by Lees (1953), must have been either a hot liquid, or an aggregation of cold solid particles subsequently compacted. If the second hypothesis is true, which seems likely, the earth must have gained heat by adiabatic compression and impact heating. Those in favour of the permanence of the oceans point out that their pattern, and the arrangement of the continents, is such that most of the land-masses are grouped in the northern hemisphere, situated around a polar ocean, while in the southern hemisphere the most continuous stretch of ocean occurs in high southern latitudes, around a south polar continent. This more or less antipodal arrangement led Vening-Meinesz (1934) to suggest that it could be due to the action of a series of convection currents. He supposed that these currents extended down to the limit of the earth's mantle at a depth of 2900 km. The probable arrangement of the currents would lead to eight cells, four rising and four sinking. Each cell would occupy one octant of the globe, as shown in figure 2.7. The light material would tend to accumulate at those places where the currents sink, eventually consolidating to form the continents. On the assumption that two cells are arranged along the axis of rotation, the sinking cells would be situated symmetrically around the north pole, and at the opposite end of the axis, round the south pole. The rising cells, now the oceans, would have been situated at the north pole

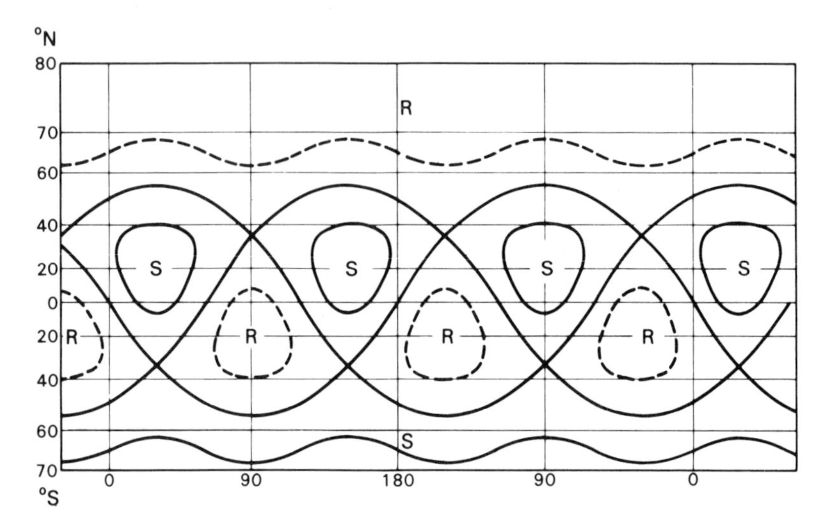

Figure 2.7 Theoretical arrangement of convection cells. R indicates rising currents and S sinking currents.

and symmetrically around the southern continent. In this way the basic arrangement of land and sea could be accounted for. More problems, however, are raised than solved by these speculations.

While the very different structure of the oceans and continents does point to their permanent role on the earth, this does not necessarily imply that their positions relative to one another must have been static all the time.

2.2 Theories of drifting continents and changing oceans

The original theory of Wegener, in which he proposed the idea of the drifting continents, was put forward to explain the distribution of the deposits of the Permo-Carboniferous Ice Age, which interested him as a meteorologist. He assumed that all the continents were grouped together as one unit, which he called Pangaea. This land-mass was centred around the south pole, then situated off South Africa. This pattern explains the distribution of glacial tillites in South Africa, India, South America, and Australia.

The climatic evidence is in some ways the most difficult to explain by other hypotheses, particularly the distribution of the glacial deposits already mentioned. Ice ages are rare phenomena in the history of the earth. One occurred in late pre-Cambrian time, but only two have occurred in the period since the Cambrian—in the Permo-Carboniferous and the Pleistocene. They exert a powerful influence on the world climate and oceanography.

The similarity of structures on either side of the Atlantic Ocean is striking. The similarity of the shape of the coastlines is also impressive. Bullard (1969) has fitted the shelf edges on either side of the Atlantic by computer and the best-fit connection shows a very high degree of correspondence (figure 2.8). His results suggest independent movement of North and South America relative to Europe and Africa at an early stage in the splitting process.

Perhaps the strongest evidence put forward in favour of the theory of continental drift is derived from palaeomagnetic studies. These have already been mentioned and it has been suggested that an abrupt change in the magnetic field took place between the steady state of the Permo-Carboniferous period and that of the Tertiary. This change could be interpreted as the result of continental drift, although this is not the only possibility. Collinson and Runcorn (1960) have assembled palaeomagnetic evidence which they feel shows some evidence in favour of drifting. They show that the position of the pole relative to America has been moving about in the Pacific. In the late pre-Cambrian it was situated in the tropical central Pacific; it then moved across the tropical Pacific westwards towards Asia in the late Palaeozoic and early Mesozoic, and thence across the north Pacific to its present position. This route is displaced relative to that worked out from observations in Europe. The displacement of America relative to Europe has been about 30° of longitude since the beginning of the Mesozoic.

One difficulty about the continental drift theory, as proposed by Wegener, is the fact that it must have taken place fairly suddenly during the Mesozoic and subsequently, while no similar movement took place during earlier geological time.

Other theories, which do not strictly come into the permanency category include that of Hills (1947). His theory is thought to be possible by Jeffreys (1950). Hills suggests that the continents were formed at an early stage in the history of an originally molten earth. Hot material was thought to be carried towards the poles by convection currents, where it would tend to concentrate as floating crystals of lighter material. As the currents slowed down due to increasing viscosity, the light material would tend to drift away from the poles towards the equator. Two continents of about equal size, called Laurasia and Gondwanaland, would result. Hills considered that the Atlantic was a rift ocean formed by the splitting of Laurasia.

Dietz (1961) initiated the idea of the new global tectonics. He suggested that many features of the ocean floor and its borderland could be explained by the action of convection currents, rising under the centre of the oceans and sinking under the continental edge. This process would allow continental drift as the continental blocks were dragged by the currents on the moving mantle in which the currents were operating. He has suggested that the ocean floor could be the outcropping mantle, moving slowly at a few centimetres a year towards the margin, leaving ridges in the centres of the oceans and adding to the continents at their edges. If the continental blocks move along with the ocean floor, a stable Atlantic type margin is formed, but if the ocean floor slips under the continental block, a Pacific type margin, with deep trenches, island arcs, and deep focus earthquakes, is formed.

3 Major structural processes: the new global tectonics

The discussion of the major structural characteristics of the oceans in section 1 hinted at the processes that may have formed the features. Further evidence for these processes will

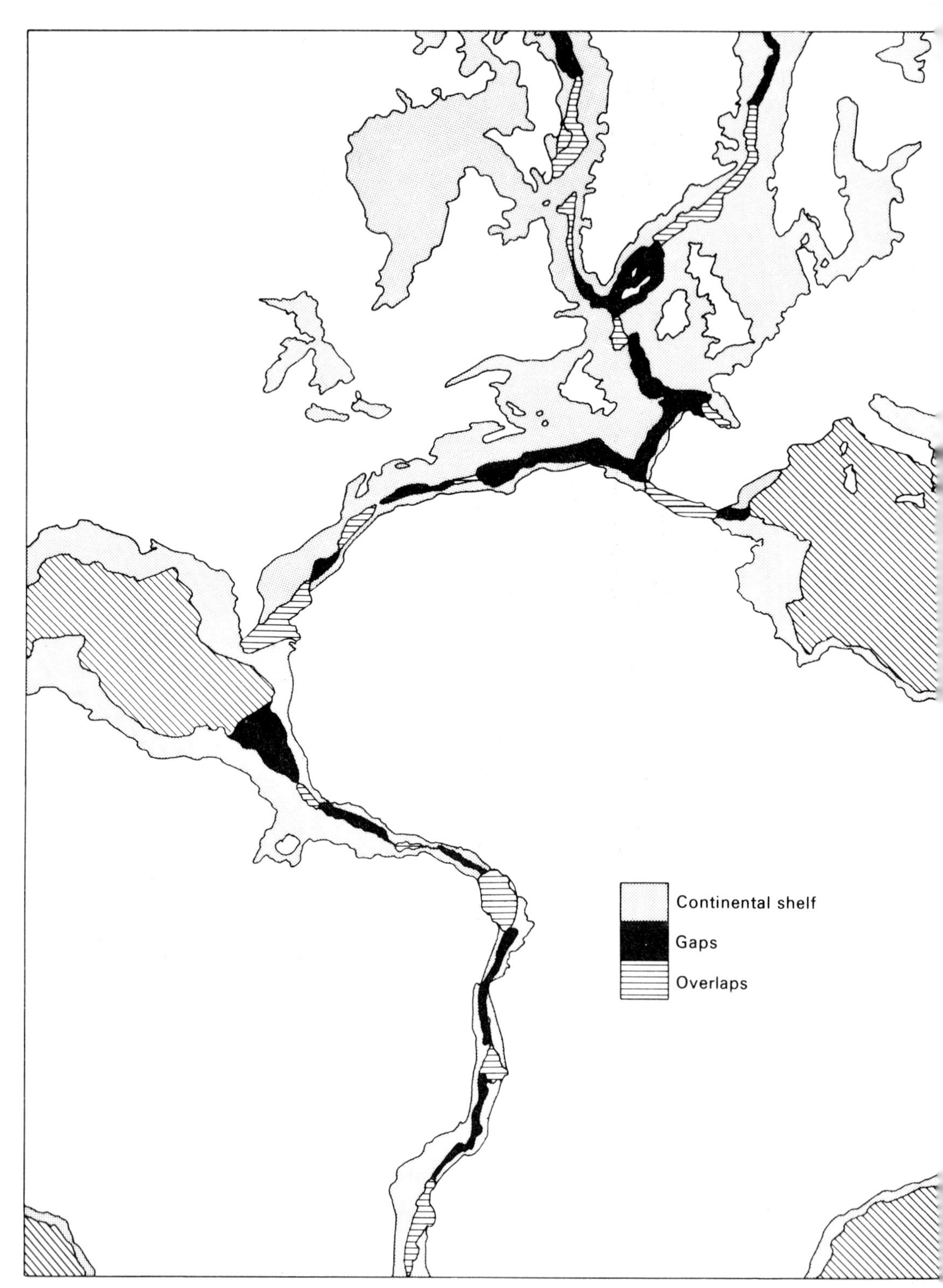

Figure 2.8 Computer aided analysis of the probable fit of the continents before rifting. The dotted area is the continental shelf, at the margin of which the fit was made. The solid black indicates gaps, and the dashed areas are overlaps. (*After Bullard, 1969.*)

be given in the next chapter, in which the different morphological features of the oceans are considered in rather more detail. In this section interest is focused on the operation of the processes as a world-wide system. The clues on which the theory has been built are derived from the geophysical observations that have been accumulating with increasing speed. Magnetic surveys, which revealed the offsetting of strips of alternately magnetized material about the crests of the oceanic ridge system, have been particularly important.

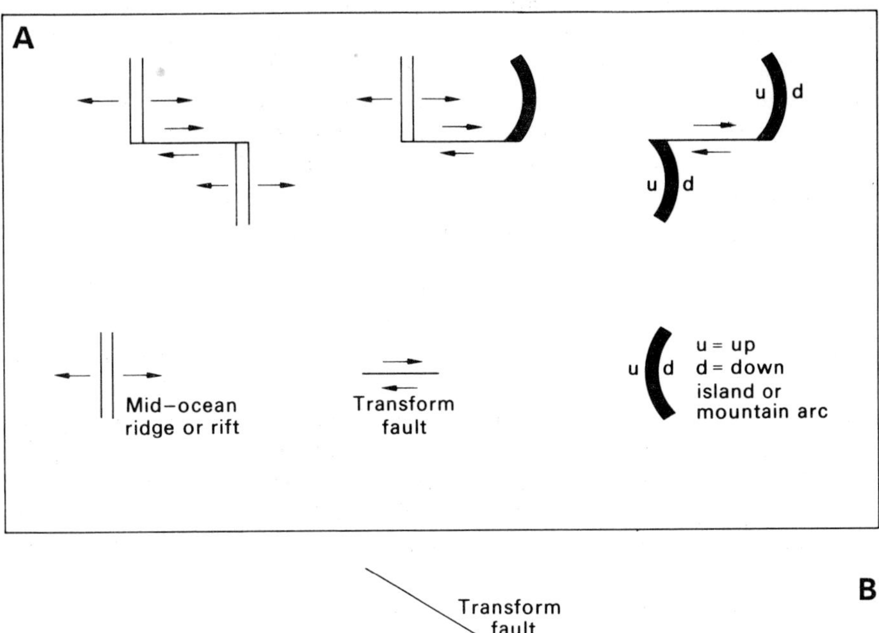

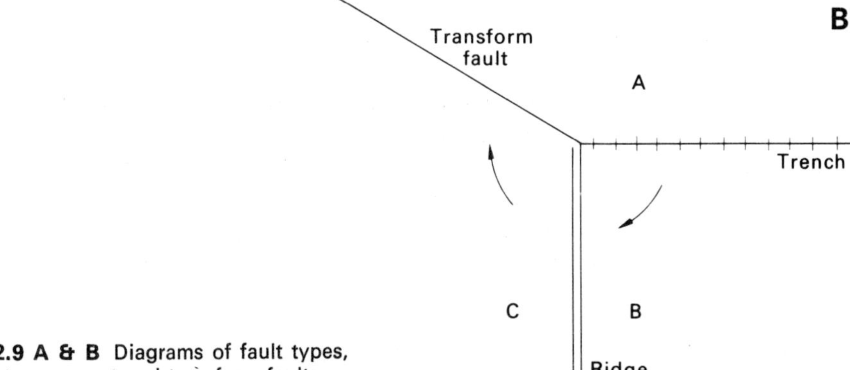

Figure 2.9 A & B Diagrams of fault types, showing transcurrent and transform faults.

The essential element of the global tectonics is the spreading of the sea floor from the ridges. The spreading is associated with expansion of the crust on either side of the ridge, as new crustal material is formed from the upwelling mantle at the ridge crest. The ridge crests are offset at intervals by faults running normal to the ridge crest. These faults have been termed transform faults. The material on either side of the fault moves laterally in opposite directions between the axes of spreading on either side of the crest as shown in figure 2.9A. Normal faults on the other hand occur along the crest section of the ridges, forming the central rift-valley. Both locations are zones of tensional tectonics. The trenches at the active oceanic margins are zones of compressional tectonics, where spreading oceanic crust sinks down as it comes into contact with another crustal slab, while the fracture zones are areas of shear tectonics. All these structural features have their morphological response. Many of the fracture zones of the Pacific can be recognized both by their relief and by the offsetting of magnetic anomalies, while the relief of the ridge crests in different parts of the ocean shows the effect of rifting. The pattern of movement can be related to the heat flow variations, the magnetic anomaly pattern, the seismic activity, and the deep-seated structure. The tectonic processes are operating on a sphere, and this fact determines the character of some of the resulting features.

McKenzie and Parker (1967) use the north Pacific as an example of how tectonics operate on a sphere. The transform faults are always pure slip movement and are parallel to the relative velocity vector between two moving plates or slabs. Figure 2.9B shows how a ridge and trench can meet to form a transform fault. This term has been applied to that section of the fractures between the two offset crests of the ridge, where the movement is in opposite directions. The transform fault is a type of transcurrent fault in which the reverse direction of movement is confined to the section that lies between the position of the ridge crest on either side of the fault. In a normal transcurrent fault the offset movement is in the reverse direction throughout the length of the fault. The relative movement of the two plates on a sphere is a rotation about some axis, and the amount of movement depends on the position relative to the axis. Where three plates meet, if two velocity vectors are given, the third can be determined. In the north Pacific there are two principal plates. Their junction is indicated by the earthquake zone that extends from the Gulf of California to central Japan. The trend of this junction varies widely and rapidly. The position of the pole for the movement of these plates is at 50°N 85°W, a position determined partly by the strike of the San Andreas Rift. The variation in the trend of the faults controls the distribution of trenches, andesitic volcanoes, and the intermediate and deep focus earthquakes. These features occur in Mexico, Alaska, eastern Aleutians, Kamchatka, and Japan, but they do not occur where the faults are of a transform nature. The transform faults in the southeast Pacific run east–west, indicating that the ocean is moving east between the East Pacific Rise and South America, relative to the main Pacific plate. The Atlantic plate is moving southeast relative to the Pacific one. If the rate of the latter is less than that of the former, then the crust must be consumed in the Chile Trench. The direction of spread of the Pacific seems to have changed in the north but not in the south.

3.1 Rates of spreading

The rates of spreading have been worked out by matching the magnetic reversal pattern on either side of the ridge crest and noting the distance between the particular dated band and the crest. It is assumed that the material becomes magnetized as it cools below the Curie point on reaching the upper levels of the crust. The age of the rock when it cools and becomes magnetized can be determined by the Potassium–Argon method. Some of the spreading patterns are remarkably symmetrical and consistent. Vine (in Phinney, editor, 1968) has shown that the Juan de Fuca Ridge off Vancouver Island in western North America shows a very symmetrical pattern, indicating a spread of 3 cm/year. Another ridge showing a consistent pattern is the Reykjanes Ridge southwest of Iceland in the

Atlantic, where the spreading rate is also 3 cm/year. The Juan de Fuca Ridge is separated from the Gorda Ridge to the south by an active transform fault, along which movement is occurring at a mean rate of 3 cm/year and which has moved in the last 5 million years. Bullard (1968) has assembled much of the evidence from reversal of magnetism data and gives the following rates for different ridge spreading (table 2.4).

Table 2.4 Rates of ocean floor spreading

Atlantic Ocean	varies from 1·0 to 2·25 cm/year, including the Reykjanes Ridge
Indian Ocean	varies from 1·0 to 2·4 cm/year, the lower value applying to the Red Sea and the higher to the Chagos-St Paul Ridge
Pacific Ocean	5·1 cm/year at 40 and 45°S 112°W 6·0 cm/year at 17°S 113°W, both rates occurring on the East Pacific Rise
All rates apply to the last 3·6 million years	

The rates of spreading vary both with place and time. The average value for the Pacific is 4 cm/year for 50 million years. All the profiles show changes in speed about 25 million years ago. This change is associated with increasing frequency of magnetic reversals since this date. The south Atlantic is 6400 km wide and, if spreading is assumed to have occurred at 1·5 cm/year, it would have started to form about 210 million years ago, in mid-Triassic times. The pattern of magnetic anomalies in the Bay of Biscay suggests that Spain has rotated relative to France as the Bay has split open. The axis of the pattern bisects the Bay. There are similar patterns in the Red Sea, the Gulf of Aden, and the Labrador Sea, all of which have formed by splitting fairly recently.

Heat differences could be important in explaining the rate of movement caused by the spreading. There is a difference in the heat flow under the continents and oceans. The high oceanic heat flow has a deeper source and the oceans are hotter by 200°C than the continents at the same level. There is, therefore, a tendency for the crust to move towards the continents from the oceans. The pattern of movement is controlled by the positions of the continents and oceans. The time scale of the convectional movement is long, as it takes 100 million years for conduction of heat through 40 km of rock. The pattern of continents and oceans and the convection pattern is a coupled system.

Heirtzler *et al.* (1968) summarize the magnetic anomaly data for the three oceans. Their results are presented in figure 2.10, which shows the spreading rates of the Indian and Pacific oceans relative to that of the south Atlantic going back over a period of nearly 72 million years to number 31 of the magnetic anomaly series in the Pacific Ocean, but only half as far in the Indian Ocean. The diagram indicates the relative rate of spreading in the south Pacific, which has been steady since the Cretaceous period. An early stage of movement was the spreading of the Africa–South America block about the southwest branch of the mid-Indian Ocean Ridge in Mesozoic and Permian times. This movement ceased in the Cretaceous. In the second phase, Africa and South America split and India moved north in upper Cretaceous times, with some east–west movement. At the same time New Zealand separated from the Antarctic. In the third phase the southeast and northwest branches of the mid-Indian Ocean Ridge became active at the end of the Eocene. Australia, moving north, also separated from Antarctica. For 10 million years the south

Atlantic and Pacific Oceans have been rotating about the same pole and the Indian Ocean about a different one.

The pattern of spreading has been discussed by Le Pichon (1968) in terms of the theoretical movement of slabs on a sphere and the evidence for this movement. The plates should move as units, and their movement should be parallel to the transform faults, which should be arcs of small circles about the centre of rotational movement. The angular velocity of rotation should be the same everywhere, and so should increase with the sine of the distance expressed in degrees of arc from the centre of rotation, reaching a maximum along the equator of rotation.

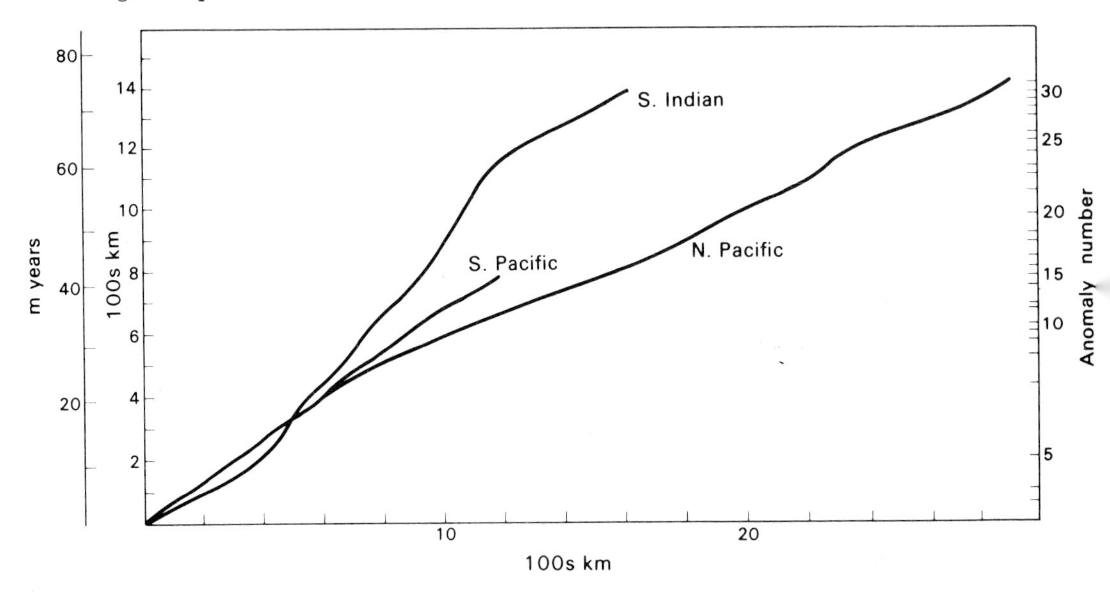

Figure 2.10 Rates of ocean spreading obtained from magnetic reversal data. The anomaly numbers are shown on the right. The data are plotted relative to the south Atlantic rate of spreading. (*After Heirtzler et al., 1968.*)

Fracture zones in the Atlantic between 30°N and 10°S are very nearly small circles centred about a point near the southern tip of Greenland. The measured rate of spreading agrees with the theoretical, so there need be no distortion of the oceanic or continental blocks as they move. The movement of the fault system on the west coast of North America, including the San Andreas system, is compatible with the rotation of the Pacific floor away from North America about a point also situated near the southern tip of Greenland. The opening of the north and south Pacific, the Atlantic, the Arctic, and the Indian Oceans can each be described by a single rotation. Six large rigid plates (shown in figure 2.11) are adopted in the model, and their relative movement can explain many results derived from geophysical observations, and including relief, seismic, and geological data.

The spreading appears to have been episodic, and can be carried back to the Mesozoic, fitting in with the orogenic phases. Spreading rates vary by a factor of 2 in the oceans, ranging between 1 and 6 cm/year, the greater rates occurring now south of the equator. There is a relatively simple pattern of opening in the Atlantic and Pacific, taking place

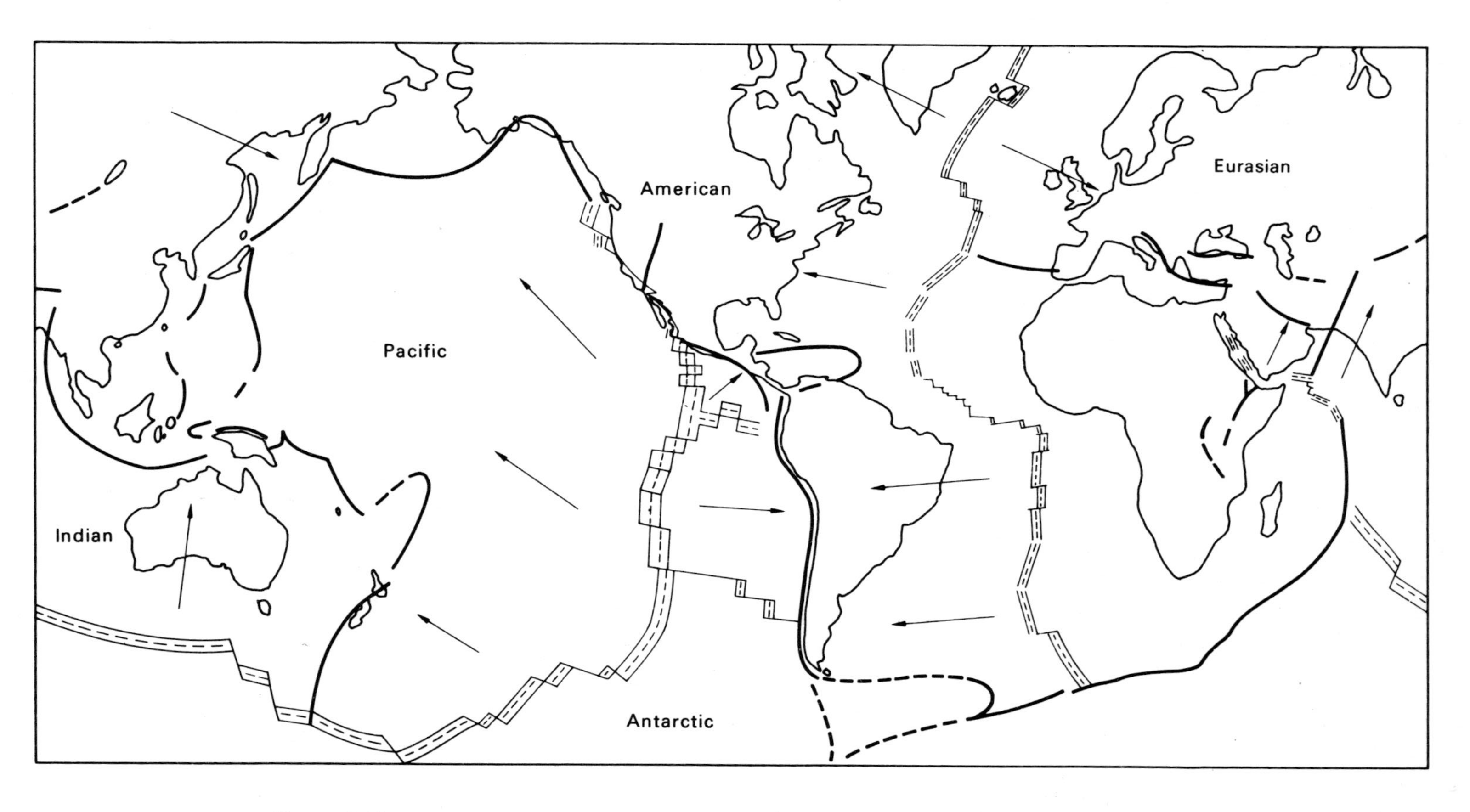

Figure 2.11 Pattern of plate movement. The six main plates are named. (*After Bullard, 1969, op. cit.*)

along the same axis from the same centre. These openings are linked by two oblique openings in the Indian and Arctic Oceans. The movement of America away from Africa is taking place at a rate of $3{\cdot}7 \times 10^{-7}$ degrees/year, from the point near the southern tip of Greenland. The split between Eurasia and Greenland and America goes from the Lena delta to the Azores and includes the Arctic Ridge. The spreading rate along this ridge is about 1 cm/year. Iceland forms a particularly active section along this zone. Features associated with this fracture zone may include areas of subsidence, such as the North Sea and the Rhine graben. The centre of this zone of spreading is at 78°N 102°E, with a standard deviation of 9·1° and a rate of rotation of $2{\cdot}8 \times 10^{-7}$ degrees/year. The Pacific is rotating away from North America at $6{\cdot}5 \times 10^{-7}$ degrees/year. The rate of slippage along the San Andreas Rift has been about 5 cm/year, while the creation of the Gulf of California has involved a movement of 300 to 450 km since the late Miocene. The Indian Ocean plate, which includes Australia, is rotating away from the African block. Spreading in association with this movement is taking place all the way from the Red Sea, through the Gulf of Aden to the south of Australia. The angular rate of opening is 4×10^{-7} around a point at 26°N 21°E.

Maxwell *et al.* (1970) have discussed the implications of the third leg of the *Glomar Challenger* drilling survey, in which the south Atlantic was studied. The evidence supports sea floor spreading. Seven holes were drilled on either side of the mid-Atlantic Ridge, one on the Rio Grande Rise and one near the Walvis Ridge. Up to 1 km thickness of sediment was drilled in water 5 km deep. The microfossils span the period back to 76 million years ago in the upper Cretaceous, with a gap in the middle Miocene. All the sediments of the mid-Atlantic Ridge are pelagic. The basalt base was reached at all 7 sites, and the age of the sediments bore a direct relationship to distance from the ridge axis. The thickness, however, bore no simple relationship to age. The sites with the youngest basement ages, which was Miocene, had the greatest thickness of sediment of 176 and 179 m, while the oldest basement age of upper Cretaceous had only 72 m of sediment. The differences could partly be accounted for by variations in the rate of sedimentation with lithology, which range from 1·8 cm/1000 years in foraminiferal chalk ooze to 0·02 cm/1000 years in Red Clay. The primary calcareous sediments have undergone different degrees of dissolution during deposition. The conclusion was reached that the south Atlantic has been spreading at a constant rate in the last 76 million years, a spreading half rate of 2 cm/year, which agrees with the magnetic studies. The sediments of upper Miocene age are richer in calcium carbonate, and suggest a rejuvenation of the mid-Atlantic Ridge, following a phase of deeper water in the late Mesozoic. The mantle in this region is abnormally light due to excessive heat flow. The continental shelves of the south Atlantic probably started to form in the early Cretaceous. The present crestal elevation of the ridge has probably never been exceeded. Table 2.5 gives the estimated age of the basement and distance from the ridge axis at the sites drilled in leg 3.

Early stages in the opening of the south Atlantic have been considered by Le Pichon and Hayes (1971), and they suggest that it opened in two phases, beginning at 140 million and 80 million years ago. The pole of rotation describing the relative motion of the South American and South African plates changed radically when the constraints imposed by adjacent continental blocks were relaxed. The early phase of opening occurred about a pole situated near one end of the line of opening, resulting in rapid differential opening

Table 2.5 South Atlantic basement age, sediment age, and distance from the ridge axis

Site	Age of basement (million years)	Palaeontological age of sediment above base (million years)		Distance from ridge axis (linear km)		Rotation at 62°N 36°W km	
16	9	11	± 1	191	± 5	221	± 20
15	21	24	1	380	10	422	20
18	—	26	1	506	20	506	20
17	34–38	33	2	643	20	718	20
14	38–39	40	1·5	727	10	745	10
19	53	49	1	990	10	1010	10
20	70–72	67	1	1270	20	1303	10
21	—	more than 76		1617	20	1686	10

to the south. In the later phase the mid-Atlantic Ridge formed and it has little variation of width over its length. The differential opening is shown by a rate of 6·2 cm/year along the Falkland fracture zone and 1·8 cm/year at the northern limit. The Rio Grande and Walvis Ridges formed at the same time as those near the equator, and are small circles about the pole of spreading, which migrated northwards between the two phases of spreading. The pole of spreading was computed relative to the Newfoundland, Bahama, and Guinea fracture zones (Le Pichon and Fox, 1971). It is situated 25° from the Newfoundland fracture zone and 50° from the Bahama one. The total opening about this pole was 46°. The basins mark the division between the early and later opening and the fracture zones do not extend beyond them.

Further south Dietz and Sproll (1970) have tested the fit of South Africa and Antarctica by computer best-fit methods. The computed fit at the 1800 m depth line was good, with a total mismatch of only 59,900 km^2 over the distance from the Weddell Sea to Princess Martha Coast, which was matched with the coast between Durban and Mozambique.

Active sea floor spreading is taking place in the southwest Indian Ocean (Bugh, 1971). The area surveyed around 39°E and 44°S is spreading at 1 cm/year in a direction 20° east of north. The evidence suggests an intersection of a fracture zone with a ridge. Two active transform faults were crossed.

W. J. Morgan (1968) has tested his model of crustal plate movement in three junction zones. He uses the strike direction of the great faults to calculate the position of the pole of spreading, in relation to spreading rates calculated from magnetic reversal data. The position of the pole of spreading between the African and American plates is given as 62°N ± 5°, 36°W ± 2°, with a maximum spreading velocity at the mid-Atlantic Ridge of 1·8 ± 0·1 cm/year. A similar study of the movement of the Pacific block relative to the North American one is made from measurements on the great faults of the northeast Pacific and the Juan de Fuca and Gorda Ridges. Much of the boundary between the two plates in this zone is faulted, along the San Andreas and Queen Charlotte Island Faults. The pole of relative rotation is determined to be at 53°N ± 3°, 53°W ± 5°. The rate of movement of the Pacific block relative to the North American is determined from movement along the San Andreas Rift, estimated at 6 cm/year ± 1 cm/year, giving a relative movement of 4·0 ± 0·6 cm/year. The Pacific block is thought to be moving relative to the Antarctic block about a pole of relative movement situated at 71°S ± 2°, 118°E ± 6°, with a maximum spreading rate of 5·7 ± 0·3 cm/year. From these estimates the movement of the African block relative to the Antarctic block is calculated. The results suggest a pole at 15°S, 15°W, with a possible spreading rate of 1·5 cm/year.

The pattern of the splits is the response of a thick lithosphere to underlying stresses, which cause breaks in zones of weakness. The pattern is not a random one and two main openings have centres within 15 degrees of each other. The axis of rotation of the spreading is at about 60°N, 50°W. There must be some compensation for the zones of spreading if the earth is not expanding. The evidence does not favour an expanding earth, so that it is necessary to consider zones where the excess crustal material is re-absorbed. The crust is lost at present in the deep trenches and has been lost in the Tertiary mountains. The main trench system, where the excess crust has been consumed, may have been in the eastern Pacific in the early Cainozoic, but it has been in the western Pacific at least since the latest Miocene. Where the rate of movement reaches 8 to 9 cm/year, two parallel trenches form. Some of the trenches have a strike-slip motion, such as the western part of the Puerto Rico Trench and the east–west part of the South Sandwich Trench. The pattern of spreading over the last 60 million years has operated in different phases that can be identified by the magnetic reversal anomaly pattern. This has been divided into 32 phases, number 5 occurring 10 million years ago and the last, 32, at the limit of the recognizable pattern 60 million years ago. During this period three phases are indicated. The first in the Mesozoic is a period of basin formation. Second, in the early Cainozoic, the mid-ocean ridges were created. During the third, the crestal regions were formed. The active drift of North and South America started about 120 million years and ended about 70 million years ago, moving at a rate of 3–4 cm/year. Most of the drift occurred in early and mid-Cretaceous times, when the maximum spreading rate was about 6–8 cm/year. Northwest America had over-ridden the trench system on its border by the Miocene, when movement ceased. The third cycle started in the late Miocene, when a new pattern started in the Pacific. The East Pacific Rise is thus a much younger feature than the mid-Atlantic Ridge.

Bott (1971) has also suggested that ocean spreading takes place in phases. There are four peaks in the observations of igneous and metamorphic age determinations at 350, 1100, 1800, and 2700 million years ago. The phases may be related to the closure of oceans as a result of drifting. In the first stage of a cycle two oceans expand and one is closing. In the second stage the shrinking ocean disappears and adjacent continents collide. In stage 3 a new convection cell develops under the joined continents, forming a new expanding ocean, and one of the other oceans now becomes a shrinking ocean. At the present rate of movement, the Pacific Ocean should close in 200 million years, bringing the present cycle to stage 2. It started 300 million years ago when the Atlantic and Indian Oceans started to expand. The driving force is thought to be cellular suboceanic convection in the upper mantle. The earth acts as an inefficient heat engine, in that about 1/1000–1/100 of the escaping heat is converted into strain energy in the lithosphere, which is released by tectonic activity concentrated along the mobile belts.

3.2 Mechanism

The mechanism responsible for the spreading process is generally thought to be associated with convection cells in the earth's mantle. The effect of temperature variations is an important aspect of the problem, and the state of the materials of the crust and mantle must also be considered. Isacks, Sykes and Oliver (1969) suggest that the upper part of the earth consists of three flat-lying layers. The uppermost is the lithosphere, consisting of the crust and upper mantle. This layer has strength to a depth of 100 km over much of the earth. The second layer is the asthenosphere, literally 'no-strength-shell', but more accurately it should be the low-hardness layer, as it is a layer of little strength extending through several hundred kilometres in thickness. Below it lies the third layer, the mesophere, which may have strength. It consists of the lower part of the mantle and is passive or inert. The lithosphere is also called the tectosphere, while the asthenosphere is the low-velocity layer of seismology, lying between two high-velocity layers.

The major zones of tectonic activity are the ocean ridges, the island arcs, and the major strike-slip faults, also known as great faults. At these sites the lithosphere is discontinuous, but elsewhere it is continuous. The lithosphere is thus broken into 6 large plates that have already been mentioned, with possibly some minor blocks. The movement between the plates is self-consistent on a world scale and can explain the pattern of vulcanism and earthquakes. At some of the oceanic–continental margins the lithosphere is healed or unbroken, while at others it plunges down below the adjacent plate. Along these planes the deep sea trenches form. At times the lithosphere plunges in different directions below the adjacent plate, the two trenches that result being linked by a transform fault (see figure 2.9). Examples of this pattern occur in the New Hebrides, Fiji and Tonga, the East Pacific Rise, and western South America. In order to compensate for the spreading of the crust from the ridges, there must be a counter-movement deep in the lithosphere. The movement of the surface plates may control the movement in the lithosphere, rather than convection cells. Where the lithosphere sinks down it may drag the upper crust with it, thus causing the spreading of the ocean floor. Thus the basic cause of the diving down of the crust may be gravitational instability, resulting from the surface cooling and hence high density of the near-surface mantle material. It is not yet certain whether the lithosphere or the asthenosphere is the active element of the system. The convective zone has probably thinned through time.

The global tectonic scheme is indicated by the pattern of earthquakes. These outline the plates. Although the pattern of plates has partially been determined from the pattern of earthquakes, it does nevertheless fit well with other independent evidence. Between 1961 and 1967, 29,000 earthquakes were reported. Their pattern forms a continuous linear arrangement. They are most common in zones of convergence, but also occur in zones of divergence and lateral slip. The zones are broader on the continents than in the oceans, owing to more complex structural control and processes.

Only two plates are involved in the north Pacific, with movement occurring between them along the great transform faults of the San Andreas Rift, Queen Charlotte Islands, and the Gulf of California. Dip-slip motion occurs in western USA, Canada and Alaska, and the eastern Aleutians. The plate is underthrusting the Kurile, Kamchatka, and Japanese arcs. The great east–west fracture zones of the northeast Pacific were formed about 10 million years ago, when the direction of movement changed from east–west to northwest–southeast. The rate of underthrusting decreases from north to south, and is associated with a decrease of volcanic activity. There is greater seismic and volcanic activity as the slip vectors change from strike-slip to dip-slip motion. Tsunami activity is generated mainly in the high rate dip-slip zones, rather than in the strike-slip zones. The heat flow anomalies are compatible with a simple model of cooling lithosphere, which is 50 km thick, but it fits less well with convection cells. The size of earthquakes may be related to the thickness of the lithosphere. Island arcs contribute more than 90 per cent of world shallow earthquake energy and nearly all the deep earthquake energy. Many different methods of computation suggest that the lithosphere is about 100 km thick. These include isostasy, heat flow anomalies, absence of earthquakes, gravity anomalies, and the propagation of S_n waves. There is much evidence of lateral variability of the outer few hundred kilometres of the earth's lithosphere and asthenosphere.

Nearly all the evidence favours the major findings of the new geophysical interpretation of crustal processes, but there are nevertheless many problems to be solved and details to be filled in. It is worthwhile, therefore, to consider other possible evidence and to keep an open mind. Some of this evidence has been assembled by Meyerhoff (1970), who makes the point that palaeomagnetic studies have led to a bewildering variety of former pole positions and interpretations of continental movement. He asserts that this evidence cannot be used to show movement between the continents. He used marine evaporites, coal, desert eolian sandstone, and tillite as indicators of palaeoclimatic conditions. Maps of the distribution of these deposits over geological time from Proterozoic onwards show that the belts have been consistent throughout. The width of the belts fluctuate through time, but they maintain their correct relationships. Evaporite maximums indicate exceptionally warm phases during which coal belts are narrow. Glaciation occurs, on the other hand, when the coal belts are wide. The major point of interest in the analysis is that 95 per cent of all ancient evaporites are in areas that today receive less than 100 cm annual precipitation.

The conclusion drawn is that the planetary wind and ocean current systems have undergone little change since the middle of Proterozoic time, with the corollary that the rotational pole, ocean basins, and continents have maintained their present positions relative to the poles during this period of 800 to 1000 million years. It is further suggested that the Gulf Stream and North Atlantic Drift have been in existence all the time, to account for the tectonic history in relation to the evaporite deposition of the Arctic and North America.

Meyerhoff and Teichert (1971) have further suggested that large-scale glaciation and coal formation in the upper Palaeozoic could not have taken place without adequate water supply. They could, therefore, not have occurred in the centre of a large land-mass, such as the reconstructed Pangea or Laurasia and Gondwanaland. Shallow seas would not have allowed suitable oceanic circulation. They conclude that reconstructions based on plate tectonics cannot account for all the observations and require reconsideration.

Hammond (1971) has fully accepted the evidence of sea floor spreading and considers that now the main aim should be an elucidation of the mechanism. This could be due to radioactive decay generating heat, phase changes associated with the core, tidal forces from the moon, or other gravitational forces. Thermal convection could work in two ways. Either heated fluid could rise from below, due to phase changes, or sinking cold material in the marginal trenches could pull the plates with it. A third possibility is deep convection in the lower mantle.

The majority opinion at present favours the global tectonics associated with the movement of plates. Bott (1971) summarizes the position by stating that plate tectonics can show how ocean floor spreading and continental drift can occur on a nearly spherical earth without deforming the ocean floor or continents, except in well-known mobile belts. The process is generally agreed to be some form of convection. This is likely to occur in the upper mantle and to be an effective mechanism of heat transfer. Convection through the whole depth of the mantle or in the lower mantle may be inhibited by the high viscosity that is likely to occur at the base of the mantle. The pattern of convection is unlikely to be affected by the rotation of the earth and turbulent flow is also unlikely to occur. The whole process is probably complex for several reasons. These include the rheological structure, the presence of heat sources within and above the convection cells (as well as below them), the strong lithosphere near the surface, and the convection occurring far from the outer margin. It is likely that the convection occurs in the uppermost 500 km. These complexities mean that an analytical solution may well be impossible. Nevertheless, the large body of evidence collected from many sources in favour of the theory is impressive.

Further reading

DIETZ, R. S. 1961: Continents and ocean basins evolution by sea floor spreading. *Nature* **190**, 854–7. (One of the first published accounts of the new global tectonics.)

GASKELL, T. F. 1960: *Under the deep oceans*. London: Eyre and Spottiswoode. (A general account of geophysical techniques.)

HEIRTZLER, J. R. 1968: Sea-floor spreading. *Scientific American* **219,** 60–70.

HESS, H. H. 1965: Mid-ocean ridges and tectonics of the sea floor. In Whittard, W. F. and Bradshaw, R. (editors), *Submarine geology and geophysics. Proc. 17th Colston Res. Soc. Bristol.* London: Butterworth, pp. 317–32.

HILL, M. N. (editor). 1963: *The sea.* Vol. **III**. *The earth beneath the sea.* Section 1. Geophysical exploration. New York: Wiley, pp. 2–232. (A detailed account of geophysical methods and results.)

LE PICHON, X. 1968: Sea-floor spreading and continental drift. *J. Geophys. Res.* **73,** 3661–97. (A technical account of the new global tectonics.)

MAXWELL, A. E. (editor). 1971: *The sea.* Vol. **IV,** Parts 1 and 2. New York: Wiley. (A technical account of recent developments in global tectonics and geophysics of the ocean basins, bringing Hill, M. N. (editor), *The sea,* Vol. **III,** up-to-date.)

PHINNEY, R. A. (editor). 1968: *History of the earth's crust.* Princeton, NJ: Princeton Univ. Press. (A collection of contributions on the new global tectonics.)

TALWANI, M. 1964: A review of marine geophysics. *Mar. Geol.* **2,** 29–80. (A review article on the state of marine geophysics at the time.)

WEGENER, A. 1922: The origin of continents and oceans. (One of the first suggestions of mobile continents, which is of renewed interest in connection with the new global tectonics.) 1966 Reprint, translation of 4th revised German edition by J. Biram. New York: Dover Pubs., 246 pp.

The above list is intended to give examples of various stages in the ideas of the development of the new global tectonics, some at a general level and others more specialized.

3 Continental margins

The continental margins are of particular significance in marine geomorphology because they cover the zone linking the continents with the deep ocean basins, two fundamentally different structural zones in the earth's crust. These marginal zones also have a much wider importance in many fields of activity both scientific, economic, social, political, and strategic. In the scientific field the continental margins are of interest from the geological and geophysical points of view. They play an important part also in physical oceanography, meteorology, and biology.

Geomorphologically the continental margins can be subdivided into the coastal zone, the continental shelf, the continental slope, and the continental rise, extending from the land towards the deep sea basins. These distinct units exhibit a wide range of geological structures (examined in more detail in subsequent sections). The sediments of the different zones have also proved a fruitful field for geological enquiry concerning the processes that operate in these zones. Geophysically the continental margins are of particular interest in that they cover the zone across which the Mohorovicic discontinuity rises from the low continental level of around 50 km to the high oceanic level of only about 6 km below the sea floor. This fact confirms the fundamental difference of the two crustal types.

From the point of view of physical oceanography the continental margins are the areas where the interaction of land and sea is most marked. Waves, for example, change characteristics as they enter shallow water, and in turn they modify the coastal zone, where their energy is eventually lost in bringing about changes that produce the wide variety of coastal landforms. The marginal zones of the ocean exert a strong effect on the nature of the major ocean currents, producing the well-developed western intensification of the current system and also being responsible for the eastern ocean zones of upwelling, which are so important biologically. Meteorology is also closely affected by the continental–oceanic border zone, producing on a small scale such features as land and sea breezes, and on a larger scale, the monsoon system that exerts an influence throughout the Indian Ocean north of the equator on all aspects of the character and life of the area. The marginal zones of the ocean provide the widest range of habitats and the most favourable conditions for life in the oceans. Thus biologically their flora and fauna are the richest and most varied in the sea. The flora include the attached seaweeds that can only grow in shallow water, owing to the necessity for light, while the fauna are most prolific in the generally more fertile waters of the marginal zone. The range of ecological systems is much wider in these zones of variable relief, rock and sediment type, and salinity and temperature conditions. The benthic fauna in particular are especially varied and abundant in the marginal zones.

The biological wealth of the marginal zones is the basis for the major economic importance of these areas from the point of view of the fishing industry, much of which is concentrated on the coastal waters, particularly where the shelves are wide in high latitudes, and in the rich upwelling areas of the eastern oceanic margins in middle and lower latitudes. The attached seaweeds are also of interest commercially on some rocky coasts, while the artificial culture of some benthic forms is an approach to sea farming that is growing in some areas, notably Japan. As the knowledge of the geology of the continental margins grows, their economic importance from the point of view of exploitation of non-organic resources, including oil and gas, is ever increasing. Eventually they will probably yield the major portion of oil and gas reserves and will become increasingly used for the exploitation of sand and gravel, as well as for more valuable minerals of a wide range of types.

The development and expanded use of marginal ocean resources will lead to many legal and political issues, some of which have already been met. The problem of defining the continental shelf from the legal aspect is an example, while legal problems regarding pollution and waste disposal in marginal zones are becoming pressing and require much national and international cooperation. The strategic aspects of marginal zones are also of significance, as even narrow zones of coastal waters can be of major importance from this point of view.

The coast is the zone in which the interaction of land and sea is most direct. The variety of coastal scenery depends on many variables, of which the most important include the nature of the land against which the sea rests, the marine processes that are operating in the area, and the time span over which they have been operating. The coasts of the world have been differentiated in terms of the global tectonic system into three main types: the trailing margins, the collision margins, and the marginal sea coasts. An example of a trailing margin is the coast of western Europe and Africa, along which the rift of the Atlantic Ocean has spread. A more recent example is found in the Red Sea and Gulf of Aden. The coast of eastern Asia provides an example of a marginal sea coast, while western South America is a collision margin, where two plates are coming into contact and crust is being consumed. The western coast of North America provides an example of a coast along which plates are moving laterally to a certain extent.

The trailing edge coasts may be subdivided into a Neo-type, Afro-type, and Amero-type, the Red Sea exemplifying the former, the west coasts of Africa and Europe and the east coast of Africa the Afro-type, and most of the eastern coast of North and South America the third type. The Afro-type is defined by both coasts being trailing on either side of the ocean, hence the situation is such that erosion is limited. In the Amero-type the opposite coast is a collision coast, and as a result mountainous, so that erosion is enhanced. The collision coasts can be divided into two types—the continental as exemplified by western South America and the island arc type, exemplified in the western Pacific. The marginal sea coast is one protected behind an island arc, such as the mainland coast of eastern Asia (Inman and Nordstrom, 1971).

Coasts have been broadly classified on a morphological basis according to the nature of the continental margin by Inman and Nordstrom (1971). They suggest three broad divisions on this basis: mountain coasts, coasts with narrow shelves, and coasts with wide shelves. The last two groups can be further subdivided, according to the hinterland, into hilly and plain coasts. Both the tectonic and morphological groupings produce very broad categories, which include a wide diversity of form and structure.

The diversity can partly be accounted for by variations in the operation of coastal processes, of which waves and tides are most important. Coasts can be subdivided into geographically based wave groups, such as storm wave coasts, swell wave coasts, and low-energy coasts, according to the climatic influence on the wave type. Another very important effect on coastal geomorphology is concerned with the rapid changes of sea level that have taken place during the last few tens of thousands of years. These changes, due to variations in the ice volume on land, have been exceptionally great and rapid in terms of normal sea-level changes through geological time. For this reason all the coasts of the world are young, and few have reached a stage of equilibrium under present conditions. The coastal zone is, therefore, one of active change and great variety, with erosion or accretion active along many stretches.

In some respects the continental shelves also reflect the marked changes of sea level that have affected the coastal zone, and many of the shelf sediments are out of phase with present conditions. The general form of the continental shelf, however, has been developing over a much longer time interval in many areas than the coasts. Time of shelf development must be counted in tens or hundreds of million years rather than the thousands of years that apply to coasts, considering that sea level only reached its present height about 5000 years ago.

Shelves vary in character widely around the margins of the world oceans. These types will be considered in more detail in the appropriate section, but it is worth noting that the major types of shelves, like coasts, can be related to the global tectonic system. There is a marked contrast between the continental shelves off the coasts of the Atlantic Ocean, for example, and those around the western Pacific. The Atlantic shelves are built of large piles of sediment collected from the ruptured continents as they moved apart, while the shelves of the Pacific are in general much narrower. Emery (1970) has stressed the importance of dams in defining the characteristics of many continental shelves, but those in which sedimentation has gone on a long time have now overflowed their dams in some instances (figure 3.1). It is in these zones particularly that the continental rise is a conspicuous element of the continental margin (figure 3.2). These areas include the margins off Africa, India, eastern North and South America, Antarctica, and south and west Australia.

Figure 3.1 (*opposite*) Classification and generalized types of continental shelves and their distribution. (*After Emery, 1970.*)

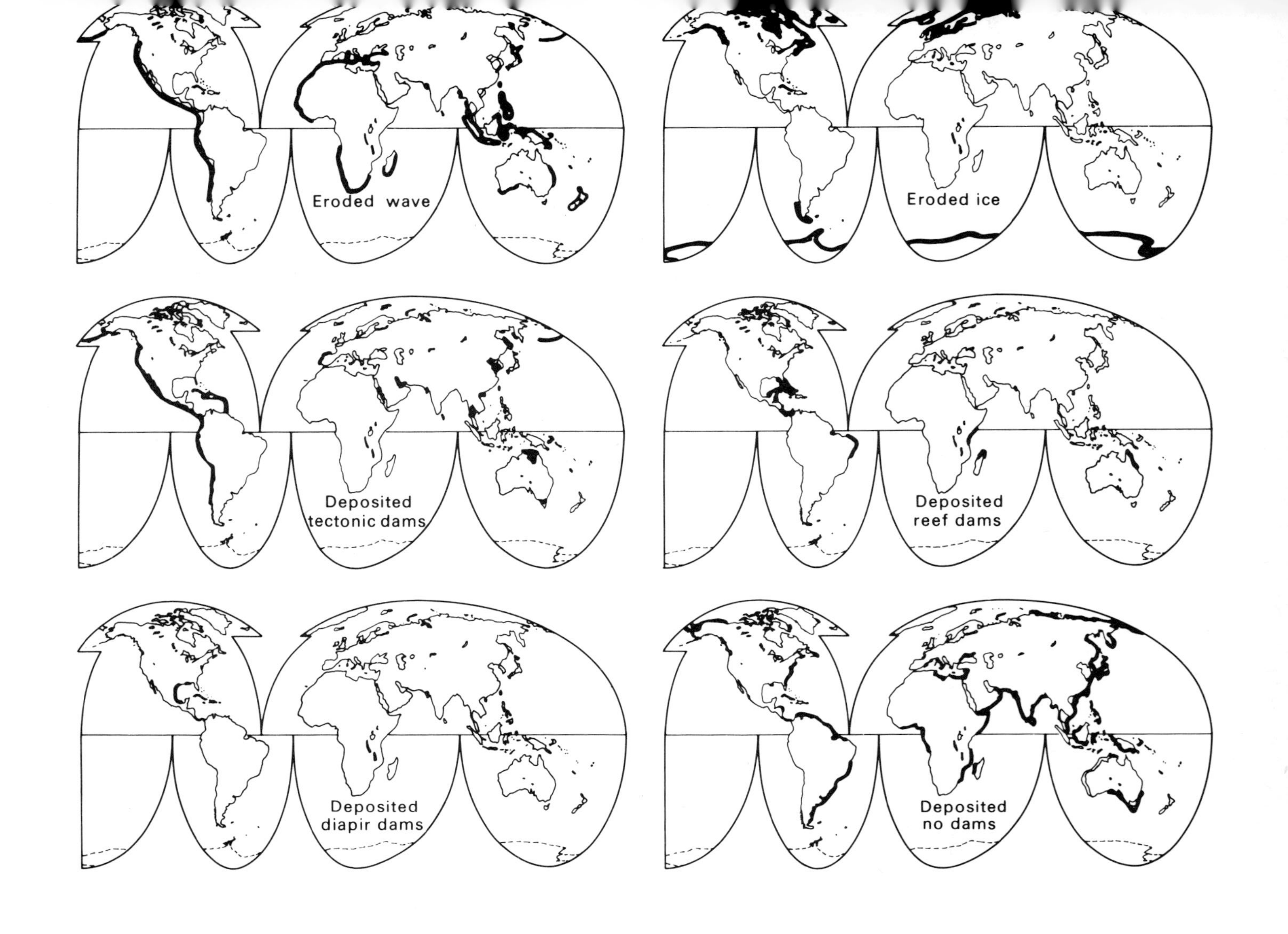

Eroded wave
Eroded ice
Deposited tectonic dams
Deposited reef dams
Deposited diapir dams
Deposited no dams

The continental rise is separated from the continental shelf by the continental slope (which forms the major relief feature of the earth's surface), in that it links the two major levels, the land surface and the deep ocean basins. The shelves usually drop off abruptly to deeper water at the upper edge of the slope, so that the slope forms an important boundary from many points of view, oceanographically and biologically. The Gulf Stream, for example, flows along the upper part of the continental slope, separating the very much cooler shelf water moving south, from the warmer water of the Gulf Stream and Sargasso Sea. Structurally the continental shelf cannot be separated from the slope, in that the two zones form two margins of one feature, although they can be differentiated on the basis of depth and slope.

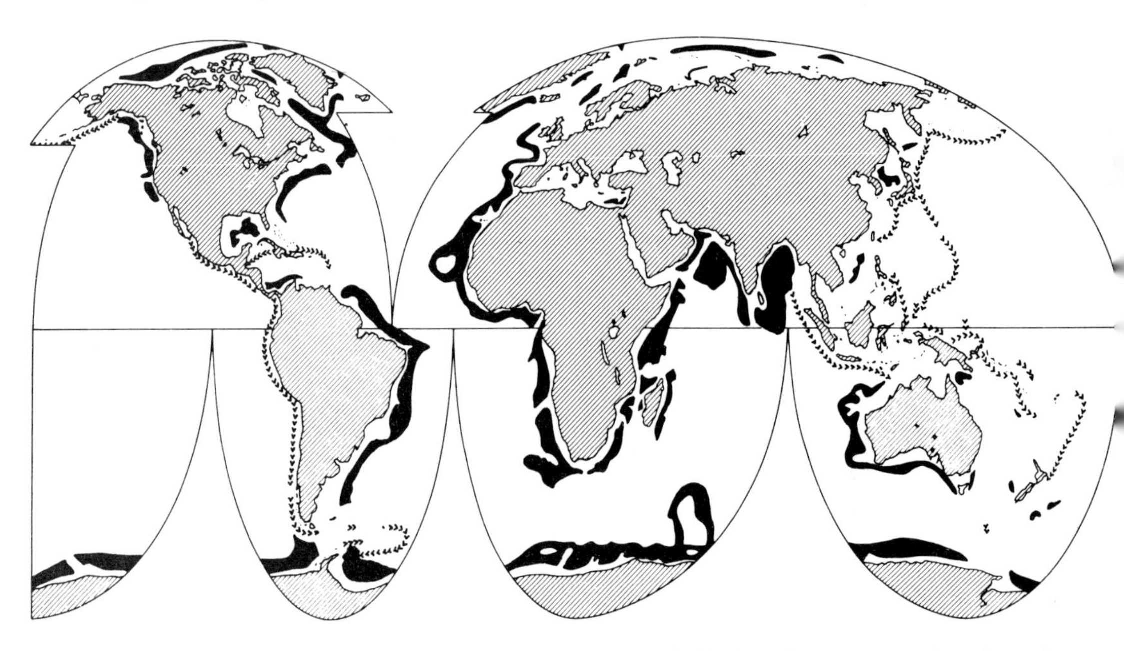

Figure 3.2 Distribution of continental rises, which are shown in black, and deep sea trenches shown by chevrons. (*After Emery* et al., *1970*.)

The shelf and slope form the zone in which some intriguing morphological features and processes occur. The presence of deep canyons has been known for a long time. As more detailed surveys have been made with precision echo-sounders, and equipment that can penetrate the shallow sedimentary layers to reveal the underlying structure, the canyons have become better known. They have been revealed around all the coasts of the world from the Arctic Ocean to the mouths of equatorial rivers and extend at times almost to the beach, while others can be traced to the bottom of the deeper abyssal basins. They are particularly important as the route along which the sediment from the land can reach the deep ocean basins. The process by which this movement takes place is by turbidity currents. A great deal of work, both theoretical and field work, has added greatly to our knowledge of this process, and its importance in a discussion of the detailed morphology of the continental shelves and slopes. Field work has included direct observations of the nature of the submarine canyon walls and floors and movement of sediment down the canyons has been observed in action in some instances. The morphological evidence concerning canyon character and formation is discussed in this chapter, while a study of the sediment associated with canyons is mentioned in chapter 5. Their influence on the morphology of the deep ocean is discussed in chapter 4, in which abyssal plains will be considered. The material moving down the canyons plays an important part in shaping the abyssal plains to which they have access.

1 Coastal geomorphology

1.1 Coastal variety and general considerations

Mention has already been made of three possible ways in which coasts can be classified. Firstly, classification can be made in relation to the pattern of plate movement associated with global tectonics, giving the trailing edge coasts, the collision coasts, and the marginal sea coasts. The second possibility mentioned was a morphological classification, which divided coasts up basically according to the relief and shelf width. The third was the dynamic classification in terms of the dominant wave regime, into storm wave coasts, west and east coast swell wave coasts, and low-energy coasts (J. L. Davies, 1964). All these classifications are very broad and within each category a wide range of coastal types is included. They do, however, provide useful criteria that are important in describing coastal landforms.

Several attempts have been made to classify coastal landforms on a genetic basis. These also bring out several variables that are significant in assessing the processes at work along the coastline. One of the best known and oldest of these classifications is that of D. W. Johnson (1919), who divided coasts up into those affected by submergence, emerged coasts, neutral and complex coasts. This classification emphasizes, in its main criterion for dichotomy, the recent and rapid changes of sea level that have affected all coasts. Another well-known classification is that of Shepard (1963a), who uses as his main classifying criterion the major process that gives the coast its character. His classification is between coasts shaped largely by terrestrial process and those shaped largely by marine processes. He thus emphasizes the dynamic and changing character of coasts.

The classification of Valentin (1952) also emphasizes change along the coast, but he recognizes that the present form of the coast may not reflect current processes. To allow for this fact he proposes two classifications: one is a classification of coastal configuration and the other is a classification of present coastal dynamics. The basic dichotomy of both classifications is between coasts that are building out or have built out, and coasts that are retreating or have retreated in the dynamic and configuration classifications respectively. The outbuilding or retreat of a coast can be brought about in two basic ways: the coast may build out as a result of excessive deposition of material, or alternatively, a fall of sea level leads to a retreat seawards of the coastal outline, and hence a gain of land. The retreat of a coast can result from coastal erosion on the one hand, and on the other, from a rise of sea level, which will drown the land. These four processes are used in Valentin's classification as four axes in relation to which any coast can be specified (figure 3.3). His latest version of the classification also allows for an equilibrium class in which the effects of a rising sea level are counteracted by outbuilding, or a falling sea level is counteracted by coastal erosion, in each instance leading to a stable position for the coast.

Further subdivision of the broad groups in most of the classifications mentioned is based on the nature of the most important process operating. Thus Valentin distinguishes between biogenic landforms, fluvial, glacial, marine, submarine, tectonic, aeolian, volcanic, karst, thermokarst, and human landforms, each of which can include the actual landform, which may be created by erosion or deposition. One important feature is that the classification can allow for a polygenetic coastal form. This is the group described as compound in Johnson's classification, and which in Shepard's may show evidence of both his major subdivisions (terrestrial and marine processes).

The main point of this brief discussion of coastal classifications is that the coastline of the world is extremely varied; it is usually compound, having been influenced by a variety of processes, the effects of which can still be discerned; and as a result few coastlines are

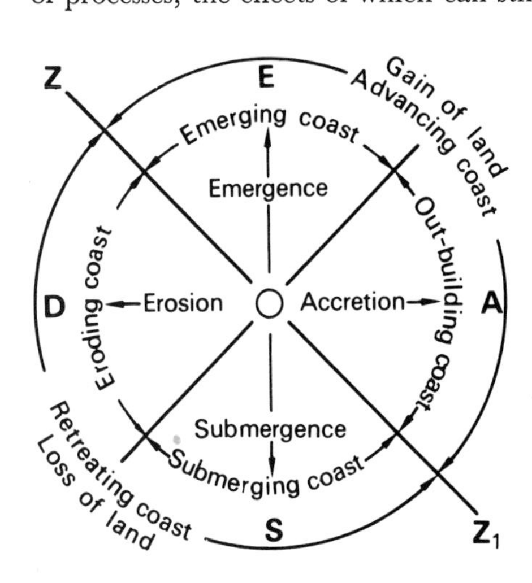

Figure 3.3 Diagram to illustrate Valentin's coastal classification, based on dynamic considerations. (*After Valentin, 1952.*)

in equilibrium with present conditions. The youth and immaturity of nearly all coasts is the result of the rapid and large fluctuations of sea level that have taken place during the Pleistocene and recent periods. Sea level only reached close to its present level about 4000–5000 years ago, since when it has been fluctuating slightly (further details are considered in chapter 6). Thus none of the present coasts of the world has had more than about 5000 years to reach equilibrium and this is a very short period of time geologically. Coasts are dynamic landforms along which change can be rapid in some circumstances, but even those that are changing rapidly can only have been modified to a relatively small degree. The coast of Holderness, for example, is retreating through erosion more rapidly than practically any other coast over a comparable stretch. It is losing ground at the rate of about 2 m/year, which over a period of 4000 years would amount to only 8 km. It is the youthfulness of present coastlines that is at least partly responsible for the great variety of forms and the compound nature of most coasts, because their earlier form has yet to be obliterated by currently acting processes.

Glaciation has played a large part in accounting for the variety of coastal forms. The growth and decay of the major ice-sheets is the most important cause of fluctuations in sea level, while the ice has directly affected the relief and landforms of long stretches of the coast in high latitudes, giving for example, the impressive fjord coasts of Norway, east Baffin Island, western Canada and Alaska, southwestern New Zealand, parts of Labrador, southern South America, and parts of Antarctica. Many stretches of coast have also been directly or indirectly affected by glacial deposition. There are the drowned drumlin coasts of western Ireland, morainic and outwash coastlines (the latter for example in south Iceland). The indirect effects of glacial deposition are important in providing distinctive beach materials. It has been suggested by Davies that the relative abundance of shingle in high latitudes is the result of glacially derived coarse sediment. Sand is certainly the predominant beach deposit of lower latitudes. Another indirect affect of the presence of ice on the coastlines is its influence on the climate. The great range in temperatures between the poles, with their ice cover, and the low latitudes gives rise to a vigorous atmospheric and oceanic circulation. This in turn induces the present pattern of wave generation, the importance of which on the coastal dynamics has been stressed by Davies (1964) in his morphogenetic classification of coastlines, in which he distinguishes storm wave and swell wave coasts among others (figure 3.4).

The coastal zone can be divided into a number of elements which determine its character. These are arranged in bands parallel to the coastline and include the backshore zone, the foreshore zone, and the nearshore zone from land to sea. The backshore zone is above the reach of the highest tide, but is nevertheless influenced by and influences the character of the whole coast. The foreshore zone is the one across which the tide ebbs and flows; it is, therefore, wide in the macrotidal environment, as around Britain, but narrow in the microtidal areas, which include the Mediterranean and the Gulf of Mexico. The nearshore zone includes the swash zone, the surf zone in which the waves break, and the area to seaward of the surf zone and break-point of the waves in which the waves and tide can exert a considerable effect on the bottom. The seaward limit of this zone is sometimes described as the surf base, or the wave base, although the terms are somewhat ambiguous and are not used consistently. The surf base will be considered as the depth beyond which waves can effect relatively little modification on rock or sediment. Its depth is probably of the order of 10–15 m. Beyond this depth, waves and tides can move material on the shelf to considerably greater depths, particularly under extreme storm conditions, so that wave base can be defined as the depth to which this movement extends, and it may be several times the depth of the surf base.

The most important distinction in the backshore zone is whether it is formed of solid rock or drift cliffs, on the one hand, or whether on the other, it is formed of deposits (such as sand dunes or salt-marsh, for example). These two major types of coast can be referred to as high coasts and low coasts respectively, although the distinction is not always clear, as some drift cliffs may be lower than some sand dune coasts. The two types do, however, reflect a difference in the coastal processes involved in their formation, in that the cliffed coasts are often eroding while the low coasts are usually accreting. There is also often a difference in the general outline of these two types of coast, and the pattern of a coastline is of considerable importance from several points of view.

A cliffed coast, particularly when it is formed of solid rock, is usually irregular in outline. The irregularity often results from variations in resistance to marine erosion of the rock types outcropping along the coast. The softer rocks are differentially eroded to give a

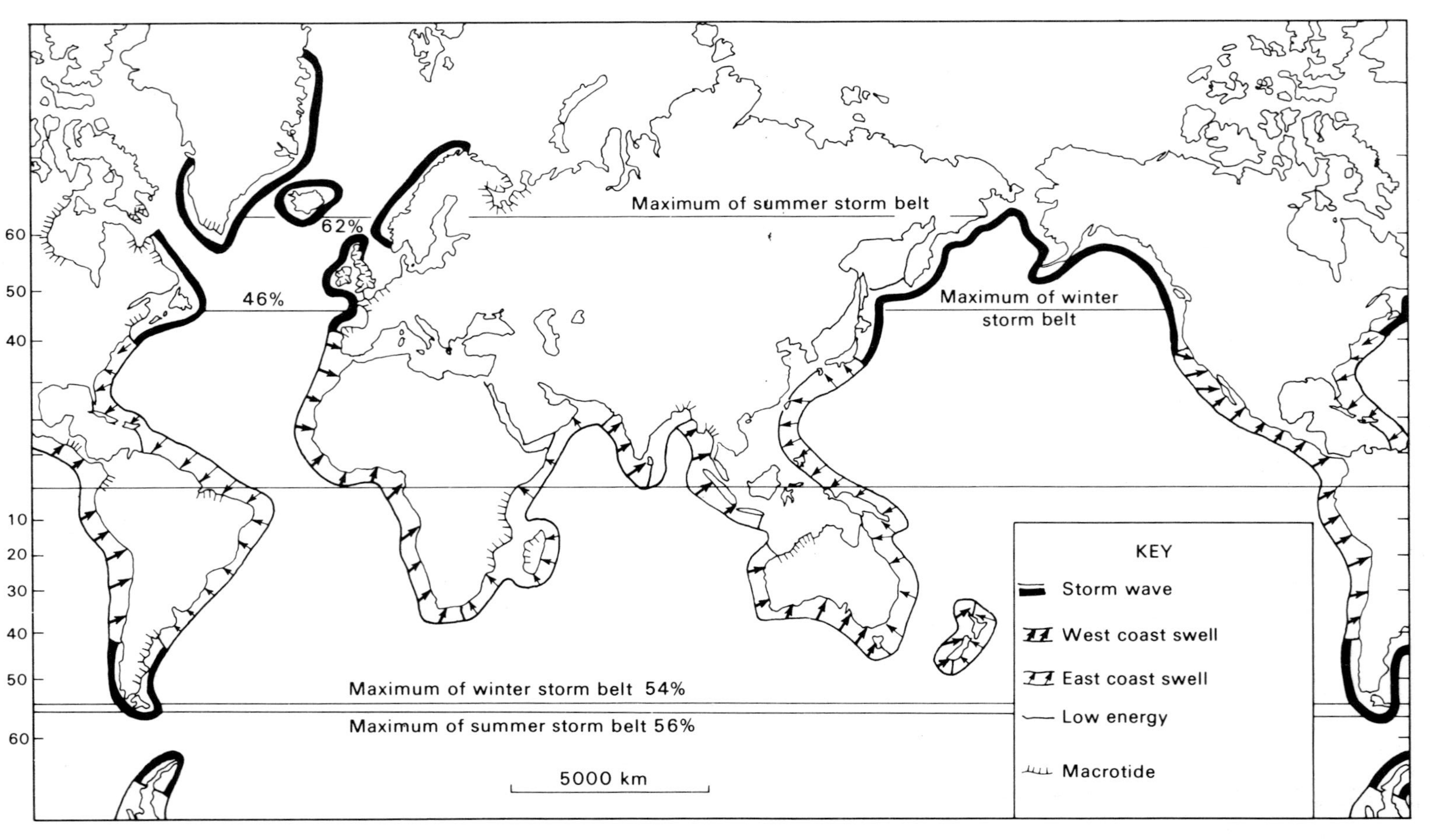

Figure 3.4 World distribution of wave types, and macrotidal zones. (*After Davies, 1964.*)

crenulate coastline, where the irregularities are on a fairly small scale, such as the coast of southwest England. The harder rocks form headlands while the softer are etched out to form bays, in which bay head beaches accumulate, these often forming the only considerable deposits on such a coast. Where the structure is such that hard rocks and soft rocks alternate on a larger scale then the headlands are more impressive and the bays deeper, often running up into estuaries, where rivers follow the softer outcrops. The ria type of coastline of southwest Ireland illustrates this type, which is typical of trailing edge coasts in many instances, where the structures run out normal or at an angle to the general trend of the coast. This type of coast is more compartmentalized from the point of view of ease of transport of material along it. Frequently different types of beach material are found in neighbouring bays and each forms a unit in the coastal system. This is exemplified on the west coast of Ireland where neighbouring bays may contain either large boulders, granitic quartz sand, or organic sand, indicating a lack of transport between them. Heavy mineral analyses have also illustrated this point in Brittany.

The low coasts that are backed by dunes or marsh are usually much straighter over longer distances so that sand can move freely along them over a considerable area. This type of coast is more common in the swell coast areas of low latitudes than in the storm wave environments of higher latitudes on the whole. The reason for this is that the swells are more often constructive, in that they move material landwards, and where this is mainly sand it can be blown inland by the wind to form dunes; marshes can develop in the lee of the dunes where the level is low enough. This occurs fairly widely owing to the recent large-scale rise of sea level, which is one of the variables that helps to account for the very common coastal type consisting of a barrier island separated from the shore by a lagoon in which sediment can accumulate. Thus the barrier island coast is typical of the low marsh and dune coast of middle to low latitudes, although such coasts are not confined only to these latitudes. They are, however, less common in areas that have undergone rapid isostatic recovery in previously glaciated areas. The barrier island coastal type is well exemplified along the eastern coast of the USA, the Gulf of Mexico, and considerable stretches of the southeast coast of Australia, as well as long stretches of the coast of India and eastern South America and the Baltic. Many of these areas have relatively low tidal ranges. This type of coast covers about 13 per cent of the coastline of the world, according to Gierloff-Emden (1961). The barriers are usually long and straight so that material can move unimpeded along them apart from tidal inlets.

1.2 Sediment movement

The movement of sediment in the nearshore and foreshore zones is one of the most important aspects of the dynamics of coastal processes. Sediment can move both alongshore and normal to the coast, but the longshore movement of sediment is of greater importance in explaining the incidence of coastal erosion and coastal accretion in many areas. The best protection for a coast from erosion is a wide high beach. The beach material absorbs the wave energy, which is expended on moving the material rather than attacking the solid rocks or drift of the backshore zone to which the waves cannot reach when the beach is high. The importance of longshore drift is apparent in the nature of many coast defence works. In many areas these consist of groynes, the aim of which is to trap material moving along the coast to raise the level of the beach and thus protect the backshore zone. Sand

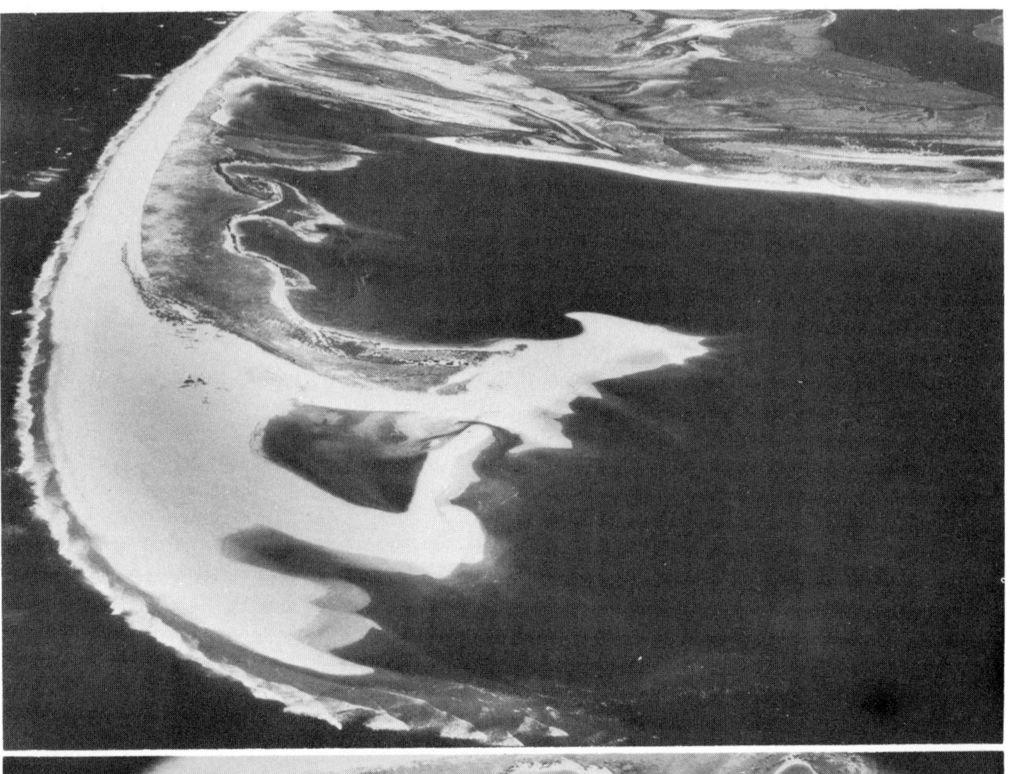

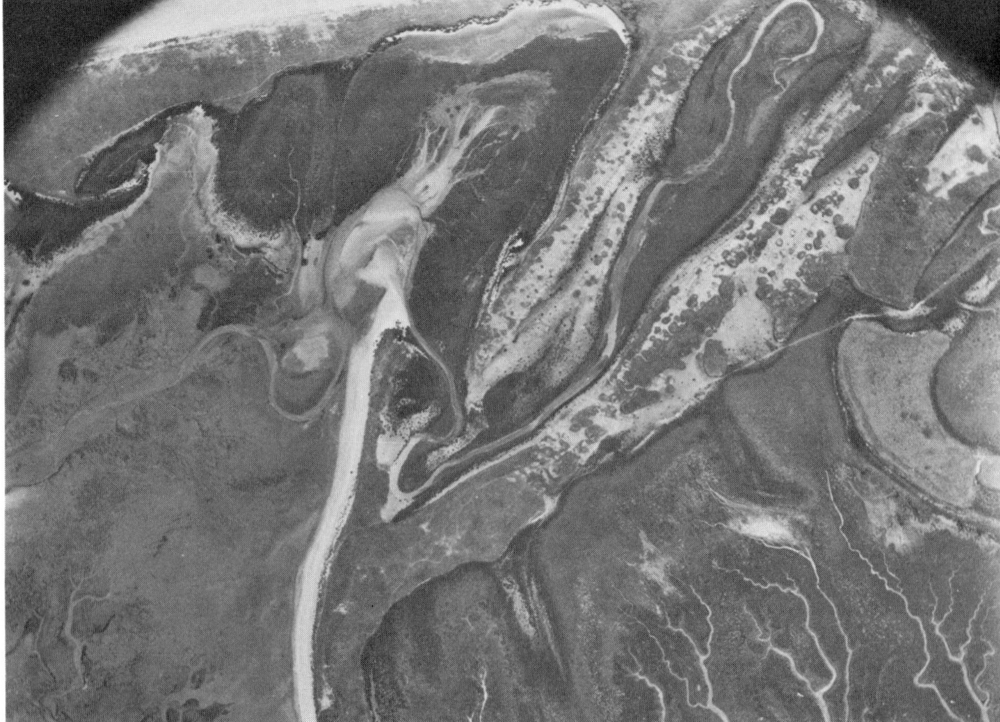

dunes are also valuable from this point of view because they contain a reserve of sand, on which the waves can expend their energy when the normal sea level is raised in a storm surge, or when destructive wave action removes sand from the foreshore. The method now considered best for beach restoration and defence in many areas consists of building up the beach by the emplacement of artificial beach fill derived either from offshore or from a suitable source on land near the eroding beach.

Beach sediment can be derived from one or more of four different sources. It can come 1) from inland via the rivers, glaciers, or wind, the former being the most important; 2) from erosion of the cliffs in the backshore zone; 3) from offshore; or 4) alongshore. The material moving alongshore must itself have come from one of the other sources. Most of the material moving onto the coast from offshore also comes originally from the land, with the exception of the organic sands, consisting of foraminifera, and either calcareous or siliceous remains of other organisms, including corals and shells.

The proportion derived from these different sources varies greatly from place to place, and now in some areas the beaches are probably living on capital with relatively little new material reaching them from any source. An example is the great shingle structure of Chesil beach on the south coast of England. The extremely well-sorted nature of this beach confirms its relict character. Owing to the recent rise of sea level many rivers are no longer bringing coarse material to the sea, which would help to account for the relict nature of many shingle beaches. Some material is probably still derived from suitable submarine outcrops, its movement towards the shore being facilitated in some instances by transport on seaweed. Another reason why some of the material brought down by the rivers cannot replenish the beaches is that where there are barrier islands, much of the sediment coming into the sea via the rivers is trapped in the lagoons. Not until these are full will material be able to by-pass them to reach the open beaches. Only the sand, however, will be of value in building up the beaches in most areas, while the mud will be carried out to deeper and quieter water where it can settle. Most cliffs probably provide relatively little beach sediment. The best cliffs for beach building are the drift cliffs. They provide material ready ground up to small size and they erode at a much greater rate than solid rock cliffs. The offshore zone can only supply a limited amount of material to the beaches because of the limited depth to which waves can move sand in large amounts. The waves are more effective where the offshore gradient is very flat because they can then move sand inland over a wide stretch, but in doing so the depth is increased and the supply eventually is reduced, unless sea level is rising. Calculations made by Van Straaten (1965) for the coast of the Netherlands indicate that about half the sand that has built the extensive barrier system along this coast was derived from the sea floor very gently stretching into the North Sea, while the other half was supplied by the Rhine and Meuse, from the estuary of which rivers it moves northwards along the coast to feed the Dutch barrier islands. A calculation made by Valentin suggests that only 3 per cent of the material eroded from the Holderness cliffs moves southwards to build up Spurn Point, the spit that prolongs this coast and extends partly across the Humber estuary (figure 3.5). The remaining material is spread over the sea floor and eventually some moves south to feed the accreting beaches of north and south Lincolnshire and the Wash.

Plate 4 Scolt Head Island, Norfolk (*opposite*). The oblique air view (*above*) shows the sandy ridges, under whose shelter salt-marsh has accumulated. The refraction of the waves around the tip of the spit can be seen. The vertical air view (*below*) shows the pattern of sandy ridges, covered by dune vegetation, with foreshore ridges and mounds. Sand waves can be seen on the tidal banks in the lower part of the picture. Tidal creeks drain the salt-marsh developed in the shelter of the ridges. (*J. K. St Joseph. University of Cambridge, Committee for Aerial Photography.*)

On a world scale, sand made of quartz grains is the most common beach material. The quartz grains are extremely resistant to further attrition and weathering and for this reason they make up a large proportion of beach sand. Basalt sand wears down much more rapidly, as do the softer organic materials. Granite and sandstone are also common rocks providing the large amount of quartz grains that make up many of the world's beaches.

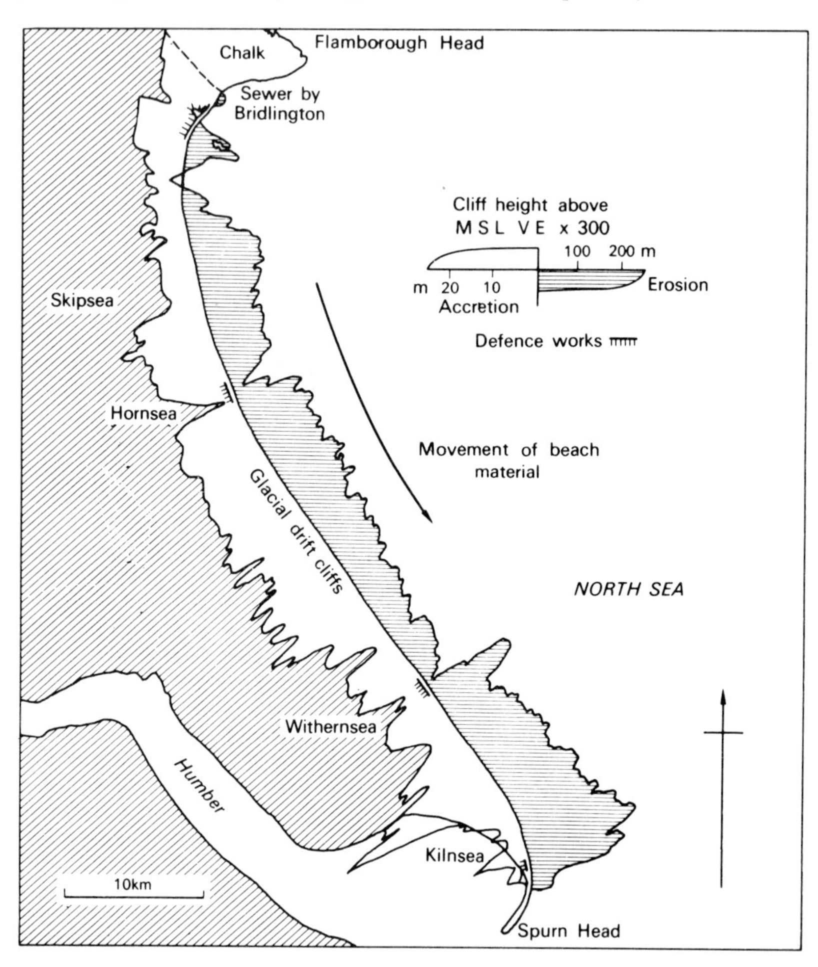

Figure 3.5 The coast of Holderness and Spurn Head, east Yorkshire, showing erosion between 1852 and 1952. (*After Valentin, 1954.*)

Movement of material normal to the shore depends on the nature of the waves that are most important in bringing about this movement. Model experiments on the movement of sand normal to the shore can be made in a narrow wave tank in which traps can provide an estimate of the net amount of material moving. A series of such experiments showed that the movement of material can be divided into two distinct zones. The area outside the break-point of the waves in deeper water was characterized by a shoreward movement of sediment under nearly all conditions. The amount of movement increased as the waves became higher, as they became longer, and as their energy increased. It also increased with decreasing beach gradient and as the water became shallower as the break-point was approached. More material moved when the bottom was rippled than when it was smooth. The only variable that adversely affected the landward movement of sand outside the break-point was the onshore wind velocity. The occurrence of a strong onshore wind set up a seaward current that reduced or reversed the landward movement of sand outside the break-point.

Inside the break-point on the other hand, the movement was less consistent. In this zone the direction of movement changed with the wave steepness, or height-to-length ratio. Steep waves, particularly when they were accompanied by a strong onshore wind, were destructive inside the break-point, moving sand offshore to the break-point, where it accumulated. These waves are similar to storm waves on a natural beach. Long, low waves that have the characteristics of swell waves were shown to move material landwards in all depths in front of the break-point so that it was built up into a bar at the limit of the swash of the waves. Thus there appears to be a fundamental difference in the effect of different wave types in the movement of sediment in the shallow water of the surf zone.

Observations made on natural beaches have shown that similar effects take place under the more complex natural conditions. It is well known that storm waves comb down beach material from the upper beach and deposit it at or beyond the breaker zone. The long swells that often follow the steep storm waves then usually return the material to the upper part of the beach. This process is cyclic on some coasts where there is a strong contrast between conditions in summer and winter. On the coast of California, for example, the beach is lowered and narrowed during the winter storms, with a submarine bar forming in the breaker zone, while in summer the constructive waves move sand up onto the foreshore and the beach becomes wider and higher. The maximum depth to which these movements take place is about 5 m in most seasons, below the lowest tide level. Large volumes of material, therefore, are only moved in relatively shallow depths in a direction normal to the shore, and such movements are rarely continuous in direction. The amount of material probably remains relatively constant in the absence of longshore movement when other conditions are also in equilibrium, such as the level of the sea and the wave pattern.

The movement of material alongshore is of considerably greater importance in many areas because it is often such that there is a net movement in one direction over a long period of time. The cause of longshore movement of material is usually the oblique approach of waves to the shore so that an alongshore component of wave energy is present. The effect of tidal streams is also significant in some areas where there is a residual movement in one direction. This occurs in areas where the flood or ebb tide is dominant; several examples occur off the east coast of Lincolnshire and East Anglia, where tidal morphology and tidal streams are such that ebb and flood tidal channels are well developed.

On most mesotidal and microtidal coasts the oblique incidence of waves is more important in causing longshore movement of material. The amount of material moving alongshore is likely to be greatest along those coasts where waves are fairly short, because short waves suffer less refraction than long ones and so reach the coast at a greater angle, resulting in a stronger longshore movement. The longer waves are greatly refracted and tend to reach the coast at a small angle, but because of the marked refraction there is likely to be more variation in the distribution of energy along the shore. This variation will result in the generation of longshore currents from zones of high waves and high energy to zones of low waves and low energy. This pattern will normally result in less consistent longshore movement of material. The movement will be more prolonged on coasts that are simple in outline, because material can move unimpeded by headlands for considerable distances in such areas. Thus longshore movement will be most consistent in direction and greatest in those areas where waves are short and approach obliquely from one dominant direction, and where the coast is relatively simple in plan.

Equations to determine the amount of material moving alongshore have been derived theoretically, in terms of wave dimensions and angle of approach, as well as from empirical observations made both in model wave tanks and full-scale observations. Model studies have indicated that when the waves are steep most of the material moves alongshore in the breaker zone. When the waves are flat the bulk of the longshore movement takes place in the swash–backwash zone, by the process of beach drifting. The optimum direction of wave approach to maximize longshore transport was found to be an angle of 30 degrees.

Measurement of longshore drift in the field can be attempted by several methods. One method is to survey a stretch of beach before an obstruction, such as a breakwater is built across it and then to resurvey the area afterwards. The beaches on the updrift side of the obstruction will accrete while those on the downdrift side will erode. An example of this method has been carried out at Santa Barbara in California, where a breakwater was built in 1929. Several surveys have been made and the rate of longshore drift assessed as 214,360 m^3/year. The erosion resulting from the building of the breakwater extended 16 km downdrift in only a few years. The volume of sand moving along the coast at Atlantic City on the east coast of USA was estimated to be 305,500 m^3/year. Another method of measuring longshore drift over a short time span is to follow marked material. This method has the advantage that the movement (because it is measured over a short time span) can be related to the waves, wind, and tidal conditions at the time. There is, however, the disadvantage that the short period of measurement may not be typical of that area. The former method is therefore more suitable for obtaining the amount of longshore transport at a particular site, while the second is better for obtaining relationships between the amount of longshore transport and the processes that cause it, especially the waves and wind.

Tracer experiments have been carried out widely, using both radioactive markers and more recently fluorescent tracers. The experiments have been carried out with both sand and shingle and have shown that sand disperses very rapidly under vigorous wave action. Ingle (1966) has made many observations on the exposed beaches of western California. His results showed that the movement was complex and very variable, with a mean rate of 305 m^3/day, but with a range from a maximum of 1430 m^3/day to a minimum of 56·6 m^3/day. The rates generally increased with the increase in breaker angle to the shore and were greatest when the wave steepness was between 0·003 and 0·0045. Size of material affected the rate of movement, and movement was only typical of the actual beach material when the tracer was in equilibrium with the actual beach material. Experiments made with shingle also showed that different sized stones moved at different rates and sometimes in different directions. In nearly all tests the scatter of points has been very variable, indicating that many variables are involved in accounting for the volume and direction of longshore movement of beach material.

On most coasts longshore movement can take place in both directions according to the direction from which the waves approach. One direction is, however, likely to be dominant. The dominant direction of movement may be determined by the pattern of wave attack, while in some areas the variation of fetch in different directions may ensure that one set of waves is more important than others. On the south coast of England, for example, both the maximum fetch and the prevalent waves come from the west so that material tends to move predominantly east along this coast. On the east coast of England the situation is not so simple because the maximum fetch is to the north and east, while the prevalent wind is from the southwest. Under these conditions the balance between movements in either direction may be closer, although in general movement is south along the coast, aided by the tide in places.

Where the coastline is in the form of broad bays, in time the longshore movement of material is likely to change the outline of the coast by transfering material downdrift and changing the alignment of the coast until it is in adjustment with the direction of wave approach. Where the waves approach from one dominant direction, the form of the beaches in the bays will tend to approach the form of a logarithmic spiral, with the tighter part of the curve in the shelter of the updrift headland and the straighter part on the downdrift side. By recognizing this pattern and its cause, it is possible to ascertain from map evidence the dominant longshore drift direction of any coast that shows the logarithmic spiral form. This is a stable form along which longshore movement is in adjustment and is reduced to a minimum.

1.3 Coastal accretion

Coastal accretion takes place where more sediment reaches a coast than leaves it. The addition of material can only take place over a long period of time where the material is moving into the area from alongshore when the sea level is static. A wide variety of forms can be produced as material builds up at the coast. The form depends both on the type of material of which it is composed and on the processes that cause the accretion.

Shingle often forms ridges built up by storm waves that can throw material above the reach of normal waves. The shingle, consisting of large particles, has a very high percolation rate so that all the swash at the top of the beach sinks into the shingle; it cannot, therefore, be carried down the beach by the backwash. Thus storm waves are responsible for the most permanent features on a shingle beach. The shingle can be thrown to considerable heights by the storm waves, a feature that is well illustrated by Chesil beach in Dorset. This shingle structure is a tombolo that ties the island of Portland to the mainland, which is separated from the shingle bank by the shallow Fleet for half its length of nearly 30 km. At the Portland end where the shingle is largest the beach crest is 13·3 m above the normal high water. The shingle has a size between about 5 and 7 cm in this section. At Abbotsbury where the bank reaches the mainland, the pebbles are about 1 cm in diameter and the bank extends about 7 m above the high water level. The extremely well-graded nature of the shingle on this beach is an indication that it is no longer receiving shingle, but is a relict feature that has probably been driven slowly landwards as sea level has risen, becoming orientated to face the storm waves and achieving the high degree of sorting in the process. The greater height of the structure at the Portland end is probably due to the greater wave energy that can reach this part of the beach, which is the most exposed and has the deepest

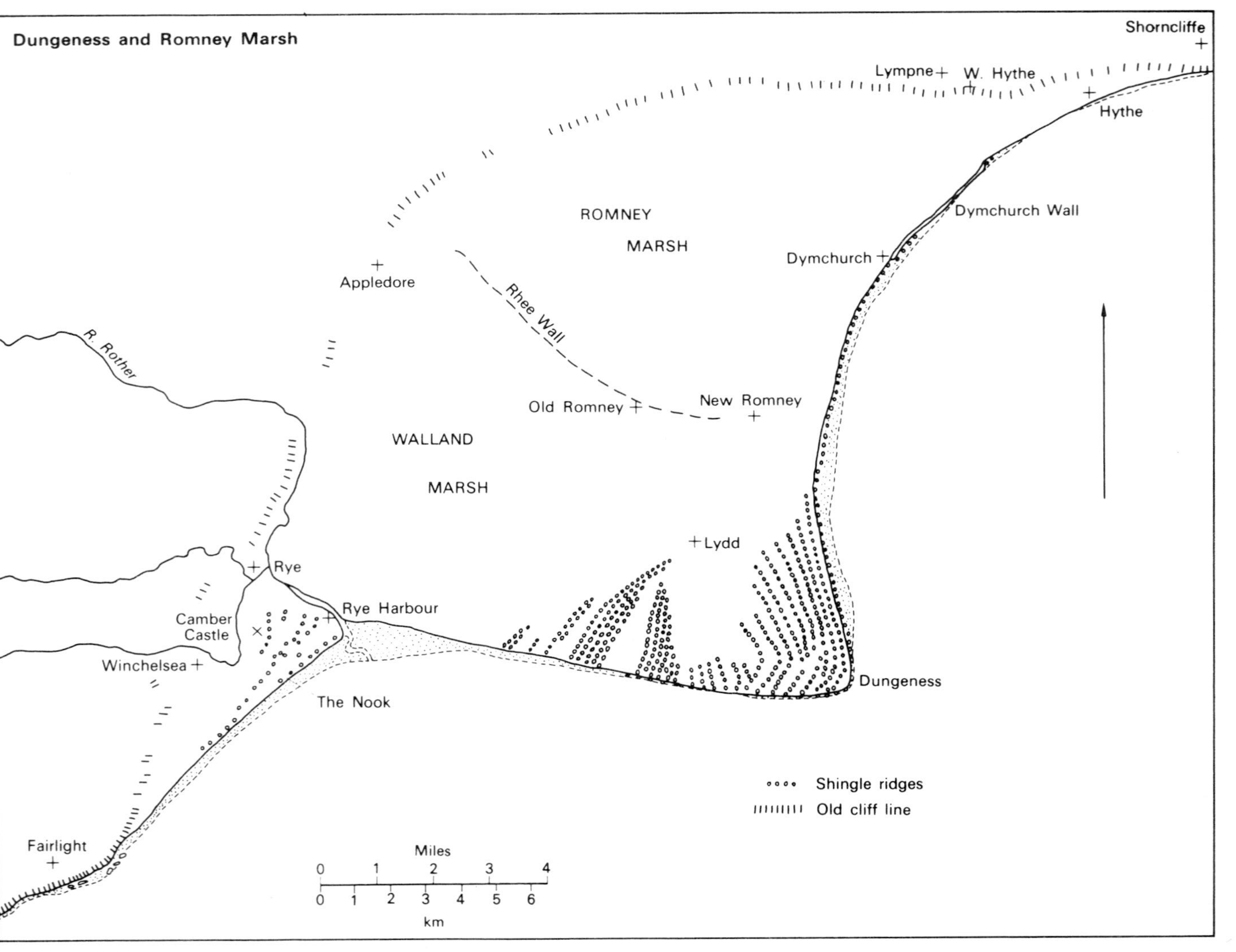

Figure 3.6 Map of Dungeness and Romney Marsh area showing alignment of shingle ridges and the old cliff line

water offshore. Thus in this position the shingle is coarsest, as coarse material is usually associated with higher energy conditions on any one beach.

Another shingle structure that is also formed mainly of storm ridges is Dungeness (figure 3.6). This feature, which is situated on the southeast coast of England, illustrates another point concerning the pattern of coastal accretion. Accretion has filled in a broad bay in this area and has subsequently built out a series of beach ridges to form a cuspate foreland. This is a triangular-shaped zone of accretion that has become increasingly angular as it has developed. The changing alignment of the shingle ridges as the feature has developed can be readily seen on air photographs. The most recent ridges lie parallel to the eastern-facing side of the ness, while the side that faces south between Rye and the point of Dungeness truncates the old ridges that run approximately from southwest to northeast. The sharpness of the point is due to the close presence of the French coast to the southeast that prevents effective waves coming from this direction. The south-facing side of the ness is shaped by the waves coming from southwest or further west up the English channel, while the east-facing side is built up by the addition of ridges thrown up by waves coming from the east across the southern North Sea. Material moves eastwards along the south-facing coast and around the point. This movement has necessitated the artificial transference of shingle in the opposite direction to prevent erosion at the site of the atomic power station built near the tip of the ness on the south side. Dungeness has built out steadily as shingle has reached it from the west. Between 1600 and 1800 progradation took place at a mean rate of just over 5 m/year. The general form of this cuspate foreland is such that it has made the coast more irregular since it started to build out from the original fairly straight bay mouth barrier form. This type of form occurs in areas where the direction of wave approach is limited to specific directions.

Hurst Castle spit on the south coast of England, which is partially protected by the Isle of Wight from waves from the southeast, also illustrates the importance of the direction of wave approach in establishing the form of the feature. This spit continues the line of the mainland coast where this turns abruptly northeast, running in a southeasterly direction for about 1·5 km before turning gradually more west to east, then it turns sharply to the northwest. The distal part of the spit consists of a series of ridges facing northeast. These ridges have been built by the action of waves approaching from the northeast along the Solent, which provides a reasonable stretch of open water. The southeasterly trending part of the spit has been built by storm waves approaching from the southwest along the English Channel, which become refracted to cause the bend along the length of the main part of the spit. The material of which the spit is formed is carried along the feature by the waves from a westerly quarter that approach the spit at an oblique angle, and can therefore effect considerable longshore movement.

Thus shingle forms are likely to be complex in outline where the direction of wave approach is limited to certain directions by the indented nature of the coastal outline and where waves from different directions can reach the beach. The features also illustrate the importance of oblique waves in carrying material into the area, while on shingle it is the storm waves that build up the highest and most permanent ridges and which determine the orientation of these features. These forms are built up normal to the direction from which the dominant storm waves approach. There are, however, other features that build up parallel to the direction from which the dominant waves come. These features are exemplified by the 'comet's tail' shingle banks that form in the shelter of an island and which are elongated downwave of the island to form an elongated bank of shingle parallel to the wave direction of approach. A number of features of this type have developed on the intricate island-studded coast of north Brittany in an area where granite supplies suitable sized shingle.

Accretion of sand on the foreshore can take place in a number of different ways, producing a wide variety of forms. Along indented coasts, and where waves are short and therefore not very refracted, complex spit forms can develop. Similar factors affect their form as

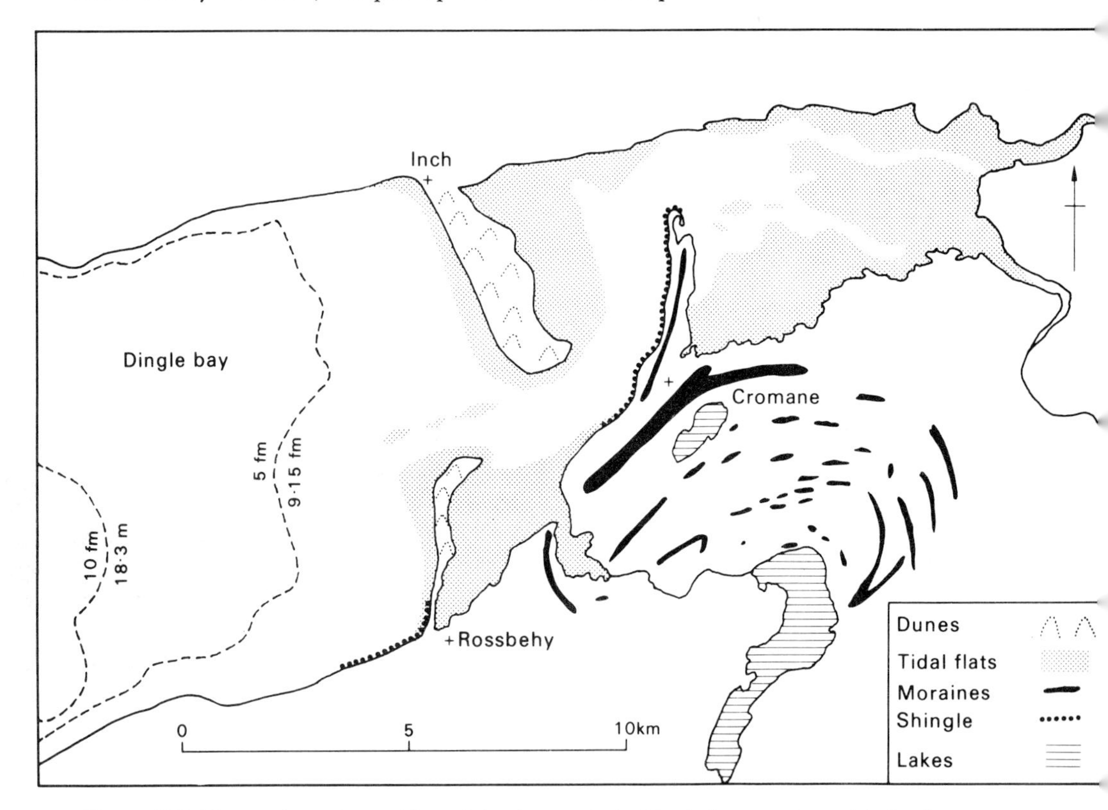

Figure 3.7 Map to show the position of the three spits in Dingle Bay, southwest Ireland.

those that build shingle features, with the important difference that it is the long swells that are the dominant waves on sand beaches rather than the storm waves. The sandy spits are, therefore, orientated to face the direction from which the long swells come. These waves are constructive on sand beaches building up a berm at the back of the foreshore owing to the continuous landward movement of sand under the action of long, low constructive waves. Because the waves are long they tend to suffer more refraction than the shorter, steeper storm waves that form the shingle storm ridges. Thus sandy spits are aligned normal to the direction of approach of long, constructive swells. The waves generally refract around the tip of the growing spit, thus producing the smoothly rounded recurves that terminate many sandy spits. These recurves often have the same logarithmic spiral form as was noted in the shape of sandy bays between headlands. Sandy Hook in New Jersey shows this form well, and it has also been recorded along the spit at Gibraltar Point in Lincolnshire.

The form of sandy spits depends in part on the exposure in relation to wave approach. The two outer spits in Dingle Bay in southwest Ireland are sandy features that run out into the bay at right angles to its long straight sides (figure 3.7). These features are formed of sand which was washed into the bay as glacial outwash, forming a gently sloping flattish sea floor. The gradient of the nearshore zone was too flat to be in equilibrium with the waves, having a slope out to about 20 m of only 1 in 400, while the equilibrium beach gradient is about 1 in 80. The waves therefore built up a berm, which has since been widened and raised by the accumulation of wind-blown sand and the development of well-vegetated sand dunes.

These spits illustrate a characteristic of many spits—a narrowing at the proximal end of the spit near where it joins the mainland. This feature is probably the result of a slow turning round of the spit to face more nearly the direction from which the long swells come, as material moves along the spit towards its distal end. The increasing width of the foreshore in this direction is evidence of this movement. The spits in Dingle Bay are unlike many spits which continue the line of the mainland coast where this turns abruptly inland. Dawlish Warren, across the mouth of the Exe estuary, and Spurn Head in Yorkshire which partly crosses the Humber estuary, provide two examples. It is quite common to find spits growing from both sides of an estuary towards the centre. In some instances there is cartographic evidence that double spits of this type formed as the result of breaching a single spit, but in some instances (for example, the two spits at the mouth of the Taw and Torridge estuaries in north Devon) the spits have grown independently in opposite directions, each receiving sediment when waves approach from a suitable direction. This form can develop where the direction of longshore drift varies under different conditions, and this is common.

Vegetation plays an important part in aiding coastal accretion in suitable areas. On sandy coasts as wind blows sand above the reach of the waves onto the backshore zone, sand-loving plants can colonise the loose sand, and help to bind it and trap further sand. The first plants to colonize growing dunes are the annuals, *Cakile maritima* and *Salsola kali* being common around British coasts. The more important plants are the dune grasses of which marram, *Ammophila arenaria*, is the most important in many areas. This grass only flourishes where sand is being actively blown around it, and then it grows up through the sand, stabilizing and raising the level of the foredunes. The grass is used artificially in many areas to encourage the growth of dunes, which provide a very valuable protection to low coasts, such as those of parts of eastern England, the Netherlands. Where fine material can accumulate in the lee of a spit, barrier, or some other structure, or where the foreshore is very sheltered inside an estuary, the plants that form salt-marsh vegetation can take root and grow. These plants are adapted to withstand immersion in salt water for part of the tidal cycle and they, like the dune plants, help to trap further silt and mud. A considerable variety of plants grow at different levels in a salt-marsh. One of the most important in British marshes at present is the rice grass, *Spartina townsendii*. This plant only developed as a hybrid in the middle of the nineteenth century in the south coast marshes. Since then it has spread widely around the south and east coasts of England and it is being increasingly used to assist marsh growth in areas that are being raised prior to reclamation. This grass grows vigorously in soft mud and is a very efficient trapper of silt, raising the marsh at up to 10 cm/year under optimal conditions, such as occur in the macrotidal, heavily silty water of Bridgwater Bay in the Severn estuary. Normal rates are considerable lower, and they fall off as the marsh level rises above that at which the plant grows best. *Spartina* has a relatively small latitudinal range of between 48 and $57\frac{1}{2}$°N and 35 to 46°S. In low latitudes the normal salt-marsh of the middle and high latitudes is replaced by mangrove swamps, which also provide useful silt trappers, although they are not so suitable for later reclamation of marsh land for agricultural purposes as the salt-marsh areas.

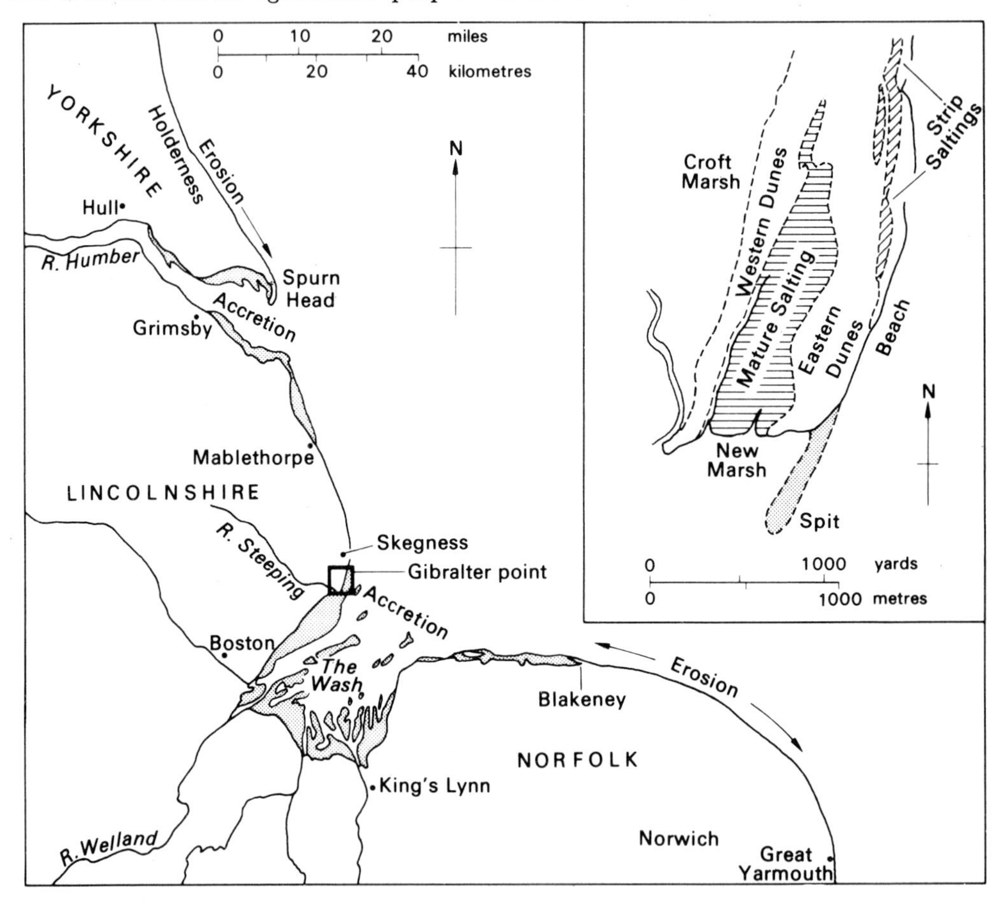

Figure 3.8 Map of the Lincolnshire coast, with inset to show details of features near Gibraltar Point.

The coast of south Lincolnshire provides a good example of coastal accretion by outbuilding eastwards (figure 3.8). The accretion takes place by a process of ridge building on the lower foreshore. The material of which the ridges are formed is brought up onto the lower foreshore by the residual movement of the tidal streams. There are many banks offshore and between them tidal channels are used as routes by which sediment passes around them and up onto the foreshore where the flood channels run towards the shore. At these points nesses of accumulation build out, creating a very flat foreshore gradient. On this low gradient the waves build up ridges of sand, which diverge slightly southwards away from the shore. The waves gradually move the ridges southwards, as the longest fetch is to the north from which direction the long, constructive ridge-building waves come. The ridges, therefore, move landwards and as new ones develop to seaward they become stabilized, while their distal ends swing inland, giving arcuate features, convex to the sea. The ridges develop sand dunes as they grow in height by the addition of wind-blown sand, and salt-marsh plants help to raise the level in the runnels that fill with fine sediment in the shelter of the ridge to seaward. By this process the coastline has built out 244 m between 1959 and 1969, where the accretion is most rapid. Over a period of about 150 years the mean accretion has been nearly 1 km, with different stretches of the coast building out at different times as the tidal channels have moved in position, thus feeding different stretches of the coast at different times.

The processes taking place on the south Lincolnshire coast illustrate on a small scale the development of the much larger barrier islands that form an important type of coastal accretion in many areas (figure 3.9). The major difference between the two types of accretion is the way in which the sediment reaches the coast. The tidal streams are more important on the Lincolnshire coast, where the tidal range is about 6 m, than on the generally low-range tidal coasts where barriers are best developed. Three possible formations of barrier islands have been advocated in the past: 1) the upbuilding of offshore bars, 2) the cutting of inlets through spits, and 3) the submergence of ridge-like coastal features (Schwartz, 1971). There are difficulties in the theory that suggest that the barriers have developed from submarine bars, as these features are normally formed at the break-point of storm waves and do not emerge above water level. Hoyt (1967) points out that if this theory was correct there should be beach deposits on the landward side of the lagoon behind the barrier, but they are not found in this position. This point is emphasized by Hoyt who advocates the importance of submergence in the formation of barrier islands. He suggests that as sea level rises, material is moved landwards by the waves to form a berm, where the gradient is flatter than the equilibrium beach gradient, and the berm will be raised by dune formation. As sea level rises the land behind the growing barrier will be flooded to form the lagoon. The rate of submergence must, however, lie within critical limits if a barrier island is to be formed and maintained during the transgression. In this view much of the material that builds up the barrier comes from the sea floor, although it is also recognized that some can move alongshore, so that in some instances barriers can form from spits. The spit hypothesis of barrier formation is supported by J. J. Fisher (1968), while Otvos (1907) advocates the offshore bar origin. His examples of the upwards growth of bars to emerge and form barriers are drawn from the coast of the Gulf of Mexico. Most of the barriers along this coast have formed in the last 3500–5000 years, during which submergence has not been at appropriate rates. The theory of Hoyt operates best when sea level is rising slowly, while the breached spit theory will work at a steady sea-level period, and the emergent offshore bar could be best achieved when sea level was falling slowly. Thus the three main theories each supposes a different type of sea-level change. Schwartz (1971), therefore, has proposed a multiple causality for the formation of barrier islands. The primary type he suggests formed as engulfed beach ridges, the secondary type either by the breaching of spits or the emergence of offshore bars, and the third by composite origin, in which a combination of the previous three processes takes place. There does appear to be evidence, in different places where sea-level changes have been dissimilar, for the

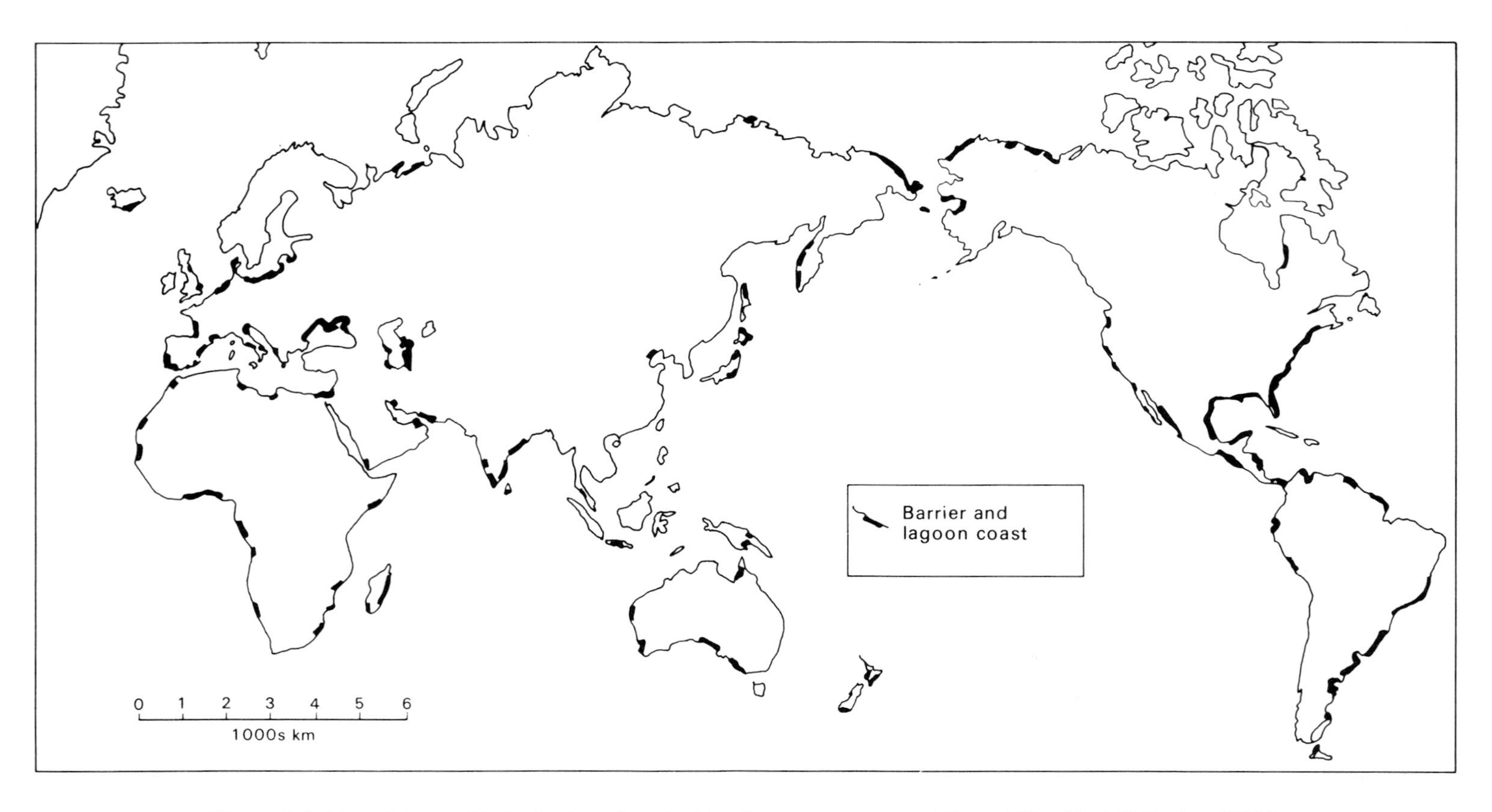

Figure 3.9 Map of the world distribution of barrier islands and lagoon coastal lines. (*After Gierloff-Emden, 1961.*)

formation of these barriers by the various processes suggested. Both longshore movement and movement from offshore have probably brought material to form the barriers. A low offshore gradient is also probably necessary for their development. In the tropics, barriers tend to be narrow, owing to the paucity of dunes in these areas. Barriers are common on the coast of southeast Australia, where much sand was left on the shelf during the low glacial sea level. This sand has been formed into an extensive barrier system by the waves of the advancing Flandrian sea. The large amount of available sand has resulted in high dunes, which exceed 150 m in height in some areas.

1.4 Coastal erosion and cliffs

Erosion of the solid rocks of the coast can only take place where waves can attack the backshore zone, or at least act as transporting agents for material that has been moved down the backshore zone onto the upper foreshore by subaerial processes, among which mass movement is most important. This situation only occurs in those areas where the beach lacks adequate material to absorb the wave energy and prevent the waves reaching the rocks. Lack of beach material will occur in those areas from which more material is moved by longshore transport than is carried into the area by the waves and tidal streams. These areas will include headlands, which are usually sites at which wave energy is concentrated by refraction, and therefore where the waves are highest and have the greatest energy. The wave-generated currents tend to flow away from these areas to those of lowest energy and smallest waves, carrying any available beach material to these low-energy zones. As material cannot reach the headland from alongshore, the rocks of these areas are open to attack. On the other hand, the headlands are usually formed of the most resistant rocks; this factor has enabled them to resist erosion in the past and thus to become outstanding zones along the coast. For this reason they often continue to resist attack. Erosion will be more severe in areas of less resistant rock or drift cliffs where adeqate beach material is lacking. The erosion of the Holderness coast has already been mentioned. It is particularly severe in this area because material moving alongshore cannot pass around the headland of Flamborough to the north of the Holderness coast, which has an open, straight form along which material can move readily. Thus material eroded from the cliffs is drifted south and does not afford protection for the drift cliffs.

Coastal erosion will be enhanced by a rising sea level. Bruun has shown that if sea level rises and the equilibrium profile is maintained, material eroded from the upper beach will be deposited on the nearshore bottom down to the limit of wave action. The result is a shoreward displacement of the upper beach as material is moved seawards. The volume of material eroded is added to the nearshore bottom, the elevation of which will equal the rise of sea level, thus maintaining the same offshore depths.

The solid rocks of the backshore zone and foreshore can only be attacked directly when they are not protected by beach material. Experiments have shown that a relatively thin cover of material is sufficient to protect the shore platform beneath the beach, as long as the material is not carried seaward by destructive waves. The experiments suggested that the depth to which sand can be disturbed is of the order of 1 cm for each 30 cm of wave height, the depth increasing as the material gets coarser. Later experiments showed that in some areas the disturbance can be nearly twice as great. Even with large waves, therefore, only a fairly thin layer of beach material will provide adequate protection. Destructive waves,

however, are likely to remove much beach material from the upper foreshore and carry it to deeper water. Under these conditions the cliffs and platform may be exposed to erosion, which is enhanced by the rasping action of the beach material carried across the platform by the waves. The cliffs are also attacked in other ways by the waves.

There are four main processes by which the waves can attack solid rocks: 1) chemical weathering processes termed corrosion, are particularly effective on limestones; 2) corrasion on the other hand, is the process whereby the rocks are directly worn away by the material thrown against them by the waves, the larger and more angular the particles the greater the damage they can do provided that the waves are powerful enough to throw them against the rocks of the cliff or drag them over the platform; 3) the process of attrition by which the material carried by the waves is itself broken down and rounded by being rubbed together and against the cliffs and platform; 4) the process of hydraulic action, probably the most effective. It takes place when waves break directly against the cliff or a sea wall. Air in cracks is compressed by the pressure of the water and then expands as the wave recedes, causing force to be exerted on the rock. Another process associated with hydraulic processes is the setting up of shock pressures.

Shock pressures are very local in occurrence and short in duration as they only occur when a pocket of air is enclosed as the wave breaks directly against a rock cliff or sea wall. If the water is too deep the wave does not break and its energy is reflected seaward to produce a clapotis, in the form of a seaward-moving wave which interacts with the incoming waves to set up a standing oscillation under ideal conditions. If the water is too shallow the wave breaks before it reaches the cliffs and again damaging shock pressures cannot be set up. Long waves of high energy that break in the optimum way to generate shock pressures can do considerable damage to cliffs by this process, despite the very short-lived and local nature of the shock, which usually only lasts for about 0·01 seconds at high intensity.

The features formed by the attack of waves on the solid rocks of the coast consist of two main elements—the cliffs of the backshore zone and the platform of the foreshore zone. There has been considerable discussion of the processes that are responsible for the formation of the shore platform. On tidal beaches the platform is a gently sloping rock surface, terminating sometimes in a low tide secondary cliff. The platform at Scarborough in Yorkshire consists of a rock slope with a gradient of about 1 in 90, with a slightly concave upward profile, extending over nearly 800 feet (250 m) where the tidal range is about 6 m. The character of shore platforms varies with the rocks across which they are cut and with the tidal conditions. Many of those along the south coast of England, along which platforms occupy 30 per cent of the coastal length, are concave in the upper part and convex in the lower portion; gradients vary from 1 in 14 to 1 in 90. Where the rocks dip between 8 and 10 degrees the platform often has the same gradient. There is evidence along much of the coast that the platform is a multiple feature, which reflects changes of sea level during the Pleistocene and recent periods, as some parts of the platforms appear to be of considerable antiquity. These platforms have been formed mainly by the processes of corrasion and attrition associated with wave action.

In tropical areas, wave-formed benches have been described by several observers. These

Plate 5 Anderby Creek, Lincolnshire (*above opposite*), showing erosion of the sand dunes by the waves generated by the storm surge of 1953. Mablethorpe Gap (*below opposite*): a breach caused by storm waves at high water level on the same occasion. The gap is being sealed by sand bags. The low level of the land behind the sea defences can be seen. (*T. Bailey Forman Newspapers Ltd.*)

features occur as relatively narrow horizontal benches at various levels up to about 6 m above sea level on microtidal coasts. The term 'water-layer' weathering has been used to describe the process that creates these benches. The flat features form as a result of chemical changes brought about by alternate wetting and drying by salt water according to some authorities, although others suggest that wave erosion is a more important process. It is likely that both processes operate in different areas and under different conditions, according to the exposure to wave attack and the type of rock. Storm erosion is probably concentrated at the back of the beach and serves to widen the platform, while water-layer weathering probably becomes more important as the platform becomes wider and flatter, a process that is achieved by the scouring action of debris-laden waves. Corrosion forms are particularly well developed on limestones and occur in both cold and warm conditions. Large pinnacles are formed in the storm wave and spray zones and overhanging vizors are characteristic of the high tide zone, below which a bench often slopes steeply seaward to low tide level at about 15 to 20 degrees, below which a low tide cliff occurs with surge caves developed below it.

Cliffs form some of the most spectacular coastal scenery, yet from the geomorphological point of view they are relatively little studied. Their variety and character depend mainly on the type of material of which they are formed, but their development is also important. The glacial and recent changes in sea level have affected the cliffs as they have all other coastal forms, with the result that only a few cliffs that are undergoing rapid retreat are entirely the product of currently acting processes. These are mainly the drift cliffs in areas where other processes allow the waves to maintain their attack on the coast. Examples have already been cited, including the drift cliffs of Holderness in Yorkshire. There are other examples on the coast of East Anglia, where erosion is very rapid in places, as at Happisburgh. The erosion of these drift cliffs and of cliffs formed in soft clays is accomplished mainly by mass movement, with the waves acting as a transporting agent to carry away the material moved onto the upper foreshore by landsliding and slumping. The latter process takes place particularly where clays underlie more coherent and permeable strata such as chalk. The rotational shear slumps that have taken place at the Warren, Folkestone in Kent illustrate this point. Marine erosion at the foot of the cliffs renders the slope unstable, and slumping occurs along circular slip planes, leading to backward tilting blocks of clay, capped with chalk. The slump blocks move down the cliff face and raise a toe of material on the beach. This toe temporarily stabilizes the cliff, until further marine removal takes away the toe material. Mudflows are common on clay cliffs where much water is available, such as at Osmington in Dorset.

The hard rock cliffs of southwest England and western Ireland retreat at a much slower pace, and these cliffs usually have a composite profile. Their form indicates a complex development during part of which the cliffs have been abandoned by the sea when low glacial sea levels occurred. During these stages the cliffs have been modified by subaerial processes, including the formation of head deposits under periglacial conditions; in some instances the cliffs have been mantled by glacial drift. Only since sea level rose to its present height about 4000–5000 years ago have the cliffs been directly affected again by the sea. During this period the sea has freshened the lower part of the cliff, forming the generally short vertical section near the base of the cliff. The upper part of the cliff profile is at a lesser angle, and this part of the profile is often referred to as the coastal bevel. All stages in the removal of the fossil section of the cliffs can be identified around the coasts of Devon and southern Ireland. At Start Point, for example, the fossil cliff is the dominant feature, while where the rocks are softer or the exposure greater, the fossil part of the profile has been mainly removed by active erosion.

Some of the cliffs of western Ireland reach great elevations. These megacliffs, as they have been called by Guilcher (1966), exceed 500 m. The cliff at Croaghan in Achil Island exceeds 666 m. Guilcher suggests that marine processes influence these cliffs to a height of nearly 400 m. One process that could help in the erosion of these cliffs is the development of joints by pressure release as the cliff retreats. This would lead to more effective weathering or periglacial removal, and hence further retreat by positive feedback. Rockfalls and landslides would also be facilitated by this process. These very high cliffs have probably been forming for a very long time, although they will only have been directly affected by the sea during periods of high sea level. They could well have been initiated during high sea-level times in the later Tertiary period. Thus some cliffs are probably among the oldest of coastal landforms.

1.5 The marine cycle

The youth and immaturity of many coastal features has been stressed, particularly the constructive features, most of which have formed since sea level reached its present height about 5000 years ago. These features contrast with the cliffs of hard rock, on the profiles of which there is often evidence of glacial and interglacial events. Thus coastal landforms are affected by processes that operate at very different tempos. On the one hand are the features that are formed by waves and which can reach equilibrium in a matter of hours or days, while on the other, are the major coastal features that develop over periods of thousands if not millions of years. In between are the seasonal and cyclic features, developing over annual or longer periods. The cyclic formation of Spurn Head illustrates a feature that has formed, evolved, and decayed at least three times with a cycle of about 250 years, according to the analysis by de Boer (1964). Beach features belong to the shorter time scale, while coastal landforms belong to the longer time scale. At present there is probably an unusual variety of coastal features, owing to the geologically very rapid and large oscillations in sea level, compared with those operating over most of geological time. These changes have affected not only the current coastal zone, but also large areas of the continental shelf, which is still out of adjustment with current processes and base level over at least part of its area.

If sea level maintains its current height for a considerable time the coastal features will be able to develop towards maturity. Russell (1967) has argued that if sea level were to remain stable, the supply of sediment to the shore would be likely to be reduced as the supply from inland decreased and as the shelves were steepened so that little could be moved landwards from offshore. This would result in accelerated coastal erosion. Curray (1964) has put forward the opposite view, his argument being based on an increase of material reaching the beaches as the estuaries and lagoons (created as a result of recent sea-level changes) become filled, allowing more sediment from inland to reach the open beaches, and eventually to build up the inner shelf area.

The theoretical cycle of marine erosion suggests that the coastal zone will retreat to be replaced by a wide, generally flat platform, a plain of marine denudation. D. W. Johnson

(1919) went as far as to suggest that the land could be lowered to wave base at a depth of about 200 m if sea level remained static for a sufficiently long period. Most of the theoretical cycles of erosion that have been put forward are based on the assumption of long continued stability of sea level. There is probably, however, a limit to which erosion can proceed with a static sea level. It seems unlikely that a wave-abraded platform could be reduced much below a gradient of 1 in 100, and the effective wave base for erosion in rock is probably of the order of 10–15 m. Thus a platform formed by waves with a static sea level could not exceed about 1 to 1·5 km in width. The width could, however, be extended if tidal or other submarine processes were effective, in greater depths than waves. A platform wider than 1·5 km could theoretically be formed if sea level were rising slowly, and it is only under these conditions that wide areas of marine erosion by wave action could occur. The only evidence that this process is effective in reality is found in the surfaces that underlie the major marine transgressions recorded in the geological strata, such as that underlying the Cretaceous strata in Europe. These generally very flat surfaces must have been trimmed by the waves that advanced over them as sea level gradually rose during the transgression. The lack of weathered material beneath the marine deposits supports this view.

During the immediate future, however, sea level is likely to continue to fluctuate too fast to allow coastal maturity to be reached. There is still a considerable amount of ice, which, if it melted, would raise sea level by about 46 m. On the other hand, a renewed glacial period is also a possibility, and this would result in a renewed fall of sea level. Meanwhile, increasing human activity along the coast will further modify the action of natural marine processes along the shorelines of the world.

2 Continental margins

The coastal zone forms the narrow linear link between the land and sea. Seaward of it lies the greater part of the continental marginal zone. Emery (1970) states that the continental margins cover 50 per cent of the area occupied by land, or 21 per cent of the total ocean area. The zone is divided into 27 million km^2 continental shelves, 28 million km^2 continental slope, and 19 million km^2 continental rise—a total of 74 million km^2. The shelf is mostly covered by sediment, much of which is now relict. The sediment has been influenced by changes of sea level, which was about 130 m lower 15,000 years ago. Most of the shelf sediments are retained by tectonic or reef dams, although some are held by the angle of the rest of the adjacent continental slope. The continental rise is formed mainly of turbidity current deposits and pelagic sediment. Mass movement and bottom currents have caused subsequent modification. The continental rises only occur where the continental margins lie within the crustal plates, and they do not occur where the deep sea trenches mark the convergence of two plates.

The Atlantic continental margins date from the Permian–Triassic period, when rifting started. Off eastern North America the basement contours show a ridge and bordering trench, which were probably originally deeper than seismic surveys suggest at present.

Plate 6 Muckross Head, Donegal (*above opposite*). A bay-head beach in a small bay on a crenulate coast. Rocks and a bouldery beach occur along the bay side and cliffs form the exposed coast in the background. There are conspicuous rip currents flowing through the swell breaking on the beach. (*C.A.M.K.*) Slea Head, County Kerry (*below*). A steep-cliffed coast on the exposed western side of Ireland. Note the relation between the bedding planes and the cliff face. Marine erosion is active at the base of the cliff which shows a bevelled profile in the upper section. A small pocket of sand beach has accumulated in the more sheltered cove. (*Republic of Ireland, Bord Fáilte.*)

These trenches trapped sediment (which also accumulated behind barriers and dams of various types) until the end of the Cretaceous, when the trap was filled and started to overflow the dam to build the continental rise.

All shelves show evidence of both erosion and deposition owing to glacial changes in sea level. Many shelves are underlain by thick unconsolidated sediment grading down into sedimentary rocks. Those in the Pacific are mainly held in place by long narrow dams of tectonic origin. Reef dams are common in low latitudes, and also occur off eastern United States.

The continental margin has an average thickness of sediment of 2 km, giving a volume of 150 million km^3, while the rest of the deep sea floor, with an area of 287 million km^3, has an average sediment thickness of 200 m, giving a volume of 57 million km^3. The continental margins, therefore, contain 73 per cent of the total sediment, on 21 per cent of the ocean area. The volume of all the continents above sea level is 125 million km^3, a value smaller than that of the marginal sediments. Most of the material eroded from the continents is contained in the continental margins. They are, therefore, likely to become increasingly important as oil reservoirs. Already 20 per cent of oil comes from offshore and the percentage is likely to rise. Some drills extend 150 km from the shore, in depths of 100 m. The maximum depth of an oil strike was at 400 m, while the maximum depth of an oil show occurred at 3572 m depth in the Sigsbee Deep. Rapid sedimentation, such as occurs around the Atlantic, is favourable for oil formation, but not for the slowly accumulating manganese and phosphorite nodules, with the exception of some banks.

2.1 Continental margin structure

The margins of the oceans and continents can be divided into three types. The first is the Atlantic type, which lacks a trench. The second is the Andean type, which is bordered by a trench, parts of which may be filled with sediments, although parts of it are not. The third type is the island arc type, where a small ocean lies between the arc and the continent. The first type will be considered in this chapter. The last two types are discussed in the next chapter, in the descriptions of the Peru–Chile Trench and the Tonga–Kermadec Trenches respectively.

The Atlantic type of margin structure is known in most detail off eastern USA. It also occurs elsewhere around the Atlantic and along the east coast of Africa and India. Clastic sediments form the continental shelf in this type, which is also characterized by many major deltas, including those of the Niger, Amazon, and Plate. The deltas, which owe their position to the tectonic arrangement of the continents, form wedges of thick sediment. The Atlantic continental margin contains many turbidites. The deep structure of this type of continental margin has been described by Worzel (1965). He points out the similarities between the continental margin of the eastern USA and the oceanic trenches: both have a downwarp or graben fault in the transitional region. For this reason they have similar seismic and gravity anomalies.

The structure of the east coast of the USA shows a thickening of sediment from the shelf offshore, with continental crust under the shelf and oceanic crust beyond the slope. The level of the crust–mantle boundary changes from a depth of 32 km under the continent to 12–13 km under the ocean over a distance of about 200 km. At Cape May the sediment is thicker and the slope of the basement rock is gentler. The transition zone of oceanic to continental crust is variable and difficult to decipher. At Cape Hatteras the Moho slopes steeply, changing from a level of 32 km to 14 km in a distance of 100 km.

A section southeast of Buenos Aires in South America shows a great thickening of sediment beneath the outer shelf and further seawards. The basement drops off fairly gently and is similar to the Cape May section. On the west side of America, a section near San Francisco crosses the continental shelf, on which there is little sediment. The mantle–crust boundary drops in a series of undulations, changing from 13 km to 22 km in 200 km horizontally. A similar distance covers the change from continental to oceanic crust across the continental margin near Antofagasta, Chile. A somewhat similar pattern is found across the coastal margin of Puerto Rico. The boundary rises from 30 km near the island to 15 km in 100 km. This type of change is found also around the Hawaiian Islands. Many different structural environments exhibit this same pattern of change from oceanic to continental structure.

Emery (1965) has described the structure off eastern USA, using data from geophysical surveys. The gravity anomaly is between −24 and +30 mgals at the shore, increasing to +100 mgals at the shelf break and +300 mgals over the continental rise. The change represents the seaward thinning of the continental crust and its replacement by denser suboceanic rocks. Magnetic data are also offset in the region, indicating a fault dating from Triassic times and extending through New Jersey along the continental slope. It then follows the New England sea mount chain southeast to the deep sea floor.

The seismic results show two deep trenches along the continental margin. The inner one is 3–5 km deep and the outer is 7–10 km deep beneath the sediment surface. The outer trench lies under the continental slope and the inner one under the outer part of the shelf. Most of the sediment fill is of early Cretaceous age, the material probably resting in part on a continuation of the Fall Zone peneplain, although this does probably not extend much further seawards. Detailed seismic profiling shows a series of layers of sediment, overlapping onto the continental slope as a growing alluvial fan.

2.2 Continental shelf types

The continental shelf links the continental slope to the land. It is a zone of shallow water, varying greatly in dimension from place to place. The legal and scientific definitions of the shelf differ. It is defined scientifically as the shallow area extending out to the top of the continental slope, where there is a sudden increase of gradient. The depth at which this change of gradient occurs is variable, and so is the shelf width. Shepard (1963a) gives the average width as 67 km and the mean slope as 0°07′. The form and material of the shelf, however, vary greatly.

2.2a Glaciated shelves

A characteristic type of shelf is found off coasts that have been glaciated. These shelves tend to be very irregular. The deep glacial valleys and fjords of the land extend some distance across the shelf, often becoming shallower towards the shelf edge. At their deepest the shelves exceed 180 m. They contain many basins in which mud tends to settle. There are also many hills on the outer shelf, often rising nearly to sea level. Banks of this type include the George's Bank off New England, and the Grand Banks off Newfoundland. They are important for the fishing industry, as their elevation raises them into the region of more active currents, by which nutrients reach the area. Their character is due to the selective scouring action of glaciers, working at a time when sea level was as much as 100 m lower than at present.

The enclosed basins and troughs are analogous to the lake basins in glaciated valleys on land. Ice can erode below sea level until it floats, where the depth exceeds 8/9ths of its thickness. Erosion is therefore not limited to the times of low sea level. Glacial deposition also had its effect on the shelf area; many of the banks may be morainic or consist of other types of glacial deposits. Another characteristic of the glaciated shelves is that there is no orderly grading of material across them, with diminishing size further offshore. In fact the coarser sediment is often found at the shelf edge.

The average depth at the edge of the shelf is 205 m and glaciated shelves are in general very wide, averaging 160 km. The widest known shelf is in the Arctic in the Barentz Sea between north Norway and Novaya Zemlya, stretching 1200 km offshore at a depth between 180 and 380 m, as far as Svalbard. It has troughs and banks.

2.2b Large rivers

Shelves off large rivers are also very wide and are more common off rivers that lack large deltas. They are found in the Yellow Sea, off Northern Siberia, off the Alaskan side of the Bering Sea, and in the Gulf of Siam. These shelves tend to be fairly shallow extending down to about 100 m. The inner edge of the shelf is often fairly smooth, while beyond there may be hills.

2.2c Other shelf types

Some shelves have a pattern of shallow dendritic valleys. The best example of this type is the Sunda shelf, which lies between Borneo, Java and Sumatra and the Malay Peninsula. Kuenen (1950) has described this shelf. The branching pattern of the Sunda River, flowing north between Borneo and Sumatra, can be reconstructed, while another river is indicated, flowing eastwards between Borneo and Java. They are inferred to be former river channels, extending to a depth of about 100 m, where the surrounding shelf into which they are cut is about 60–80 m deep. It is thought that this pattern can be explained by the submergence of the river valleys due to a rise in sea level of about 90 m. The channels do not contain basins, but grade smoothly offshore. They were cut during the period of glacially lowered sea level, which is of the correct order. The distribution of fresh-water fish in the western rivers of Borneo are very similar to those found in Sumatra, which supports this hypothesis.

Other characteristic shelf types include those formed of coral reefs in very shallow water at the edge of the shelf, since reef-forming coral cannot grow in deep water. The offshore slope may be very steep. The depth at the edge of this type of shelf is about 20 m. Off coasts of young mountain ranges the shelf is usually very narrow, averaging 16 km in width. An example of this shelf type is found around much of the Pacific coastline. Such shelves also tend to be fairly deep, their edge being at about 150 m.

2.3 Origin of the shelf

The shelf is complementary to the steep continental slope. Four possibilities have been suggested for their formation. First, they could be constructional features, built up by deposition from the land. Second, they could result from the planing off, by marine erosion, of offshore islands, while the intervening basins become filled with finer sediment. A third possibility is that they are submerged deltas. A Fourth idea is that they may be submerged and down-warped, wave-cut terraces.

This last method of formation has been particularly advocated by Bourcart, in his theory of continental flexure (Bourcart, 1950). He suggested that the continents may flex about the position of the present sea level, the inland area tending to rise, while the offshore zone tends to sink. There is certainly evidence for the sinking of the offshore area, as seen in the shallow water sediments (now deeply buried by later deposits) off the east coast of the USA. If this movement were to take place intermittently it would help to account for the alternating periods of regression and transgression. At times when the shelf sinks and the land is elevated inshore a regression results, rejuvenation takes place as slopes are steepened, coarse sediments are brought down, and at this stage coarse gravel and sands are laid down, perhaps as basal conglomerate.

During periods of quiet sedimentation, which follow the flexing, the land is worn down. Finer sediments are brought down to the sea, which tends to rise at this stage, resulting in a transgression. Whether the change at the coast itself is one of emergence or submergence depends on the position of the line of flexure in relation to the coast. If the line of flexure is offshore, this will cause a withdrawal of the sea as the land tilts, but if it is inland the tilt will cause drowning of the coast at the same time as rejuvenation takes place inland.

As evidence for his theory, Bourcart discusses the character of the coast off Morocco. He suggests that there is evidence for the down-warping of a succession of continental shelves, the margins of which are now at 180 m, 500 m, and 1000 m. The slope of each shelf becomes steeper as it is traced offshore. This type of warping is probably only applicable to limited areas, and even in this particular part of Morocco the evidence has been interpreted differently by other workers. It is, however, an interesting theory and the process may well have played a part in the making of the continental shelf in some areas, where there is evidence of down-warping offshore and elevation inland. The latter process can be seen in the form of elevated erosion surfaces, such as the Schooley peneplain in the Appalachians.

Emery (1970) has given some general conclusions concerning the nature of the continental shelves of the world. He shows that the shelves cover 7·5 per cent of the total area of the oceans, or an area equal to 18 per cent of the land. Only a few areas of shelf are well known scientifically, including the shelves around Britain, eastern USA, parts of western USA, including west Alaska, and small parts of northeast South America, China and Japan.

A division of shelves into those underlain by sedimentary strata, and those underlain by igneous and metamorphic strata, is proposed. In many areas the shelf is the top surface of a large prism of sediment. The sediment is held in place in a number of ways, which form another method of classification. Some shelves are held up by long, narrow fault blocks. This arrangement occurs around much of the Pacific, where deep trenches are common. Along parts of the west coast of USA, a single geological dam runs uninterrupted for thousands of kilometres along the coast. The rocks of these dams are several hundred million years old, but they were emplaced as a dam about 25 million years ago. In places the dam emerges as islands, such as the Farallon Islands off San Francisco. A similar dam occurs off the east coast of USA and in the Yellow Sea. These features range over a great age, some being as old as 500 million years. The one off the east coast USA was in position between

270 and 60 million years ago, during which period the trench to landwards was filled and overflowed across the dam. Another form of dam is the diapir dam found, for example, in the Gulf of Mexico in the form of salt domes. This dam is about 150 million years old. In the eastern Gulf of Mexico another form of dam occurs. It is an algal reef dam (130 million years old), while a younger coral reef dam occurs off Florida (25 million years old). The shelves of igneous or metamorphic rock occur on top of the rock dams, in areas where shelf sediment has not accumulated or has been eroded. These areas are mainly those that have been eroded by glacial action during periods of low sea level. In some places wave erosion may account for igneous and metamorphic shelves which are mainly in high latitudes and are considered young in comparison with the sediment shelves.

There are considerable stretches of coast where there are no dams and the sediment lies at the angle of rest. This applies to the present shelf of eastern USA, large stretches of the east coast of Africa and India and a short stretch of equatorial west Africa, much of eastern South America, western and southern Australia and east Asia, and parts of north USSR. The shelves with thick sediment prisms are mature; they indicate long periods of accumulation, probably accompanied by subsidence of the foundation, as illustrated in the Bering Sea shelf. The shelves have widened the continents by up to 800 km, through the deposition, on average, of 2 km of sediment.

The details of the sediments on the top of the shelves around the world have been greatly affected by the changes of sea level of the Pleistocene. There is no even grading of sediments, and the coarser material is often found near the shelf edge. Sea level 15,000 years ago, when it was at its lowest, lay at the seaward margin of the shelf. Since this date the sea has risen across the shelf and influenced its present superficial character. Submerged barrier beaches are found on some shelves, while others have drowned glacial deposits, such as off northern USA Atlantic shelf.

Many of the sediments on the continental shelves are not yet in equilibrium with present conditions. Swift, Stanley, and Curray (1971) consider that relict sediments can be recognized as petrographic entities, and also by their morphology. There are relict features that can be recognized as old beaches, moraines, etc. Relict fauna and flora also provide means of recognizing relict sediments which are now undergoing modification, especially in the tide-swept shelves off western Europe and off northeast North America, and on the wave-swept shelves of the Middle Atlantic Bight. The sediments that show mixed characteristics (indicating relict features as well as modern modifications) are called 'palimpsest' sediments. They are on the way to becoming modern sediments, in which all relict features have been lost. The modern processes are unmixing relict sediments and forming recognizable features, such as sand waves. Some of the ridges on the bottom of the Middle Atlantic Bight are probably submerged barriers, but others could be formed by modern storms, on the evidence of their alignment (for example, between Cape Fear and Cape Romain). It is likely that both relict forms and modern features are present.

There is great economic potential on the shelves; 90 per cent of the marine food comes from them. Natural gas and oil are becoming increasingly important in this zone, from which nearly 1/5th of world production comes. Sand and gravel will become increasingly important from this source, and is already reclaimed off the USA and from the North Sea. The shelves represent the main area where the erosional remnants of the land come to rest, although some of this material finds its way to the deep oceans also. The shelves, however, only form the upper surface of the feature that includes the continental slope, together forming the continental margin.

2.3a The shelf of the eastern USA

The most common structural feature on the shelf of the eastern USA is an unconformity on top of generally seaward-dipping strata, which are probably Pliocene shales. Above the unconformity lie 10–60 m of sediments with 4 or 5 reflecting horizons. These layers are Pleistocene glacial outwash and marine sediments. The shelf must have built upwards after a period of bevelling during low glacial sea level in the early Pleistocene. Where sediments are abundant they cover the continental slope as well. The development must have been complex, as channels, buried by later sediments, cut across the shelf in many areas.

Processes that lead to these shelf characteristics include erosional ones as well as deposition. Erosion is indicated by terraces on the shelf and by the buried channels and submarine canyons. More than half the present area of the shelf is covered by relict sediments, left by the Flandrian transgression. Tertiary sediments are well represented on the shelf, although late Cretaceous and lower Tertiary strata are truncated along the continental slope. The truncation of the older deposits suggests a period of faulting before the deposition of the younger Tertiary sediments. A fault of 600 m displacement has been located on the Florida–Hatteras slope dating from late Eocene.

The details of the Pleistocene events on the shelf off the northeastern USA have been discussed by Knott and Hoskins (1968). They have recognized five sedimentary series by sparker profiling across the shelf. These series represent phases of erosion and sedimentation related to variations of glacially controlled sea levels. There is also evidence of direct glacial action in folded and overthrust sediments, which are interpreted as due to ice pressure. Parts of the shelf edge show signs of strong erosion, such as off New Jersey. Other stretches, for example off Martha's Vineyard, show evidence of continued deposition in a 200 m layer of sediment at the shelf edge. Filled channels also occur. On George's Bank there is evidence of glacial action also, including 70 m of sediment on the southern part of the bank, part of which fills buried channels. The sediments have been modified by marine action and the unconformities in the strata on the shelf indicate a complex development. The lowest, in the New Jersey–Long Island area, is overlain by 70 m of topset structures, extending over 30 km from the 146 m depth line shorewards. The surface dips offshore at 0·1°. The second unconformity cuts off the offshore ends of the unconformable strata. The dip of the second unconformity is 2–3°. Surface 3 cuts across the shoreward part of the strata belonging to the second set. Surfaces 4 and 5 also form bench-like features overlying the older strata. The arrangement of the strata varies along the coast, so that their formation represents locally varying processes. The shelf off Martha's Vineyard shows filled channels, which were probably cut by glacial melt-waters. The sediments on the shelf here include disturbed beds, overlain by 10–20 m of undisturbed sediment. Both morainic deposits and lobate glacial outwash deposits seem to be included among the strata. The ice appears to have disturbed the bedding of some of the sediments and the junction between the disturbed and smooth areas probably indicates the ice limit.

The general form of the deposits suggests that they are mainly the result of progradation. This process has not been continuous, however, and erosion has occurred with varying intensity in the area extending from New Jersey to Nantucket Island. Both fluvial and marine forces have operated in the southern part of the area off Long Island. Well-developed unconformities indicate the variation in supply of sediment and in the transgression and regressions associated with alternating glacial and interglacial phases. The unconformities mark the glacial episodes; it is suggested that the lowest may mark the Nebraskan, while the second could be Kansan, when the deposits of the Aftonian Interglacial were truncated. An unusual amount of foreset deposition followed this event, possibly in the Yarmouth Interglacial, which appears to have been long. The deeper wave-cut bench may be an erosional feature of Illinoian time, while the overlying benches date from the Wisconsin. George's Bank, which lies further north, has been deeply eroded by ice. The evidence suggests four or five cycles during which the bank was covered by glacial outwash, and then dissected by subaerial erosion during periods of low sea level. The dissection was smoothed and the channels filled during the ensuing transgression.

2.3b Northwest USA

The continental margin of the northwest of North America around the Bering Sea forms a contrast to the eastern margin, and has been studied by Scholl *et al.* (1968). The distance between the tip of the Alaskan Peninsula and Cape Navarin in Siberia is 1300 km. The Bering shelf is shallower than 200 m, and a steep scarp separates the shelf from the deep Aleutian basin, which lies between the shelf and the Aleutian Islands. The steep scarp drops nearly 3000 m in about 25 km, in a rugged slope, scarred with canyons. It separates two very flat areas, the Bering shelf above and the Aleutian basin below. The slope gradient is about 5°, but in some parts it is 10–15° for the first 2000 m.

The seismic reflection profiles show three layers of material on the shelf and slope. The lowest is the acoustic basement of lithified rock. The middle one is a 1–1·5 km thick layer of stratified sediments, which are semi-consolidated sedimentary rock and unconsolidated sediments. The third layer only occurs below the slope and consists of material belonging to the continental rise. The top of the basement sequence may be an erosional feature, cutting across Mesozoic sediments. The main layered sequence above the older material is as much as 1500 m thick, and it covers shelf and slope alike, passing beneath the rise unit in deeper water.

The second unit has well-developed layering within it, and a seismic velocity of 1·7 km/sec. In places it represents a complex history. The strata seem to have accumulated as neritic or hemipelagic deposits of Tertiary age. Three samples that have been analysed have a mid-Tertiary marine flora and are diatomaceous siltstone and tuffaceous sandstone. In the southeastern section there is a surface-mantling unit overlying the main layered sequence unconformably; this is still accumulating rapidly and is a diatomaceous and terrigeneous deposit. Modern rates of sedimentation as high as 15 cm/1000 years have been recorded. It probably started to accumulate 2 million years ago, allowing a more rapid rate of deposition in the Pleistocene. Its base thus may mark the Pliocene–Pleistocene boundary and it can be considered the uppermost member of the main layered sequence.

The history of this coastal margin appears to fall into two main phases. First, an out-building phase was associated with faulting and subsidence of an older margin, the acoustic basement, and the accumulation of the main layered sequence. Second, a later destructional–constructional phase was associated with intense canyon cutting and the formation of submarine fans at the base of the slope, and the surface-mantling sedimentation at a higher level. The acoustic basement must have started to down-warp in the early Tertiary. Some erosion of these older rocks occurred on the outer shelf before the main layered sequence was deposited. Some faulting and folding has also affected the main layered sequence, and these movements have affected the pattern of submarine canyons. There has been considerable subsidence of the Aleutian basin and its steep northern margin during the Tertiary period. This must have amounted to at least 1·5 km in view of the thickness of the sediments in the main layer. If there has been erosion of the underlying basement, then the total subsidence must have been about 3–4 km at least. Thus crustal subsidence of the Aleutian basin would account for its northern margin. The Bering shelf is a slowly subsiding continental platform.

2.3c The southern Californian Borderland

The relief off the southern Californian coast is complex, and the continental shelf itself is narrow. It ranges from less than 1·6 km wide to 24 km, with an average width of 6·5 km (Emery, 1960). There are also shelves around the offshore islands, such as San Clemente Island, becoming wider further west. This increase may be due to greater exposure of the outer islands, which are also composed of less resistant rock. In general, however, there is no close correlation between shelf width and degree of exposure. The whole area of banks, islands, and basins forming the Californian Borderland extends over a distance of 240 km to the edge of the continental slope. The flat tops of some of the banks are up to 12 km wide, with many rock outcrops. Rock also outcrops on the islands and mainland shelves, forming an irregular bottom.

Erosional terraces are found on some of the shelves, where the deposits (usually thin) may include rounded gravel. Many of these terraces are probably related to eustatic changes in sea level, having been formed during glacial lowering of sea level. The shallowest one occurs at a depth of 150–180 m. Its Pleistocene age is indicated by the truncation of rocks of Pliocene age. Many of the terraces are wider and flatter than those which are now emerged above sea level, being up to 8 km wide and having a slope of 1°, as opposed to the maximum width of 0·8 km and 3° slope of the emerged terraces. These terraces are probably younger than the main diastrophic movements on this coast, which are responsible for some of the deeper features of the terraces. Some evidence of warping is given by the increasing depth of the terraces offshore. The difference is greater as the terraces become deeper, which shows that the deeper terraces have been down-warped to a greater degree in the offshore zone, than the shallower ones further inland.

Between the shelves and banks there are basins and troughs. The depths of the basins below their sills lie between 150 m and 88 m. The depth of the sills lie mostly within a range of 36 m. Those basins nearer the shore have flatter floors, often filling the whole area below the sill. Further offshore the floors are only about half covered below the sill. This difference is due to the greater rate of sedimentation in the basins nearest to the shore. The basins mostly trend northwest to southeast, being angular in shape in many instances and thus resembling the land basins on the adjacent coast. The nearshore basins are probably filled largely by the action of turbidity currents, but this method is not so effective in the offshore basins, where sedimentation is by organic material and by the deposition of fine detritus brought in suspension from the shore.

The continental borderland of this area is composed mainly of sedimentary rocks of Tertiary age, although volcanic and older metamorphic rocks also outcrop. An even covering of sediment of recent date overlies these solid rocks. The modern shelf deposits are often lens-shaped, thinning near the outer edge of the shelf. At times coarse material of Pleistocene age is found on the edge of the shelf and at the shore, while the intervening area is covered by a thin layer of more recent finer sediments.

The structure of this area is very strongly influenced by faulting, showing distinctly in the relief pattern. In the submarine environment, faults can be most easily recognized by the relief, which is not modified by subsequent erosion as it is on land. A fault scarp remains a conspicuous feature of the submarine relief. Seven major faults, all trending northwest to southeast, occur between the shore and the continental slope, while one trending east to west occurs near Los Angeles. These primary faults are connected by shorter faults, and in this way the basin and bank structure of this area has been formed. Movement has been both vertical and horizontal along the faults.

The depth of the Moho increases from 13 km below sea level on the abyssal plain, to 18 km at the top of the continental slope, and 32 km at the coastline at San Diego, near the Californian–Mexican boundary. This agrees with the positive Bouguer anomaly of gravity, which increases from 0 mgals along the coast to +80–+100 mgals on the upper edge of the slope. The faulting which gave rise to this complex structure took place around the Miocene period. Since the blocking out of the continental borderland in the Miocene, the outer part has been down-warped, a process that has been going on into the post-Pleistocene period. The material displaced at depth may have helped to build up the uplifted inland area, but its bulk is greater than can be accounted for in this way. The structure of this area gives some support to the idea of continental flexure.

2.3d The continental shelf and adjacent shallow basins around Britain

The character of the shallow sea floor round Britain is very different. The shelf extends for a distance of about 480 km off Land's End and the adjacent basins cover the whole of the North Sea and English Channel. The outer limit of the shelf is marked by the 200 m line.

The structure of the shelf will be considered in connection with the continental slope in this area. The shelf off Land's End consists of deep basement rocks, over which sediments have been deposited for a long period to build up the present shallow depth. Sediments are more consolidated with depth. The relief of the area covering the Western Approaches to the English Channel and Irish Sea has been discussed by Robinson (1952), who shows that to the west of Land's End a series of northeast to southwest troughs dissect the shelf and descend to 27·0–36·5 m below its surface. These banks and troughs may have been formed in the same way as the rias of southwest Ireland, being due to folding in the Armorican orogeny. Southeast of the banks and troughs, off the entrance to the English Channel, the shelf is smoother and slopes very gently seawards. It is in this region that the seismic surveys suggest a great thickness of sedimentary rocks, overlying the basement rocks. This is confirmed by the work done in the English Channel, (Hill and W. B. R. King, 1953, and W. B. R. King, 1954), which shows that there is over 915 m of New Red Sandstone over a considerable width of the western Channel. Its base rises to above sea level on the shores of the Channel.

The shelf off western Ireland and Scotland has been glaciated as shown by the deep clefts and troughs. Some of these clefts, as in the Inner Sound of Raasay, may be tectonic features, modified by glacial erosion (Robinson, 1949).

The North Sea is very shallow, very little of it exceeding 90 m, except for a few deep holes. Only the northern part of the sea can be considered as continental shelf. The shallow southern part of the sea is better described as an adjacent basin. The northern part, where the depth is rather greater, extends northwards from the Forties, in the latitude of south Scotland. The depths here vary from about 70 to about 90 m, except for the deeper Norwegian Trench, which follows the south coast of Norway and penetrates about 880 km into the shelf. It is deepest at its head, where the water is 775 m deep, while near its mouth, in the neighbourhood of Bergen, it is only 200 m deep. O. Holtedahl (1950, 1952) and H. Holtedahl (1958) have suggested that this trench is of tectonic origin, and earthquakes still take place along it. It may have originated by the tilting of Scandinavia northeast and east in the early Tertiary. More detailed surveys of the trench show it to be double. It has no doubt been altered to some extent by the ice-sheets that passed south across it.

The southern North Sea is very shallow, much of it being under 36 m deep. It also clearly shows the influence of the glacial period. During ice advances it was dry land, at times over-run by ice. Rings of stony ground, which continue the lines of the end moraines of Jutland, are probably the seaward continuation of moraines, now submerged beneath the sea. Some of the trenches cut across the North Sea floor and may be parts of the old courses of the Thames and Rhine, which at times of low sea level joined to flow north to the open sea. The latest date at which this took place was during the low sea level of the Mesolithic period, about 8000 BC.

The east to west trend of the Silver Pit, south of the Dogger Bank, has been interpreted as an old urstromthal, cut by glacial melt-water flowing along the edge of the retreating North Sea ice-sheet. The Devil's Hole, which is 238 m deep, in an area only 73 m deep, is more difficult to explain. It may have been formed by the glacial melt-water flowing beneath the ice-sheet under hydrostatic pressure. It is not being modified at present.

The Dogger Bank has been interpreted as the remains of a glacial moraine by Stride (1959a). Seismic survey failed to reveal solid rock in this vicinity. This shoal area, which is more than 18 m above its surroundings, is 96 km wide at a depth of 36 m, and is less than 18 m deep at its crest. The moraine was probably deposited by the ice-sheet of the last glaciation, and the ridge has not been much modified since then, as early post-glacial peats are still found on the top and flanks. The offshore banks, which are so characteristic of the southern North Sea, are shaped mainly by tidal streams.

The southern North Sea is structurally a geosynclinal area, which has been subsiding slowly for a very long period, as indicated by the character of the Mesozoic and later sedimentary strata of Holland. The tide gauge data show that this area is still sinking, as sea level is rising more rapidly than on the more stable parts of the coast. It is, therefore, a feature of considerable antiquity, associated with the subsiding area of the Rhine deltaic region.

From the geological evidence (W. B. R. King, 1954) and the geophysical data, it seems that the English Channel is also a very old feature. It has been subsiding geosynclinally for a long period and accumulating sediments, while the adjacent areas of southwest England and Brittany have remained relatively high. Subsidence has been particularly active in the region of the Isle of Wight, as shown in figure 3.10, where the sub-New Red Sandstone floor is estimated to be at 3660 m below the surface. Tertiary deposits are much thicker in the Isle of Wight basin than elsewhere on the Channel floor, attaining 600 m in the basin. Some folding took place during the Miocene earth movement, and Chalk now outcrops over much of the floor of the central part of the Channel.

One of the most striking characteristics of the floor of the channel is its extremely flat nature and the absence of any thickness of modern sediments over large areas. The relief varies only within one or two metres across its whole width, and only one deep trough interrupts the even floor. This is the Hurd Deep, which is 172 m deep. The channel has a very gentle slope, falling from about 36 m at the east end to about 90 m between Land's End and Brest, where the true shelf may be considered to start. All the flat area of the Channel would have been exposed during the low sea level of glacial times. Its flatness is

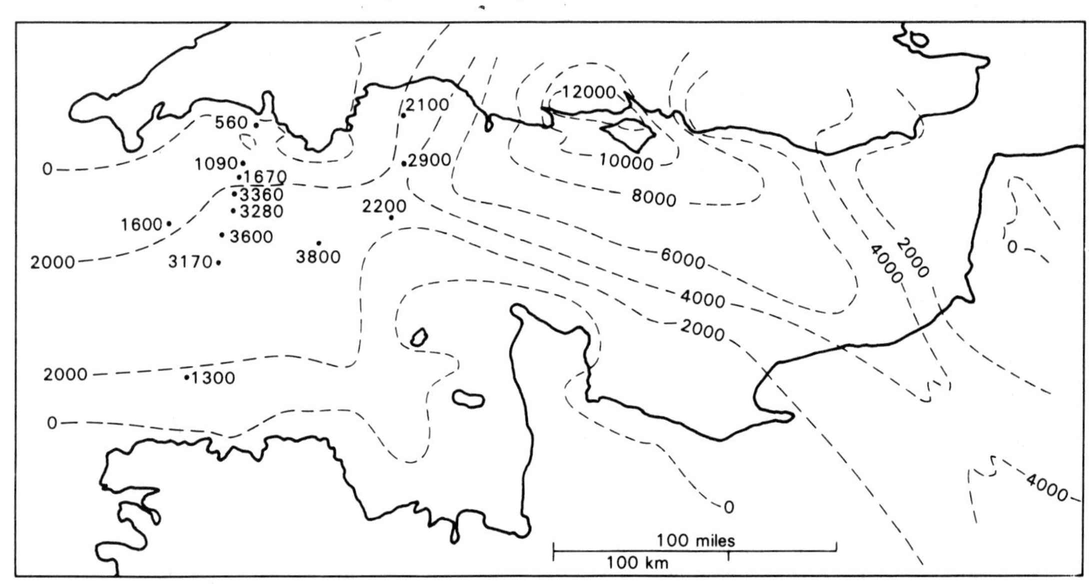

Figure 3.10 Contour map of the sub-Mesozoic—New Red Sandstone floor in the English Channel area. The depths are in feet, derived from seismic survey. (*After W. B. R. King, 1954.*)

possibly due partly to erosion by the powerful melt-water streams that flowed through the Straits of Dover and along the Channel. It would also have come under the action of waves during the Flandrian transgression, and at other times. The flat area may represent a plane of marine erosion, trimming a flat fluvio-glacial valley. Such a plane can only be cut effectively over a wide area during a period of rising sea level.

The origin of the Hurd Deep is more difficult to suggest, as it cannot have been cut by ice action, and it does not appear to be structural in origin. Shepard (1959) has suggested tidal currents as a possible erosive agent. The trench is 160 km long, but only 6½ km wide. It is asymmetrical, having a steeper northern slope. A. J. Smith and Hamilton (1970) report further observations in the area. They have mapped the Deep by 52 sparker traverses and 200 gravity stations; the results of these measurements indicate that the feature is erosional in origin. The Chalk is cut through into the underlying Jurassic strata. The trench is floored by Jurassic clays, silts, and thin limestones. A karstic origin is considered untenable for this reason. The depth of fill is variable, ranging from 60 to 90 m to zero m at the southwest end. At the northeast end three depressions form a tributary system. They are partially filled channels that can be traced to the Seine, Somme, and the Straits of Dover. The conclusion may be that the Hurd Deep is a remnant of a much larger channel, now partially filled in. It was probably eroded both by tidal and fluvial action, the latter during low glacial sea level when both melt-water and tide could scour the channel. Tidal action has maintained the Deep since sea level rose. The maximum depth is now in the zone of strongest currents.

The Irish Sea, according to the suggestion of Linton (1951), may have originated as the result of subsidence of the keystone of an arch elevated during the Tertiary, of which England, Scotland, and Wales form the eastern slope. This suggestion follows Cloos' idea of the structural history of dome-shaped uplifts. There is, however, much evidence to suggest that ancient basins of sedimentation, as in the English Channel and southern North Sea, also existed in the Irish Sea area from at least Mesozoic times onwards. Geophysical work in the Irish Sea by Bott (in Donovan, editor, 1968) shows that there are six deep sedimentary basins, especially on the east side. They are separated by areas where the basement rock is at a shallow depth. The basins are: Stranraer—Permo-Trias, 1·3–2·0 km maximum sediment thickness; Solway Firth—Carboniferous and Permo-Trias, 2·1–4·9 km thickness; east Irish Sea—same age, 2·4–6·0 km thickness; Peel—Carboniferous, +2·0 km thickness; Kish Bank, off Dublin—Permo-Trias, +2·5 km thickness; Cardigan Bay—Permo-Trias and younger, +2·0 km thickness. The crust appears to be a few kilometres thinner under the sea than under the adjacent land, with the change occurring near the coast. The basins are related to patterns in the crustal structure, shown by magnetic anomalies. They form mainly in areas of thin crust. Erosion of surrounding mountains probably caused the subsidence with flow in the upper mantle involved. The present form of the Irish Sea depends on the Caledonian or earlier crustal structure.

A considerable impetus to the exploration of the North Sea has been caused by the search for and finding of oil and gas. The North Sea forms part of the continental shelf off northwest Europe, but it differs from some shelves in its long continued tendency to subside in the southern basin portion. The basin extends into north Germany, Holland, and Denmark, but its connection with the outer shelf of northwest Europe is less certain (Donovan, editor, 1968). The rate of subsidence has varied over time and space, but the region is approximately in isostatic balance, which implies a crustal thinning of 5–7 km. The forces causing subsidence are small, and the cause appears to be a transformation of basalt into ecologite, which being dense sinks, so that the process is irreversible. The process is load governed. Heating of the sedimentary column will give rise to light melts that will accumulate in domes or ridges. These result in different movements as will salt diapir formation, a process that also operates in the North Sea. These features are important for gas and oil collection. The gas and oil are associated with Mesozoic deposits and also with the deeper Palaeozoic and lower Triassic strata. Considerable areas of the North Sea are underlain by Permian salt deposits, which form domes in places. Subsidence in parts has allowed 2440 m of Tertiary and Quaternary sediments to accumulate in the north Holland area, while 1220–1520 m have accumulated in the southeast North Sea. Gas was first discovered in commercial quantities in the North Sea late in 1965, and since then oil has been located in commercial quantities, particularly in that area.

2.3e Tidal morphology

(i) *Channels and banks:* The macrotidal environment around the British Isles and the abundance of loose sediment are conducive to the development of morphological features by tidal streams. These features will, therefore, be considered mainly in relation to the west

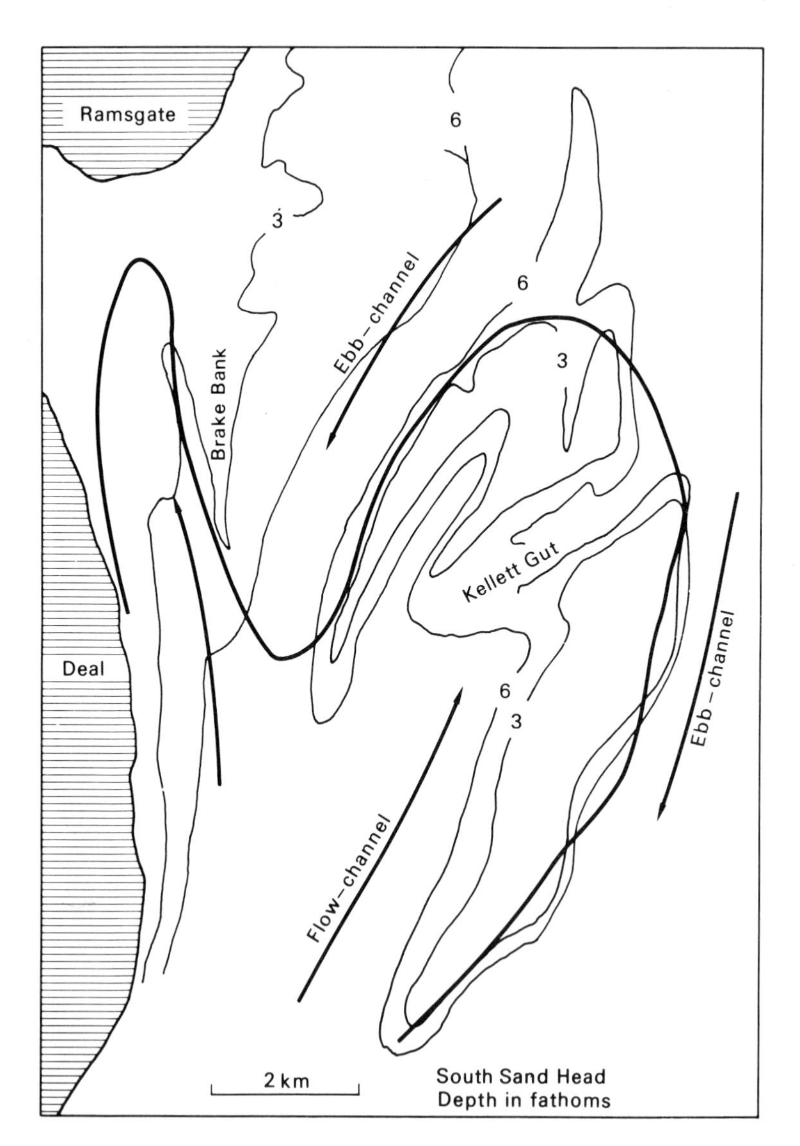

Figure 3.11 The ebb and flood-channels of the Goodwin Sands region off the south-east coast of England. (*After Cloet, 1954b.*)

European shelf and shallow seas, where they are well developed. Ebb and flood-channels form where there is plenty of loose sediment (mainly sand) on the sea floor and where the tidal streams are almost rectilinear. A completely rectilinear system of tidal streams is found off Skegness in Lincolnshire, where the currents flow either due north, with the ebb, or due south, with the flood-tide. Where the tidal currents are rectilinear, the flood-stream tends to use one channel for longer than the ebb, which uses a neighbouring parallel one for a longer period. The interdigitating channels are usually separated by elongated or curved banks of sand, deposited in the areas of quieter water between the major current channels.

This system of tidal channels and banks (first discussed by Van Veen in relation to the southern North Sea) has now been recognized wherever suitable conditions of sediment supply and tidal streams exist. The association of channels and offshore sandbanks is exemplified by the Goodwin Sand off Kent, and by the sandbanks in the Wash. In the Goodwin area there are two main northerly-directed flood-channels, between which are southerly-directed ebb-channels. The interaction of these different channels accounts for the elliptical shape of the bank, and for the channel between it and the Brake Bank, which is situated near the Kent coast. These banks are shown in figure 3.11. The major axis of the ellipse formed by the banks is aligned parallel to the direction of the main tidal streams, in a north-northeast to south-southwest direction (Cloet, 1954a).

The ebb-stream may still be running out of a river estuary, when the tide has already turned offshore. The flood-tide, therefore, tends to avoid the ebb-channel and initiates one parallel to it. There is a fairly wide variation in the pattern of the ebb and flood-channels, which is sometimes related to the configuration of the coast or estuary in which they occur (Robinson, 1960).

The tidal streams and the channels that they create are of considerable importance since they affect navigation in shallow water and in the estuaries where they occur; they are becoming increasingly important from this point of view as giant tankers are constructed of greater draught. They also affect coastal development geomorphologically particularly because the streams rarely flow equally in both directions. Nearly always either the flood or ebb is the stronger, with the result that there is a current residual and a net movement of material in one direction. This movement will influence the general direction of transport of bed material in the area. A vector diagram can be drawn to indicate the direction and size of the residual. It is drawn by plotting the hourly values of the tidal stream vectors as a continuous figure. The net or residual movement can be found by joining the initial and final points. As stronger currents can move more material in proportion to the square or some other power of the current velocity, the current can usefully be plotted as a power of their velocity in order to arrive at their net sediment transporting capacity.

There is a well-developed tidal morphology pattern off the southern Lincolnshire coast. The offshore banks and their associated tidal channels are shown in figure 3.12, which illustrates their spatial arrangement and typical profiles across them (C. A. M. King, 1964). The tidal streams are almost entirely rectilinear, and there is a southerly flood residual off Skegness of 41 cm/sec. Another tidal channel with a flood residual lies close to the coast, and flood tidal streams exceed 200 cm/sec in this channel. The flood-channel swings close into the lower foreshore near the outlet of the River Steeping. There is very coarse material on the bed of this channel, indicating the fast velocity. A flood-channel also runs along the seaward side of the Outer Dogshead and Outer southeast Knock Banks, while ebb-channels occur on both sides of the Outer Knock Banks. The latter carry sediment northwards. The pattern of banks and channels is continually altering and new banks and channels have developed recently. It is likely that the ebb-channels have increased in importance. A complex circulation probably allows material moving from both north and south to come ashore

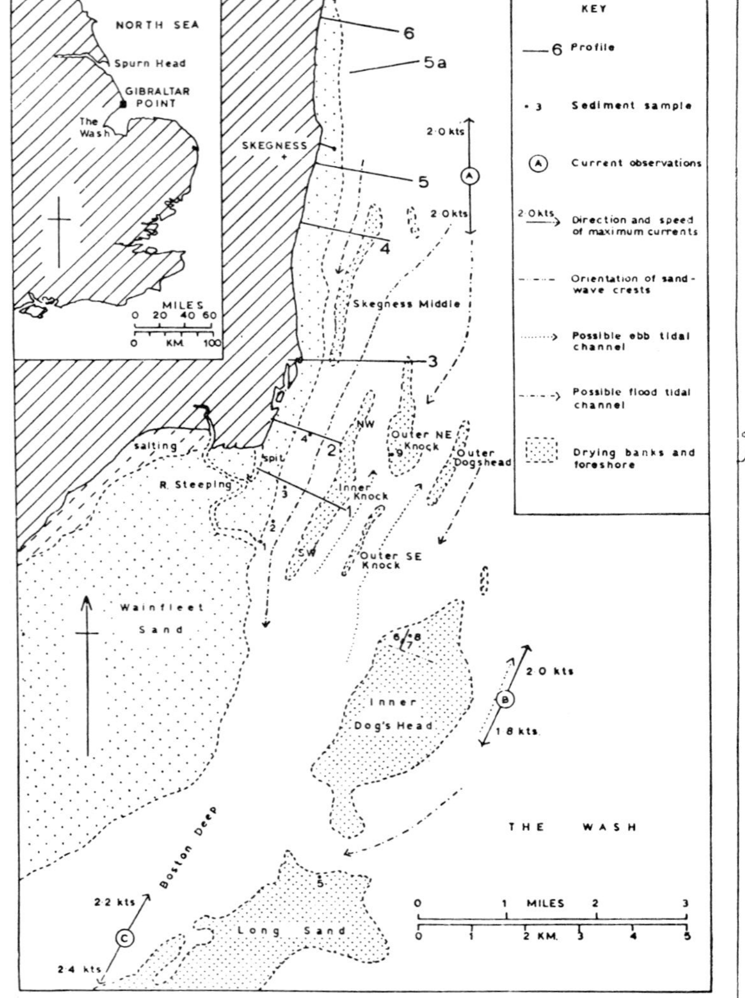

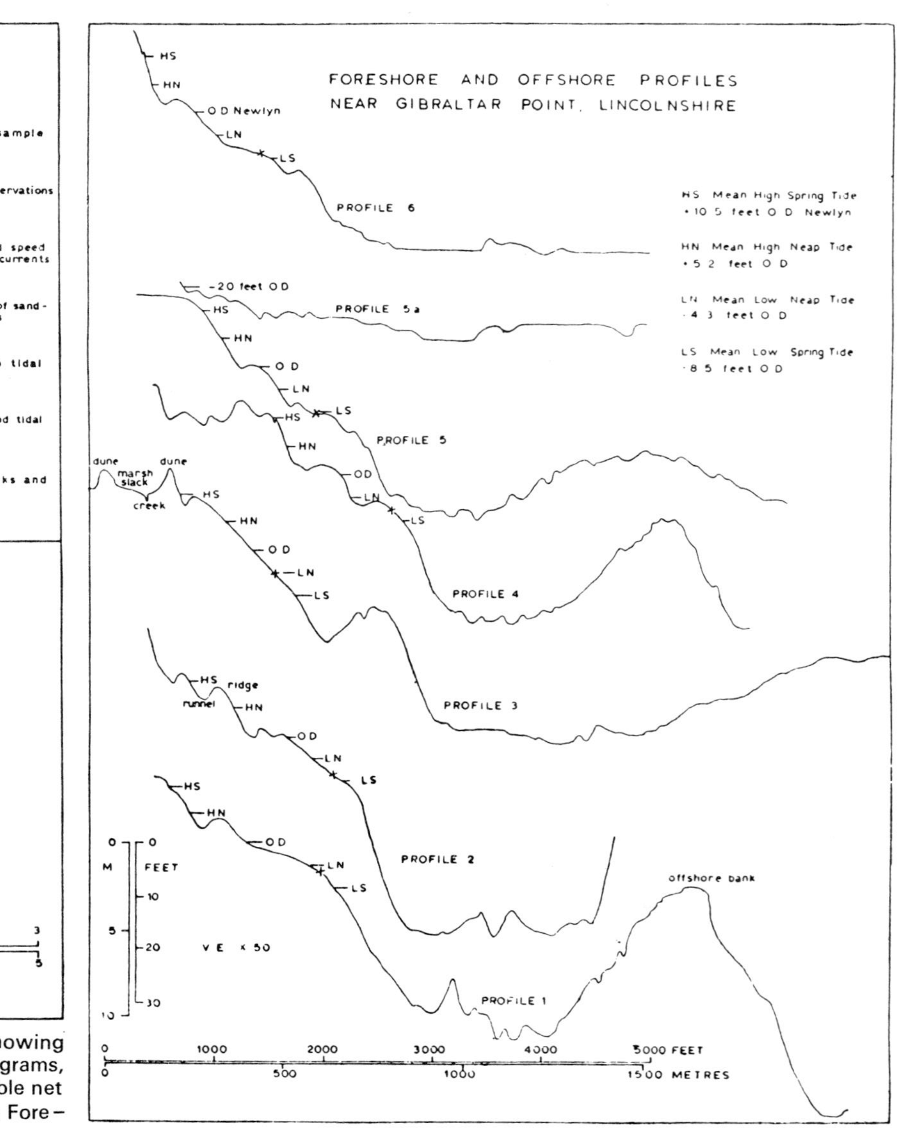

Figure 3.12 A: Map of the area around Gibraltar Point showing the nature of the offshore relief. The position of the vector diagrams, and the direction of maximum flow, are shown. The probable net direction of tidal flow in some of the channels is indicated. **B:** Foreshore and offshore surveys along the lines indicated in **A**.

in the area, accounting for the marked coastal accretion in this zone (Robinson, 1964 and King, 1973).

Sea bed drifter experiments support the morphological interpretation of the channels and banks in the Lincolnshire area. The two forms of evidence both provide information about the trajectories of material under the influence of tidal streams. The tidal morphology helps to account for the large number of recoveries from the southern Lincolnshire coast and their absence from the central section. The tidal morphology suggests that much of the material moving into the Wash remains in it, as the flood-channels are dominant, and these allow material to move into the Wash from the north. The banks between the channels are often very steep-sided and massive. Those off Gibraltar Point consist of fine, well-sorted sand.

The pattern of tidal streams and their associated banks and channels also occurs off the coast of east Anglia (Robinson, 1966). A flood-tide channel directs material south onto the northern flank of a ness of material, which forms a protuberance at several points along this stretch of coast. An ebb-channel on the southern side of the ness allows material to reach its southern flank. The growth and movement of the individual nesses can be traced and related to the development of offshore banks, which are the result of the tidal stream pattern.

A good example of the movement of the nesses is given by Benacre Ness, which has moved north in the direction contrary to the normal movement of beach material along this coast. A flood and ebb-channel system is well developed off this coast. Between 1824 and 1865 an ebb-channel was established off the position then occupied by the ness. This ebb-channel gradually moved north after 1865, followed by the ness itself, covering a distance of 2 km. The Barnard Shoal has developed in the zone where the streams are opposed to one another. The tidal streams in this area are strong, reaching 175 cm/sec 6·4 km offshore and 118 cm/sec close inshore in Yarmouth Roads. The supply of sand in the offshore zone is also plentiful.

A sandbank formed by tidal action is described by J. Dungan Smith (1969) in Vineyard Sound, off Martha's Vineyard near Cape Cod. The bank is called Middle Ground and is orientated west-southwest to east-northeast. It is about 5 km long and 0·5 km wide, and is formed entirely of sand. An anti-clockwise circulation takes place around the bank, owing to the dominance of the northeast-flowing flood tidal stream on the southern side and the southwest-flowing ebb on the northern side. The average maximum shear velocity of the tidal stream is 4·5 m/sec. The bank has been formed by the tidal streams, although its position is determined by a remnant of morainic material.

(ii) *Sand waves:* Smaller morphological features are superimposed on the large-scale tidally formed sandbanks and channels that have just been described. Sand waves are the most conspicuous of these features and are widespread where sediment is available; they are orientated with their crests normal to the direction of maximum tidal flow. They are normally asymmetrical, with the steep side on the lee.

Sand waves have been recorded both in very shallow water offshore and in fairly deep water on the edge of the continental shelf. Sand waves on the La Chapelle Bank cover an area 19 km long, ending abruptly at the edge of the continental shelf. Their lateral extent is 8 km. The average height of their crests is 7·6 m, reaching a maximum of 12·2 m. The mean wave length is 854 m, although there is considerable variation. The wave crests are orientated at 110°, parallel to the shelf edge and normal to the current. They are composed mainly of sand with some shell gravel and their asymmetry suggests westward transport of sediment, as their steeper face is to the west. The direction is confirmed by the current measurements in the area that have already been mentioned.

Stride (1963) has explored much of the shelf and shallow sea area around Britain and found that sand waves are readily identified in echo-sounder traces; their asymmetry gives useful evidence of the direction of sand transport. Stride found sand waves up to 18·3 m high and 915 m long; off the coast of East Anglia they are often 4·6 m high and 137 m long. Their dimensions are more regular where the sand supply is abundant. Wave crests tend to be straighter in the northern Irish Sea and the Bristol Channel; they are more sinuous in the southern Celtic Sea and the western English Channel. The former occur mainly where the tidal streams are rectilinear, reaching diametrically opposed peak velocities. The latter occur where there is a greater spread of tidal stream direction of flow and where the surface tidal streams are greater than 51 cm/sec. Sand waves are found in a belt 100 km wide off the east coast of England and the west coast of Germany, over much of the Irish Sea, the Bristol Channel, the English Channel, and the Celtic Sea.

Most of the waves are asymmetrical and even a difference of 5 cm/sec in the ebb and flood-peak velocities is enough to cause asymmetry and net sediment transport. In the English Channel there appears to be a circulatory pattern of movement in the bays off the south coast of England. A detailed study of the sand waves on Warts Bank, southwest of the Isle of Man in the Irish Sea by N. S. Jones *et al.* (1965) showed that the waves travelled at an average rate of 5–10 cm/day, but one crest covered 74 cm on a single flood-tide. This bank is elongated roughly east to west. The waves on the northern side moved southwest while those on the south side moved northeast, on the evidence of their asymmetry. The flood is dominant on one side of the bank and the ebb on the other. The streams over the bank reach a maximum of about 50 cm/sec and the waves occur in depths of 10–40 m.

Another set of waves has been surveyed off the Yorkshire coast by Dingle (1965), who supports the view that the tidal streams converge off Flamborough Head. Some of the waves were uniform, while others were variable in length, the latter occurring where the sediment was thicker and the former nearer the coast where the tidal streams were more rectilinear. The currents were too slow in this area for sand waves to form in water deeper than 55 m.

McCave (1971) has described sand waves in the southern North Sea off Holland. The waves reach a maximum of 7 m height, decreasing northwards and eastwards to 2 m. Wave lengths vary between 200 and 500 m and only form where the sand volume is adequate and wave action is not excessive. These factors delimit the western and eastern margins of the sand wave field. The waves also only develop where the tidal ellipse is asymmetrical and the current velocities exceed 60 cm/sec. Where the sand waves exceed 5 m in height, mega-ripples occur and these are associated with bed-load transport.

A very uniform set of small sand waves has been recorded in St Andrew Bay, Florida, by Salsman *et al.* (1966) who observed the movement of the waves for 2 years. The waves are orientated normal to the tidal currents, and are 30–60 cm high and 13–20 m long. The waves, composed of fine sand, are advancing into a muddy area and covered 11·44 m in

849 days. Their movement was most rapid when the tidal streams were at their maxima of 40 cm/sec at flood-tide. The sand waves were formed by the tidal streams in a channel behind an offshore barrier.

(iii) *Sand ribbons:* Ribbons of sand form in some areas instead of sand waves, but they also are related to the direction of sediment transport by tidal streams. Sand ribbons have been revealed by acoustic sounding over an extensive area off southern Britain, where the sea floor consists of patches of sand, sand and gravel, stones or shells. They trend parallel to the coast along the English Channel, the Bristol Channel, and the Irish Sea. This is also the direction of the strongest tidal streams. They vary from place to place and die away where the tidal stream velocity falls below 50 cm/sec on the surface. As the tidal streams increase above this speed the ribbons become sharply defined, and when the current exceeds 100 cm/sec their length-to-width ratio increases. They are often only a few centimetres thick, but may be up to 2½ km long, with a maximum width of 100 m in the Bristol Channel.

(iv) *Material transport by tidal streams:* An analysis of the asymmetry of sand waves suggests that there is a general northwards movement, estimated at 40 million m^3/year in a 64 km wide zone, up the coast of the Netherlands. Off East Anglia in a belt 16 km wide, movement is to the south and this direction also applies a short distance off Yorkshire and Lincolnshire. The movement along the Netherlands coast has been confirmed by radioactive tracer observations. The wave form suggests that there is little movement at the eastern end of the English Channel, where the sand waves are symmetrical. In the western English Channel the main direction of transport is to the west, changing to southwest as the shelf edge is approached. Closer inshore the movements are more complex.

Current observations and sand wave data indicate that movement of sand is northwards along the western side of the long banks off the north coast of Norfolk, and southwards on the eastern side. The net direction is indicated by the steeper northeastern slopes of the banks, which agrees with the net northerly flow both north and south of them. The banks are sufficiently far offshore to lack transverse obstructions to sand flow. They are straight and almost parallel, separated by 27–42 m depths, increasing to the west. Over the banks, depths are sometimes only 3 m. The internal structure of the banks in some cases suggests movement of the whole bank to the northeast. The linear banks may have an element of lateral flow in the form of helical spirals, with diverging flow on the floor, which would account for the lack of material in this position in areas with anti-clockwise circulation. The pattern would only occur with anti-clockwise circulation. Where the circulation is clockwise, convergence occurs near the sea floor and divergence near the surface. This situation is found off Norfolk and thus helps to account for the presence of the banks in this area. The asymmetry could be due to stronger secondary flow on the eastern side of the channel.

Tidal streams show considerable variation according to the wave conditions, and hence the volume of material transported varies widely in both space and time, according to observations reported by M. A. Johnson and Stride (1969). They calculated that the neap-tide rate of transport could be only 1/10 that of the spring tide rate under calm conditions. With storm waves, however, the neap-rate could equal the spring rate. This could occur

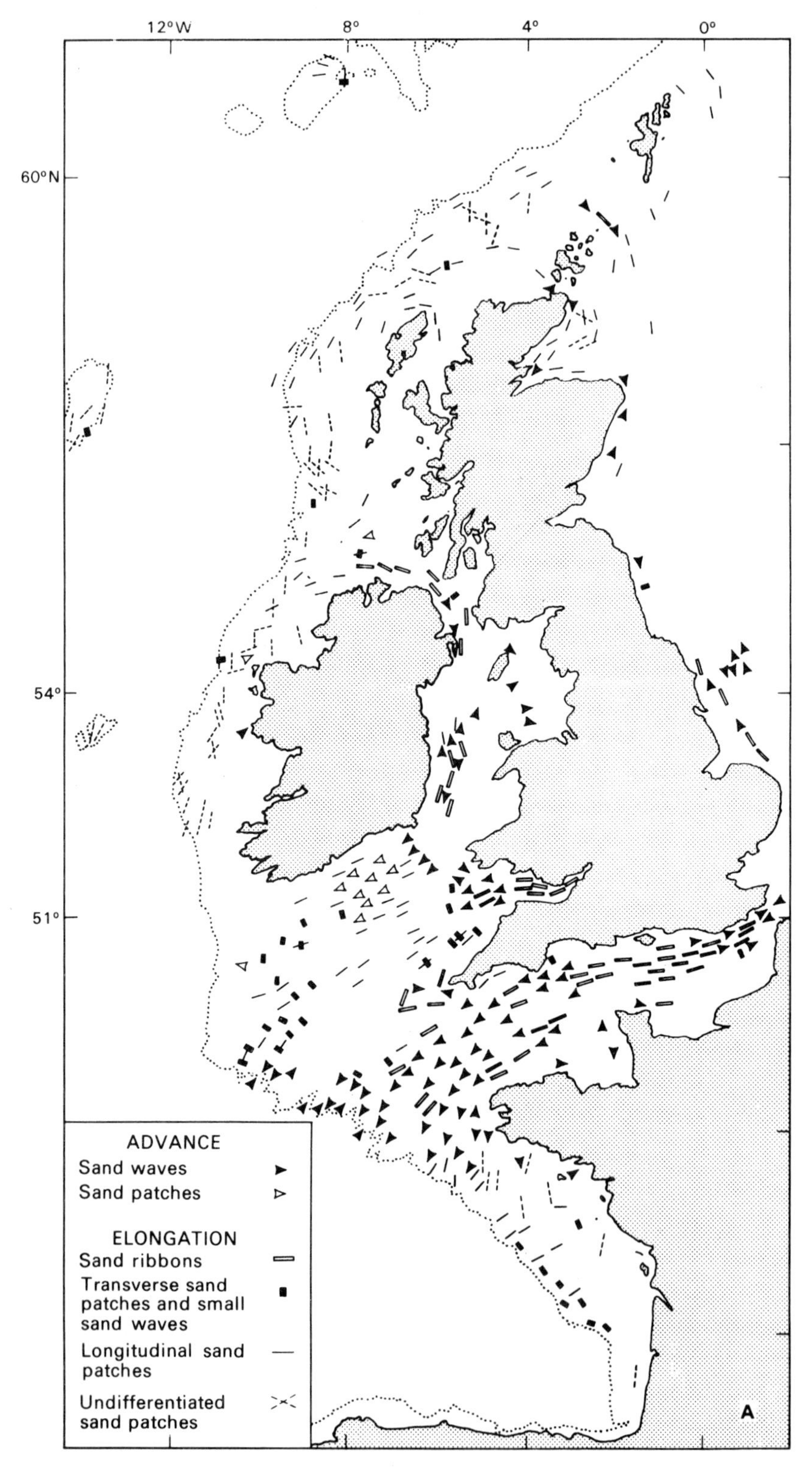

Figure 3.13 A: The elongation of different types of sand bodies and their inferred direction of advance around the British Isles and France.

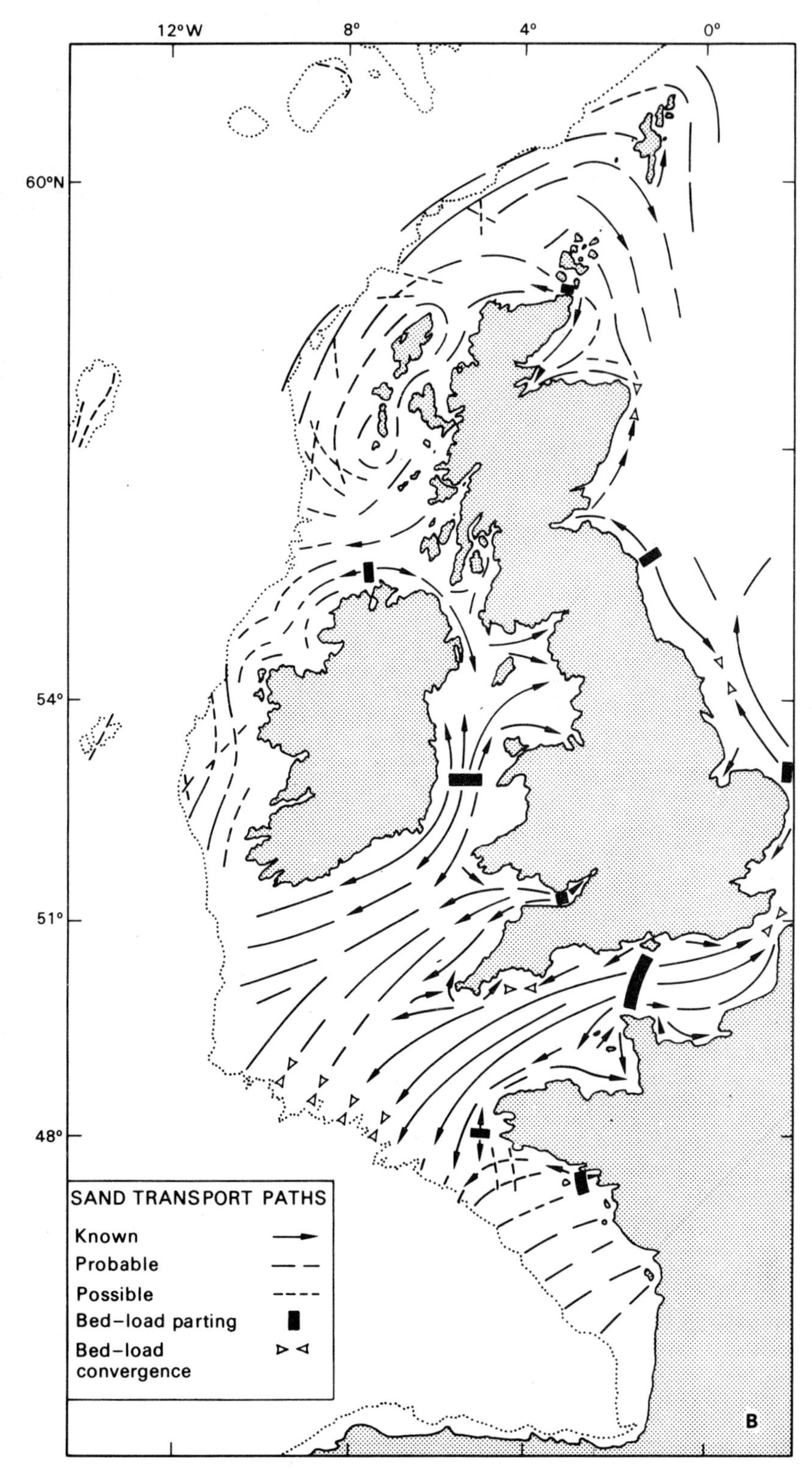

Figure 3.13 B: Generalized transport paths on the continental shelf around the British Isles and France (*After Kenyon and Stride,* 1970.)

when the neap-rate was augmented by storm waves generated by northerly winds in the Flemish Bight in the North Sea between East Anglia and Holland. At spring tide, storm waves could increase by ten times the transport rate characteristic of calm conditions and as much sand could be moved in one day as in a month of normal conditions; the movement also could be in the opposite direction.

The decrease of water temperature in winter causes a significant increase of material transport. The annual net total moved by tidal streams in a 64 km wide belt occupied by sand waves is about 10 times the annual volume of transport along the littoral zone off Holland, which has been estimated at 0·8 million m^3. The rate per unit width is, however, greater in the littoral zone.

Further confirmation of the correlation between sand wave asymmetry and current residual direction has been obtained from observations by Belderson and Stride (1969) in the Irish Sea. The evidence suggests that sand is travelling towards the Solway Firth and Liverpool Bay. There is a patch of muddy sediment off the coast of Westmorland and Lancashire south of St Bees Head. Observations showed that streams in this muddy zone were less than 44 cm/sec at spring tide, which is in keeping with the deposition of mud.

Kenyon and Stride (1970) have collected all the available data on transport of sediment by tidal streams and other currents, using the evidence of the bottom features already described. Their results are summarized in figure 3.13. The general pattern shows movement parallel to the coast off western Britain, but normal to the coast of western France. Tidal streams are the most effective transporting agent, but short-period waves and surges can occasionally be effective under extreme conditions. Wind-driven currents move mainly fine sediment in suspension. The tidal streams are strong enough to move sediment of sand size most of the time. Locally near headlands and in straits the tidal stream velocities exceed 258 cm/sec, while over large areas streams of 100–150 cm/sec occur, with values falling to 25–50 cm/sec in the middle and outer shelf area. The storm waves mainly cause an increase in the movement by tidal streams.

Sand ribbons are associated with streams over 100 cm/sec at the surface, and their long axes are a reliable indication of flow direction, although a unique direction is not provided, except where thin wisps extend from the main ribbon. Small-scale forms may change with the changing current direction and are, therefore, not diagnostic of long-term movements. Sand waves occur where the streams are 65 to 115 cm/sec, and enough sand is available. Sand patches, often of elongated linguoid form, occur where streams are less than 50 cm/sec at the surface, and movement is parallel to the elongation of the patch.

A number of sediment partings occur, including those off the Isle of Wight, Lleyn Peninsula in north Wales, the Tweed, Cromer at the northeast corner of East Anglia, Malin Head in northern Ireland, and the Pentland Firth in north Scotland.

The shelf is probably losing sand to the continental slope and deep sea floor between southern Ireland and southwest France, but little sediment moves offshore off western Ireland or northwards from western Scotland. Sand may move around Scotland into the North Sea and northern Irish Sea. Some of the sand bodies off western Scotland appear to trend with the diurnal tide component rather than the semi-diurnal, which is normally greater, but they could be related to storm waves. The importance of the diurnal tidal component in this area has, however, already been mentioned.

3 Continental slope and rise

3.1 Character

The continental slope and rise form the major relief feature, linking the shelves (which are structurally marginal to the continents) to the deep ocean floor. It is the zone in which the Moho sinks down beneath the continents, as indicated in the profiles described in the last section. The slope is the steep upper portion of this marginal zone, while the rise is the gentler lower portion, which rises gradually up from the abyssal plains. Shepard (1963a) gives the average gradient of the slope as 4°17′ to a depth of 1830 m. One of the most important features of the shelf, slope, and rise zones are the submarine canyons that are best developed in the slope zone. So much has been written about these features that they will be considered separately. Apart from these features the slopes show considerable range of character and origin. The slope is rarely uniform, being diversified with valleys or basins. The gradient increases to 5°40′ off faulted coasts, but is reduced to 1°20′ off large rivers. One of the most spectacular slopes is that found off west Florida. An escarpment descends very abruptly from 1100 m to more than 2920 m. It is more than 800 km in length and the level drops at a gradient greater than 1 in 2. This very steep slope, which cuts sharply across a gentle slope, is due to a fault. Another continental slope fault scarp descends to the Bartlett Trough off Santiago. This slope has a gradient of 45° and a vertical height of 6100 m. On the whole the slopes of the continental border seem to be rather steeper in the Pacific at 5°20′, than in the Atlantic (where they average 3°05′) and the Indian Ocean (average 2°55′).

Different types of slopes can be recognized in different areas. The continental slope of the northeast USA is fairly steep, but its major characteristic is the very large number of submarine canyons that dissect it. These descend at least 1800 m. The rocks exposed in them show that the continental slope is built of sedimentary rocks varying in age from the Miocene to the Cretaceous. Further south, the wide Blake Plateau is found part way up the slope. From the continental shelf the slope drops off steeply from 90 m to 730 m to the Blake Plateau, which slopes gently down to 1100 m over a horizontal distance of about 300 km, in latitude 28° to 30°N. At the outer edge of the plateau there is another abrupt break of slope, with a gradient of 15° falling to a depth of 4600 m. The surface of the plateau is of resistant material, probably rock of Cretaceous to Miocene age.

The Gulf Stream appears to have affected the plateau. It flows very rapidly along its inner edge. Shepard suggests that the Gulf Stream has prevented the building up of recent sediments over the older deposits. Deposition of this type has taken place further north where the Gulf Stream leaves the continental slope. This explanation is more likely than the suggestion that the plateau has actually been eroded by the Gulf Stream.

A gentle slope type is found off Louisiana and Texas, but west of the Mississippi River the slope becomes much more irregular, with elongated hills and basins. The basins are about 460 m deep and 48 km long. It seems that they may be blocked remnants of former valleys, as they are elongated in the direction of the slope. Landslides have been suggested as a possible causal process. The lower parts of these slopes are in general smoother, but the hills and basins descend some way down them.

Off California the offshore areas and continental slope have been intensively studied by members of the Scripps Institute of Oceanography. Here the slope lies from 80 to 260 km from the coast (Emery, 1960). It trends N 25°W, and is broken by five gaps which are probably tectonic in origin, as they are unrelated to submarine canyons. The slope itself is about 16 km wide, dropping to a depth of 3660 m. In general its profile is straight, but minor irregularities do occur, while at its base is a shallow depression, which may be a filled-in deep sea trough. Northwards, off San Francisco, this depression is replaced by an

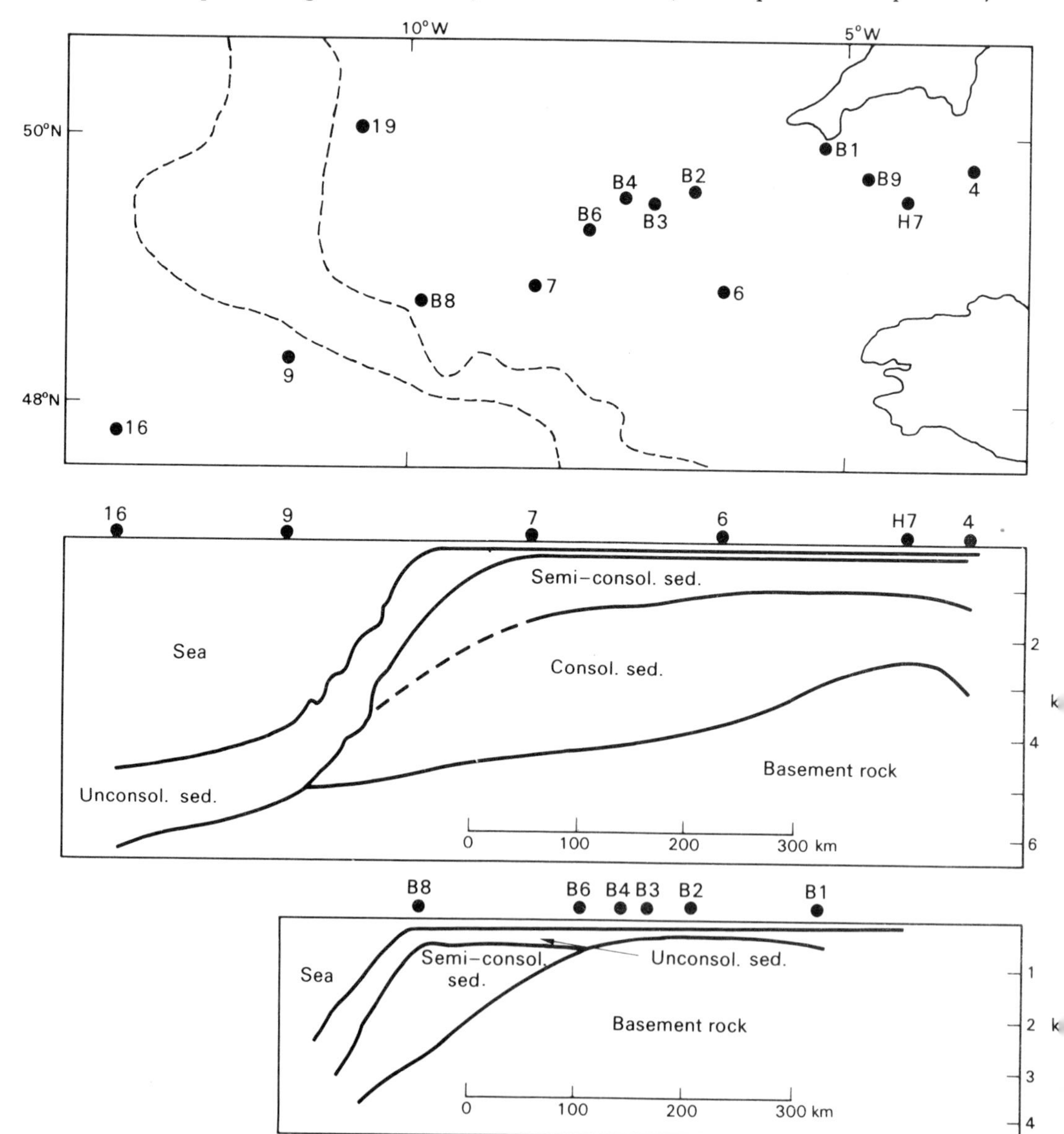

Figure 3.14 Seismic surveys of the continental shelf off southwest England. (*After Hill and Laughton, 1954.*)

apron (indicating a later stage of development), while to the south, off Mexico, the depression is deeper (indicating an earlier stage). This gradual transition northwards up the coast is probably related to the rate at which sediments are supplied to the trough. Off Mexico the dry climate prevents perennial streams, while off northern California, the streams are perennial and the hinterland is high and erosion more rapid. Off southern California the rate of trough formation at the base of the slope is about equal to the rate of sedimentation. This particular slope is not a depositional one as rock is found near the top of the slope, and the topography inshore is very complex. It seems likely that faulting has played a part in its formation.

Observations off the west coast of Britain have given useful data concerning the character and method of formation of the slope. The continental shelf is very wide and the slope drops down steeply beyond it from a depth of 200 m to 4000 m. Seismic data give evidence of the structure in this zone. Continental structure, with a thick layer of less dense, acidic type crustal rocks changes to deep ocean structure, where a thinner layer of basic rocks overlies the ultrabasic mantle. In the deep ocean beyond the slope the sediment ranged from 0 to at least 2·8 km in thickness, and probably more in places. Beneath this, at the foot of the continental slope, is a layer of continental rocks 8·1 km thick, below which the Moho is located at a depth of 13·2 km below sea level. This is transitional between the structure on the inland side of the slope and that beyond it, where the layer beneath the sediment is oceanic crustal rock with a thickness of 4·2–5 km, the Moho being only 11·4 km below sea level.

Near the edge of the shelf there are several layers, probably sediments in various stages of consolidation. The basement rock beneath these sediments slopes gently down from a depth of about 2½ km south of Plymouth to about 4½ km at 10°W, at the top of the continental slope; at its foot the basement is about 5 km below the surface. This structure is shown in figure 3.14, and is very similar to that off the northern part of eastern USA, where the basement rocks are also buried under a great thickness of later sedimentary rocks. In both areas these are mainly Mesozoic and Tertiary sediments.

Guilcher (1963) has described variations along the slope off southwestern Britain and western France. In part of the slope a terrace lies at 2000–2200 m depth. To the northwest the slope is much steeper, forming the Pendragon Scarp; elsewhere the slope is smooth and gentle, as in the Porcupine Bight. Off Morocco the slope is straight in plan but the profile steepens towards the base, which varies in depth from 2200–2400 m to 4000 m. The slope off Norway and Svalbard shows an irregular profile, with longitudinal ridges and sea mounts near its base. A similar type with sea mounts on the slope occurs off San Francisco and Nantucket Island, New England. The slope around Antarctica has been found to be very variable in character. The great variability of slopes around the world suggests a variety of origins.

3.2 Origin of the continental slope

Various theories of the possible origin of the continental slope can be suggested. As with so many other features, it is unlikely that any one explanation will account for all the continental slopes of the world. This major relief feature usually lies at the junction of the two fundamentally different structural types, with their very different levels. This fact alone would account for the presence of a major slope. The slopes off southwest Britain and eastern North America have been shown, by seismic survey, to be depositional in character. The slopes are formed of sediments in various stages of consolidation, which have been laid down on the underlying basement rock. In these areas the slope in its present form is directly the result of prolonged deposition, probably since the beginning of the Mesozoic period. The sediments off eastern North America also include Cretaceous strata. The layer of sediment is lense-shaped in places, reaching its maximum thickness on the edge of the shelf and thinning towards the shore and down the slope.

Another process goes on simultaneously with the building up of sediments, most of which accumulated in shallower water than that in which they now occur. This is the down-warping of the continental margin, a process that may well be important in the formation of the continental slope. One suggestion is that the continental slopes on both sides of the Atlantic are formed by the down-warping of Miocene peneplains. The evidence for this is not always convincing, however, and on the east coast of the USA the Cretaceous seems to be truncated by the slope, rather than bending down with it. Shepard (1948), however, argues that if down-warping were the major factor in the formation of the shelves and slopes, they should be out of isostatic adjustment, for which there is little evidence. He also points out that the straightness of many slopes does not fit in with the hypothesis of down-warping.

Shepard, therefore, suggests that faulting may be an important factor in the formation of some slopes. He comes to the conclusion that most continental slopes represent a zone of recurrent faulting where the continental and oceanic crusts meet. The faulting may be in the form of a series of step faults, or one major low-angle fault, although the latter is less likely. Examples of this type of fault slope may be cited as the 6° slope off northwest Peru, and the very steep slope leading straight down to 3660 m off the fault coast of the southwest South Island, New Zealand, where the slope drops down to more than 3660 m only 24 km offshore. Points in favour of this hypothesis include the seismic activity often associated with the slope which helps to explain the outcrop of rocks at various depths on the slope. The trenches at the foot of many slopes are also evidence in favour of faulting, according to Shepard (1973), although they can also be explained in other ways.

The most common pattern for those slopes that have been investigated most thoroughly is that of constructional processes accompanied by subsidence. This applies to the North Sea, the southwest Celtic Sea, eastern USA, and part of the Gulf of Mexico. Some of the subsidence associated with this type of continental margin is very great, indicated in the troughs found off the USA and the Bahama Banks. A drilling at Andros Island passed through 3810 m of calcareous formation. These were shallow water formations that must indicate subsidence of this order in Cretaceous and subsequent periods. The North Sea is another area of strong subsidence, with a post-Carboniferous sedimentation attaining 9000 m, probably, in places. Sediments underlying the Dogger Bank are 3400 m thick, and 600 m have accumulated in the Pleistocene in Holland. The same type of shelf and slope also seems to occur off west and northwest Australia, with sedimentation in four large basins, in which between 5420 and 12,200 m have accumulated.

Another type of continental slope is the flexured type. This has also been down-warped but it differs from the first type in the smaller volume of sediment involved. The narrow margins of the western Mediterranean exemplify the type in which the amount of sediment is small. The land has risen as the sea floor has subsided. In some areas, as off the Ivory Coast of central Africa, a large fault occurs along the continental slope at the seaward edge of a narrow shelf. A third major type consists of block-faulted slopes, giving either a series of down-faulted blocks, or a series of horsts and graben, each lower in elevation than the one nearer the land. The first of these is probably exemplified by the slope off Queensland, and the latter by the coast of southern California and its borderland. A final type is the grooved and fissured type found in regions of active glacial erosion, such as Norway, Labrador, or east Antarctica. In this final type, the fissures run parallel to the coast and represent glacially overdeepened troughs, initiated by faulting.

One of the slopes studied in most detail is that off eastern USA. Hoskins (1967) has described the slope off New England where the seismic reflection profiles show a thin layer of sediment on the outer shelf, overlying two prograded layers, the uppermost layer being only 100 m thick, in a total of 1 km. The deepest of these layers outcrops on the slope in places. These observations indicate that the continental slope is erosional in character and has not been built seaward at least since the Cretaceous. Fault zones are not indicated on the profiles. Seamounts exist on the lower part of the rise, but the slope is straight and steep, although greatly dissected in plan by submarine canyons.

3.3 The continental rise

In considering the origin of the slope it is necessary also to consider the character of the continental rise below it. The rise represents a thick apron of sediment, attaining 1·6 km thickness in places. The conformable layers of sediment involved in this great pile of material lap up onto the slope between 1200 and 2000 m below sea level. They are probably not derived from the outer shelf and upper slope, as suggested by the on-lap, because the volume involved is too great. Most of the sediment on the rise has been derived from the land, probably since Cretaceous time. The gradient of the rise varies from 1° to 6°, while the angle of rest of the sediment is probably about 1–2°. The lower portion of the exposed slope consists of upper Cretaceous material, so the rise material must date from after this period off the east coast of the USA. It is inferred that the bulk of the sediment accumulated on a Palaeozoic base during Cretaceous time to form the shelf. Its edge may then have been down-faulted, exposing the edges of the layers on the slope. Material from the edge of the slope and shelf, and continental material, have accumulated since on the rise. The edge of the shelf may have been down-warped further in the Tertiary, allowing accumulation of the sedimentary wedge on the shelf during the middle to late Tertiary period. Down-faulting continued as the trench beneath the rise filled up with thickening sediment. This down-faulting may have produced the present slope.

The currents that flow along the slope may also have played a part in its erosion. Other erosional processes include canyon cutting: slumping may also have been active in association with earthquake activity. The rise is thus a major zone of deposition on the continental margin during recent geological time and at present.

Further observations on the continental rise are given by Fox, Heezen, and Harian (1968) by means of undersea photography, showing that ripples and scour marks occur, probably created by the western boundary current beneath which they occur. There is a belt of hills 64–130 km wide on the lower part of the continental rise, which trends northeast to southwest from the Hudson Canyon to the Hatteras Canyon, a distance of 520 km in depths of 4500–5300 m. The hills occur in the area adjacent to the Hatteras abyssal plain near the Hatteras Canyon. The individual hills are 4–6 km wide and 60–100 m high with slopes of 1 : 120 to 1 : 180. The hills are partially buried by sediment. The photographs show signs of a strong southwesterly current flowing obliquely across the axial trend of the hills, parallel to the regional contours. Evidence suggests that the hills have migrated up-current and up-slope through time, in the way that anti-dunes move upstream in a river. They are interpreted as abyssal mud dunes, and it is thought that they indicate the direction of Tertiary bottom currents. They have been found also on the outer ridge of the Blake Plateau–Bahama area, in the Argentine basin, and in the South China Sea.

Further evidence of the part water movements play in the shaping of the continental rise has been presented by Heezen, Hollister, and Ruddiman (1966). Their observations relate to the rise off southeastern USA, where the rise is a broad uniformly sloping and smooth wedge of sediment 1–10 km thick and 100–1000 km wide. It lies at the base of the continental slope where major trenches are absent. It is normally covered by lutites and is probably partly composed of them. The wedge thins seawards towards the abyssal plain. Its general

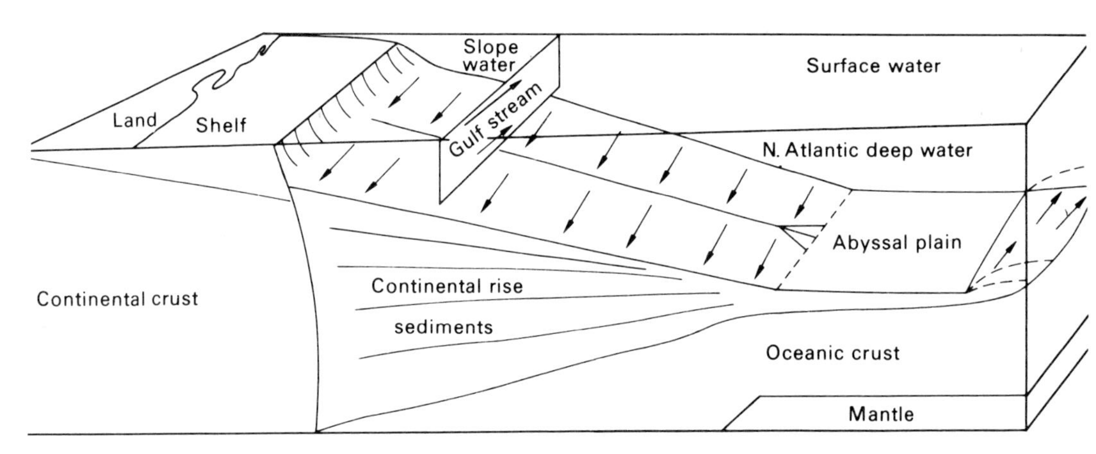

Figure 3.15 Block diagrams of the continental rise sediments and the pattern of currents. (*After Heezen, Hollister, and Ruddiman, 1966.*)

relationships are shown in figure 3·15. Most of the material composing it is terrestrial in origin, so that its rate of deposition varies with the rate of subaerial denudation. Rates are high in comparison with other parts of the ocean. The values vary between 5 and 50 cm/1000 years, and were probably higher during the Pleistocene when sea level was lower.

The situation is rather different near the Blake Plateau, an area showing evidence of elevation in the landward dip of some of its sedimentary layers. It is bounded by a steep escarpment, beyond which an outer ridge lies between the escarpment and the Hatteras abyssal plain. The Blake Plateau has only a thin layer of sediment, because deposition is prevented by the intensity of the current. Where currents flow south in the north Atlantic there should be a western intensification. Near-bottom velocities of 18 cm/sec have been observed in the south-flowing western boundary current, a speed that could transport sediment such as that found on the continental rise. Ripple marks indicate movement on the Blake–Bahama outer ridge. The volume transport of deep water moving south under the Gulf Stream in the western boundary bottom current is 4 to 12 $\times$ 10^6 m^3/sec. If the water contained only 1 $\times$ 10^{-6} g/cm^3 of sediment, and the volume passing south of Cape

Hatteras in 10,000 years could be 5×10^6 m^3/sec then the total volume of clay would be $1 \cdot 6 \times 10^{18}$ cm^3. This material is thought to have moved towards the outer ridge. Sediment transport will be greatest near the top of the continental rise, due to deflection to the right, where velocities are high enough for lutite transport, but not for erosion. Under these conditions sediment can accumulate at a uniform thickness at any given depth, the thickness being greater as the depth decreases and the volume of flow increases. The outer ridge accumulated because the ridge beyond the Blake Plateau caused an easterly deflection of the northerly-flowing current on the western flank of the ridge, while a westerly deflection occurred in the south-flowing current on the eastern flank of the ridge, as shown in the inset of figure 3.13. Sediment thus accumulated on the ridge where the two currents met. The lutite found on the continental rise cannot be accounted for by turbidity currents, particularly on the outer ridge, which is isolated from the main rise further north and so has no direct access to sediment derived from the land. The outer ridge is 1500 km long and it illustrates the importance of transport parallel to the velocity of the main stream in deep sea currents that flow parallel to the bottom contours. Thus transport of sediment is along and not across the bottom contours in these conditions. This conclusion is supported by the observations that the thickest sediments are found under the axis of the deep geostrophic currents, which are closest to the continental slope where the current is flowing south. This accounts for the wedge form of the sediment in the continental rise and the ridge form of the outer ridge associated with the Bahama Bank and its associated spur. The Gulf Stream, on the other hand, flows along the upper part of the continental slope, much nearer the land, than the deep currents associated with the continental rise and outer ridge in this area.

Elsewhere movement is normal to the contours as shown by Uchupi (1967). Slumping and gravitational sliding have modified the continental slope southeast of Long Island. Such movement could be initiated by rapid deposition or earthquake activity, and could have occurred in late Tertiary or Pleistocene times. Slump structures are shown by rotation of bedding and displacement of the sedimentary prism, with beds dipping towards the continent in places. Sometimes two slump blocks have been superimposed. The movement may have disturbed the sediments on the rise and caused them to move also. Such movement could have caused the lower continental rise hills off Cape Hatteras.

Since shallow structures can be readily identified by seismic surveys, some interesting examples of diapiric structures have been revealed. Diapirs of salt and other types are common on land, and now they have also been discovered in the deep oceans and the continental margins. Salt domes have been proved by the JOIDES drilling in the Sigsbee Knolls in the Gulf of Mexico. Seismic surveys have shown that these domes were derived from mid-Mesozoic salt. Similar features occur in the South China Sea, near the Balearic Islands, and other small basins which could have been of continental type. Deep sea diapirs have been located off the Grand Banks in the lower continental rise and Sohm Abyssal Plain, and at the base of the continental slopes off the Grand Banks and Nova Scotia. Salt is the most likely origin of these diapirs, because of the presence of the Jurassic salt layers on the Grand Banks of Permian age. Seismic surveys have revealed salt domes on the Nova Scotia shelf. Reconstructions of the pre-rifting land distribution suggest the presence of evaporite basins that at the time were continuous both to northwest Africa and to the area around the Grand Banks and Nova Scotia. Rifting has since separated the two regions by the width of the Atlantic ocean. Similar diapiric features occur in 2400 fathoms

Plate 7 Mudlump islands off the mouth of South Pass, Mississippi River delta, Louisiana. View from southwest shows four islands and surf breaking over mudlump shoals. The sand spit has resulted from wave reworking of the river mouth bar and joins a mudlump island to the South Pass jetties (see Plate 8 below). The boundary between turbid river waters and gulf waters is shown clearly in the foreground. (*James P. Morgan.*)

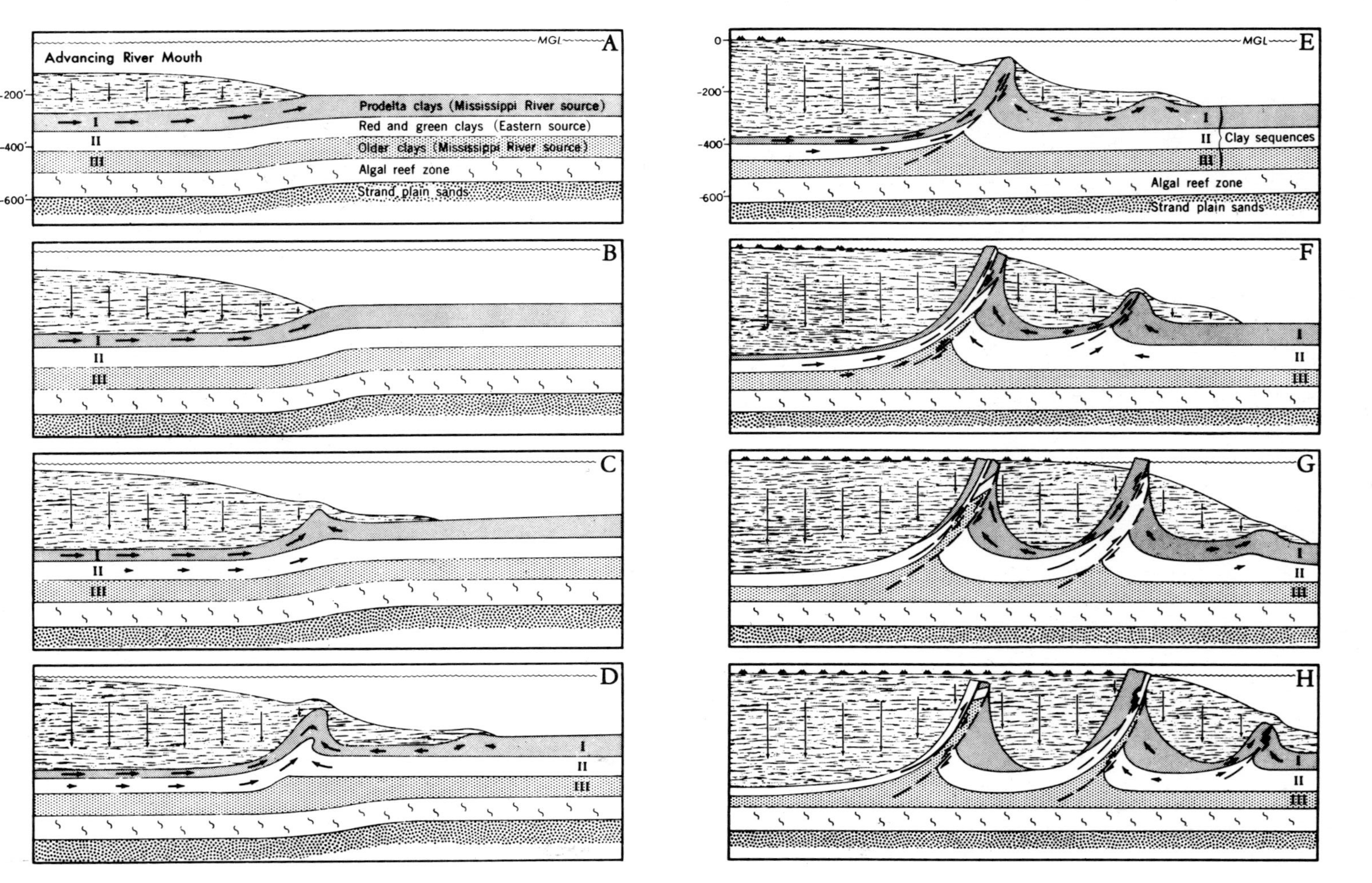

Plate 8 Diagrammatic representation of South Pass mudlump origin and development, illustrating the formation of the diapiric structures and the emergence of the offshore mudlump at stage G. (*Morgan, Coleman and Gagliano, 1968, 160–61.*)

Plate 9 Mudlumps caused by diapiric intrusion off the mouth of Southeast Pass, Mississippi River delta, Louisiana. The pattern of wave refraction around the mudlump in the right foreground of the photograph shows very clearly. (*James P. Morgan.*)

(4400 m) north of the Cape Verde Islands. These are also probably salt, as they are close to known salt deposits: they lack magnetic signature, and they closely resemble the diapirs of the Gulf of Mexico. Confusion can occur between true diapirs and similar features in the form of pseudo-diapirs, such as the partially or completely buried relief that is found on the flanks of the mid-Atlantic Ridge, for example.

It seems likely that the evaporites originated as rifting started. The exact time of opening is not certain. Triassic and Permian have been suggested, but the former seems the more likely because of evidence of tension and faulting in the Triassic; magnetic evidence also supports this date. Drainage reversals in the eastern United States also agree with this date. Cores that include sediments of the proto-Atlantic are very similar to those obtained from the developing ocean in the Red Sea. They both suggest restricted circulation, giving organic-rich sediments with black saprolitic material. The present Red Sea cores are similar to those of the early Cretaceous from the Atlantic. Salt can form in deep water where evaporation exceeds runoff and precipitation, in an originally well-oxygenated basin with normal fauna. The next stage is euxinic, with a stagnant zone below sill depth, oxygen-deficient bottom water, anaerobic benthos, normal nekton, and sapropel facies. The name, euxinic, is derived from the Black Sea, which illustrates these characteristics. The third stage is the ephemeral stage—the stagnant bottom persists; salt begins to precipitate at the surface; it dissolves at depth and fauna are sparse. The fourth stage is the permanent evaporite stage. Evaporation still exceeds precipitation and runoff. The bottom water is brine-saturated; halite and gypsum precipitate at the surface and are preserved at depth, where bottom brine is replaced by salt. This process operates in the Red Sea at present. As the Atlantic widened it became re-oxygenated by more vigorous circulation (probably aided by the shifting positions of the poles), causing colder water to sink in high latitudes. Well-developed thermohaline circulation in the Atlantic is post-Cretaceous or post-Eocene in age, prior to which period sapropelitic clays accumulated. These organic-rich deposits give rise to petroleum accumulation (Delany, editor, 1970).

4 Submarine canyons

4.1 Distribution

As the continental slope is surveyed in more detail and more extensively, so the distribution of submarine canyons is found to be wider and wider. These deep valleys, which incise the slope and some of which cut across the continental shelf, often extend far out into the abyssal plain. They are found off a wide variety of continental shelf and slope types. They occur in the Arctic (Carsola, 1954), off straight coasts in both stable and unstable areas, areas of uplift and of subsidence, off deltas, associated with estuaries, and in enclosed seas such as the Mediterranean, where they extend far below the sill depth. Many canyons are associated with the mouths of large rivers. These include the Hudson, the Mississippi, the Yukon, the Indus, the Ganges, the Columbia, the Congo, the Susquehanna, the old mouth of the Tagus, the Adour, and many others. Not all canyons, however, are connected with the mouth of a large river. Where canyons appear to be lacking, this may be due to inadequate soundings.

4.2 Types of canyons

Three broad types of canyons used to be distinguished. The first were the fairly small gorges originating fairly near the edge of the continental shelf and running down the slope to a great depth. The second type is rather similar in form, but extends right across the shelf and starts near the mouth of a large river. The third type has a branching, dendritic pattern, deeply incised into the edge of the shelf and slope.

The Oceanographer Canyon exemplifies the first type. It cuts into the New England shelf, heading 21 km inside the edge of the shelf, at a point 160 km from the land. It is the largest of a series along this shelf, many of which only head 8–16 km from its edge. The floor has an average gradient of 1 : 23 to a depth of 2000 m, while at its head its depth is 110 m. It reaches a maximum depth of 1520 m below the surrounding shelf. In cross section it is V-shaped and Cretaceous rocks outcrop on its walls, together with late Tertiary and glacial material.

The Congo Canyon falls into the second type. It extends 36 km up the estuary of the river. It is V-shaped, unlike the flat-floored Mississippi and Indus Canyons, and it can be traced for 230 km to a depth of 2300 m. Its longitudinal gradient is 0·5° and it seems to be the only canyon along a considerable stretch of shelf to north and south, where large rivers are absent. The canyon floor and walls are now being covered by mud. Although sand and silt are carried down the river and alongshore, these materials do not seem to find their way into the canyon. The canyon probably only started to be filled during the last few thousand years.

The Hudson Canyon is of particular interest because it is one that can be traced to great depths; deposition at the mouth of the canyon has been studied by Ericson, Ewing, and Heezen (1951). This canyon originates off the coast north of Boston, Massachusetts, and has been traced to a depth of about 4900 m. It is 288 km long and links with canyons coming down into the western Atlantic basin from around Greenland, the whole system having a length of 2000 km. The bottom sampling carried out by Ericson *et al.* off the mouth of the Hudson Canyon has given important information concerning the character of the sediments associated with the canyons. Cores were taken beyond the mouth of the canyon in a submarine fan at a depth of 4600 m. These showed layers of sand, varying from thin films to layers 6 m thick, which were interbedded with abyssal clays. The sand was well graded and bedded, but was not at uniform levels, while material of Miocene age from the canyon walls was included in the sediment. Even in the deep sea basin the canyon has steep sides, the floor is flat and 5–8 km wide, and from 18 to 180 m below the smooth basin floor. This evidence shows that material from near the shore or inland can be carried right out to the centre of the deep ocean basins along the submarine canyons.

The canyons off the coast of southern California illustrate the third type. The Monterey Canyon is one of the best known. It can be traced right up onto the beach and extends for 80 km offshore to where the depth is 2740 m. A profile across it is very similar to one across the Grand Canyon of Colorado. It heads near the Salinas River, but its gradient is much steeper than that of the river. It has a gradient of 1 : 10 for the first 0·8 km and 1 : 25 for the next 72 km, compared with 1 : 1666 for the lower Salinas River. It is joined at a depth of 2000 m by the Carmel Canyon. The rocks of its walls include Miocene sediments and granite at a depth of 915 m, while samples from its floor at 1460 m show graded bedding, each layer becoming progressively finer upwards. It loses its V-shaped character at a depth of 3660 m, terminating in a fan which is cut by shallow trenches, with levees on either side. Figure 3.16 illustrates these canyons.

Now that canyons can be studied directly in the field from diving saucers and with aqualungs (at least in their upper portions), they are becoming better known and theories of their

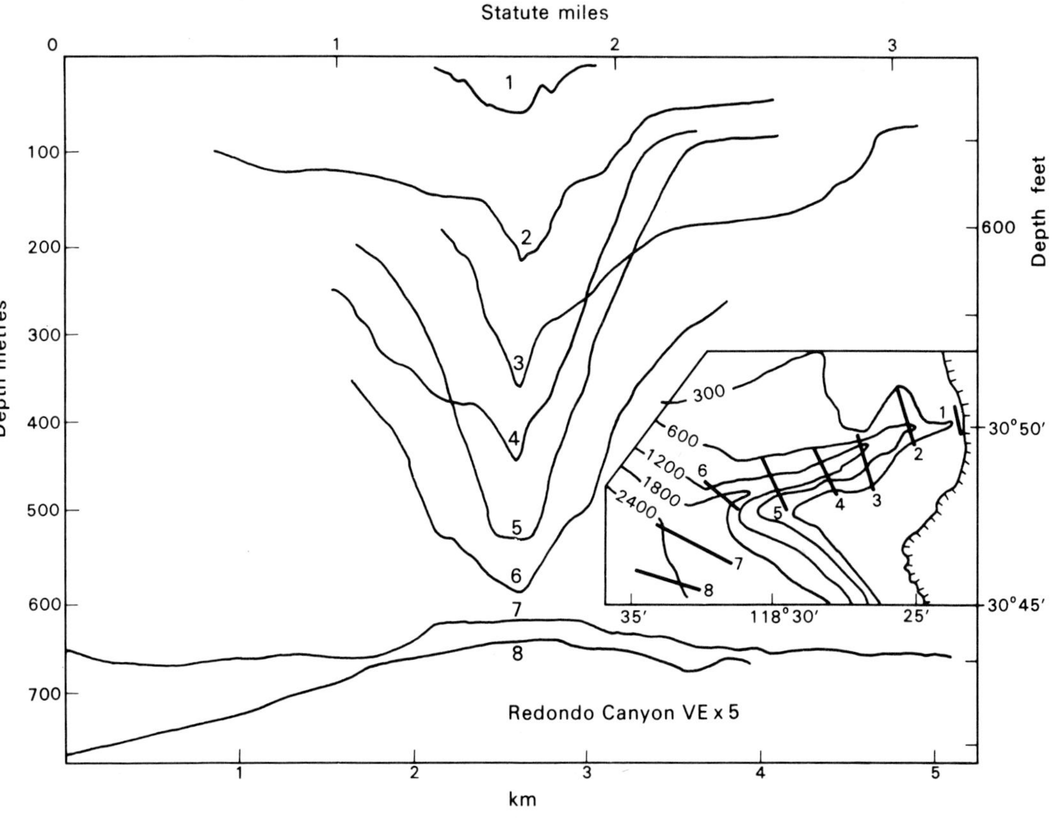

Figure 3.16 Map and profiles of Redondo Canyon, off Los Angeles, California. Profile 7 shows the submarine fan. (*After Emery, 1950.*)

formation are based on better evidence. The increase in world-wide echo-sounding has also revealed a much clearer picture of their distribution and dimensions. From these data a greater range of types has been identified, and described by Shepard and Dill (1966). Canyons occur off nearly all coasts of the world, including western North America, Bering Sea, north Alaska, eastern North America, Europe, the Mediterranean, Australia, Chile, Ceylon, Honshu, Congo, Hawaii, Bahamas, Manila, and New Zealand. Eight types of submarine valleys are distinguished by Shepard and Dill:

1 Features similar to land canyons, with V-shaped profiles, high steep walls with rock outcrops, a winding course and many tributaries. These are called submarine canyons

and are exemplified by the Monterey Canyon off California, the Hudson Canyon on the east USA coast, the Cape Breton canyons, and the Trincomalee Canyon off northwest Ceylon.

2 The second type cuts the large fans that are associated with the lower end of most large submarine valleys, and are called Fan-valleys. They are V-shaped or trough-shaped, the walls may be precipitous, but are not formed of rock and can attain 183 m height; natural levees occur on their margins. They have distributaries and can be winding.
3 The third type extends across the continental shelf and are called shelf valleys. They are rarely more than 180 m deep and do not connect with submarine canyons. The New York shelf valley is the best known. They sometimes have small basins along their length.
4 The fourth type occurs on the shelf off glaciated coasts and are called glacial troughs, although they may not be glacial in origin. They are trough-shaped, often exceeding 180 m depth and have rather large basins along their length. They have both tributaries and distributaries. The trough coming from the St Lawrence estuary is an example.
5 The fifth type is associated with deltas. They are trough-shaped with floors sloping continuously seaward down the continental slope and beyond, rarely having tributaries or hard rock outcrops. They are called delta-front troughs and occur off the Mississippi, Niger, Indus, and Ganges deltas.
6 The sixth type consists of small discontinuous valleys found on many otherwise straight slopes. They usually lack tributaries and are associated with submarine fault scarps or delta fronts. They are called slope gullies, as their relief is small and they are discontinuous.
7 The seventh type is associated with active diastrophic areas. The valleys are straight, with few tributaries and they can be related to structural trends. The San Clemente rift-valley is an example. They are called valleys and resemble grabens or rifts.
8 The eighth type is found in the deep sea. It is trough-shaped, of low relief and with few tributaries. Some of the valleys are parallel to the continental margin, others lie at a large angle to it. Some are continuations of fan channels, but others do not seem to connect with them. They are called deep sea channels. Examples occur in the mid-Atlantic abyssal basins.

Methods of studying canyons include echo-sounding to assess their relief, although various problems arise in sounding in steep areas. Sampling can provide evidence of the material on the sides and in the bottom. Remote photography and current measurements provide useful information, while scuba diving provides information directly from the heads of the canyons. The canyons must be considered in relation to their associated deep sea fans. These two features are different aspects of the same process, so that one cannot be explained without the other.

The canyons on the east coast of Australia have been described by Conolly (1968). They are large features, 80–160 km long, crossing a large amphitheatre-shaped depression that extends from the shelf break to depths of 3660 m on the continental slope east of Bass Strait. Bass Canyon has three major tributaries at 110–165 m depth and is 160 km long. Over 64 km its floor is wide and flat, 3·2–8 km wide, with 1280 m high walls in the central part of the continental slope. The Flinders Canyon to the south is a narrow gorge, with two major tributaries, extending to depths of 3840 m. The canyons formed after the deposition of 1500–3000 m of sediment in the late Jurassic to late Eocene. Canyon erosion and sedimentation has occurred in three phrases—in the late Eocene–early Oligocene, and the Miocene; the present stage started in the Pliocene. The shelf is narrow and the slope steep in this area, so conditions are suitable for canyon formation. Sediment can reach the area via the Bass Straight from Victoria and Tasmania, especially at times of low sea level. Current and wave activity is strong in the area, so sediment can move to considerable depths. The material carried through the canyons has produced a pile 3000 m thick and 80–110 km wide in front of the continental slope, thus prograding it seawards by this amount. The volume of sediment involved amounts to 83,000 km^3 since the late Eocene.

Large canyons have also been recorded off southern Australia by Von der Borch (1968). These canyons cut across the continental slope and are restricted to areas of shallow basement rock, being absent where Tertiary basins occur. The large canyons may be related to former drainage patterns of early or pre-Tertiary age, relics of which can still be identified. If this relationship is genuine, then the canyons must be old, having been initiated in early Tertiary time at the latest. At this time the heads of the large canyons may have been cut by subaerial processes. There are smaller features on the slope in the Tertiary basins, and these may date from low Pleistocene sea levels. The Perth Canyon heads 48 km offshore and has two tributaries. It extends 64 km to a depth of 2740 m. The large Esperance Canyon off the south coast has a vertical relief of 1830 m and heads 32 km offshore at 110 m. Sounding intensity is not sufficient to give much detail of the relief.

Probably the largest canyons in the world have been located in the Bering Sea off Alaska and have been described by Scholl *et al.* (1970). Three of these are exceptionally large when compared with other canyons. They are the Bering, Pribilof, and Zhemchug Canyons. The Bering is the longest and the Zhemchug has the greatest volume. The Bering Canyon is nearly 400 km long. It heads in a large amphitheatre-shaped scar on the outer shelf near Unimak Island. It is formed of a number of tributaries coming from the Aleutian Islands to the south. The canyon is winding and eventually turns from southwest to north, flanking the Umnak Plateau, which is overlain by over 3000 m of the main layered sequence previously mentioned. The canyon has an asymmetrical cross profile and flat floor and ends on the continental rise at 3200 m where it is 11·5 km wide and flanked by levees. The volume of the canyon is 4300 km^3, while most other canyons have volumes less than 500 km^3. The lower part of the canyon is a fan-valley.

The Zhemchug Canyon is considerably larger (figure 3.17), having a volume of 8500 km^3, compared with 450 for the Monterey Canyon. The Zhemchug has a double head behind the face of the continental slope, forming an elongated trough 160 km long, 25–30 km wide, and 2600 m deep. The seaward flank is steep at 20°, while the landward flank only slopes at 6–7°. It is 100 km wide but only 200 m deep where it cuts through the continental slope, although it has a central gorge 30 km wide and more than 1500 m deep. The canyons extend to a depth of 3400 m, ending in a fan-valley 4 km wide and 100 m deep. The feature is superimposed on a structural depression as is the Pribilof Canyon. Its alignment and that of the other canyons appear to be structurally controlled.

The sources of sediment associated with the canyons are provided by Alaskan rivers, including the Yukon. Glaciation helped to increase the sediment supply. Several cycles of canyon cutting are indicated by unconformities in the strata that dip down into the canyon,

A

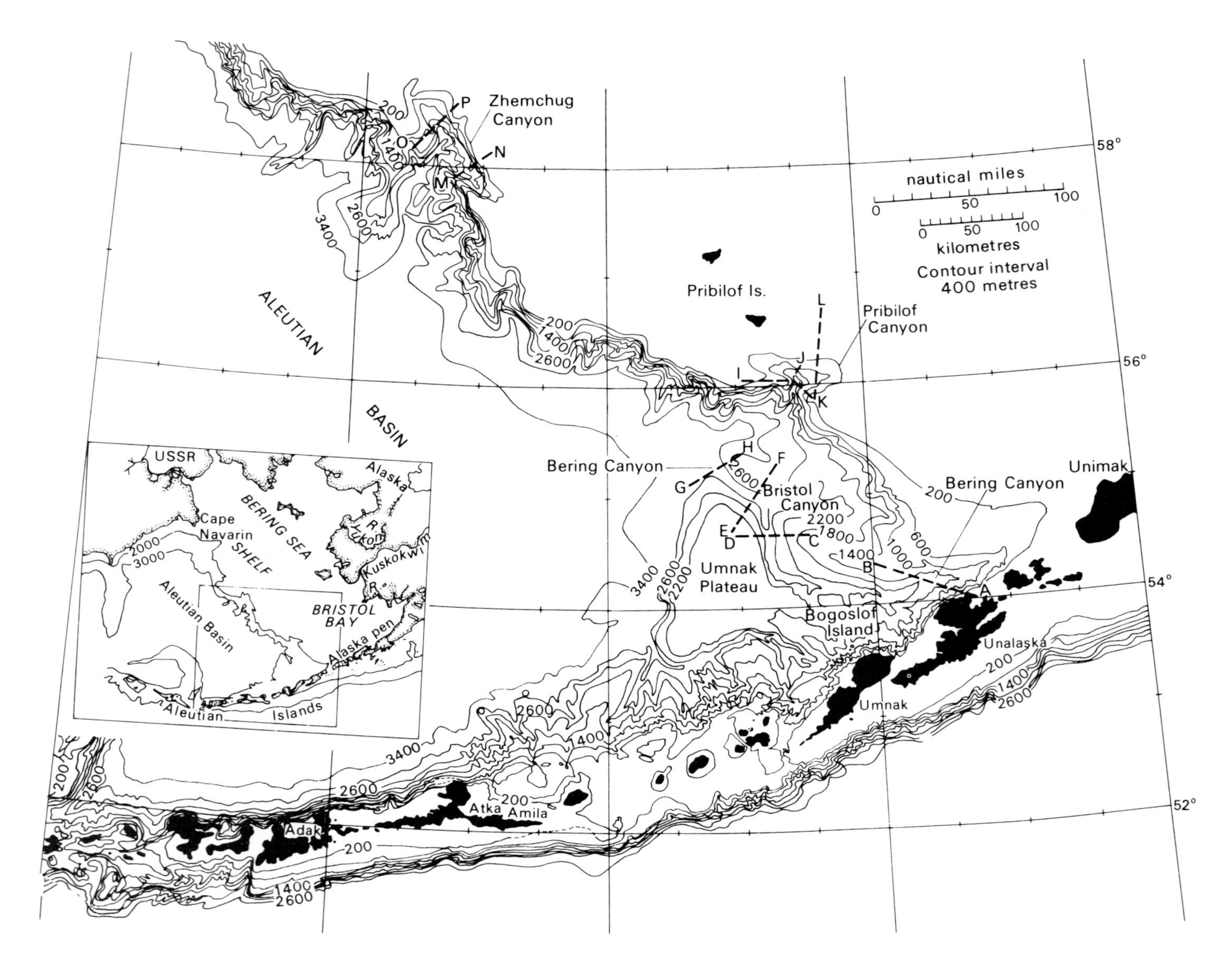

Zhemchug Canyon
ALEUTIAN
BASIN
nautical miles
kilometres
Contour interval 400 metres
Pribilof Is.
Pribilof Canyon
Bering Canyon
Bristol Canyon
Umnak Plateau
Bogoslof Island
Unimak
Unalaska
Umnak
Atka
Amila
Adak
USSR
Alaska
BERING SEA
SHELF
Cape Navarin
R. Yukon
Kuskokwim R.
BRISTOL BAY
Alaska pen.
Aleutian Basin
Aleutian Islands
58°
56°
54°
52°

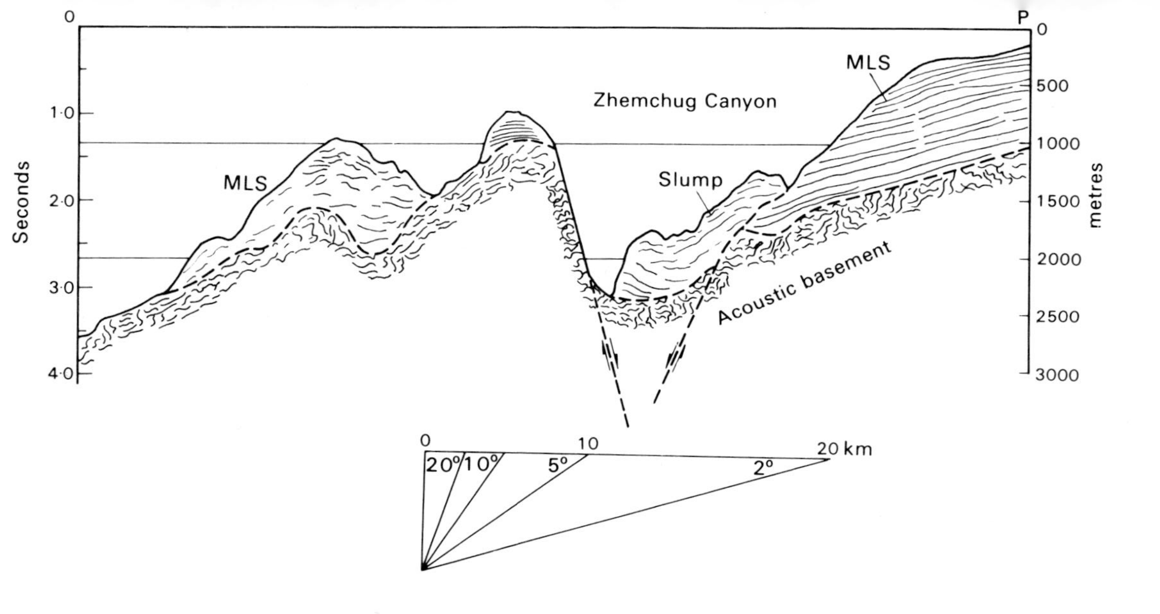

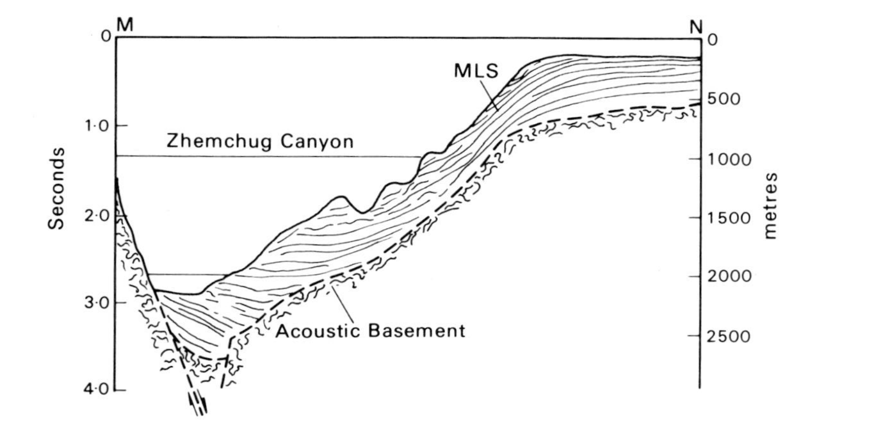

Figure 3.17 Diagrams to illustrate the character of the Zhemchug Canyon in the Aleutian basin. **A** (*above*) : The setting of the canyon and the Pribilof and Bering Canyons is shown. **B** : The geological interpretation of the acoustic profiles across the Zhemchug Canyon. MLS stands for the main layered sequence. **C** (*see over*) : Geomorphological diagram of the Zhemchug Canyon. (*After Scholl* et al., *1968.*)

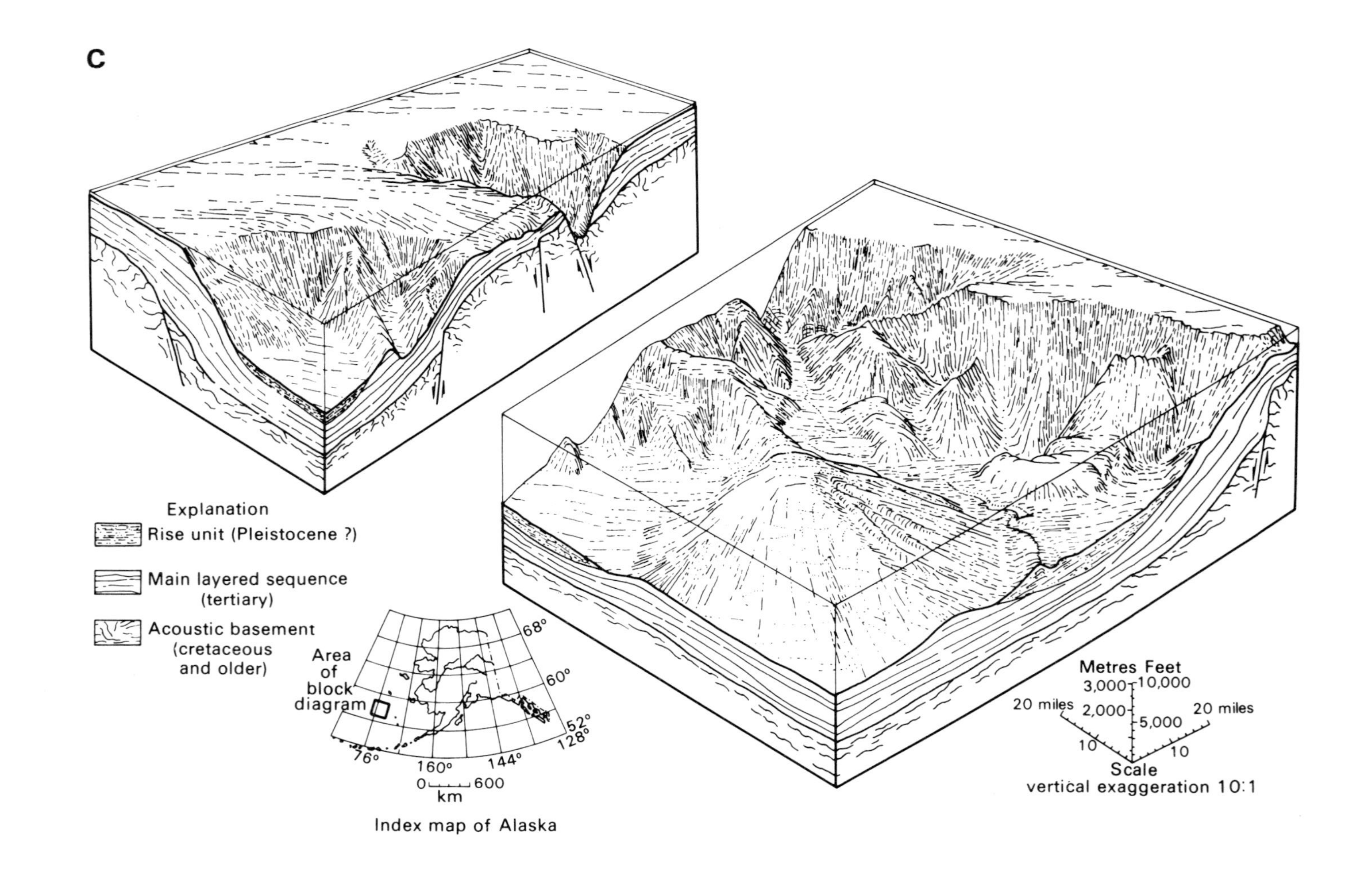

Figure 3.17C. Geomorphological diagram of the Zhemchug Canyon.

with intervening periods of filling. The Bering Canyon was probably initiated before the end of the Tertiary, with the first initial filling taking place in late Tertiary and early Quaternary. The two subsequent cutting episodes probably took place at times of glacially lowered sea level in the Pleistocene. The volume of material passing down the Zhemchug Canyon during 1·5 million years has been estimated at 225×10^{12} metric tons from the Yukon. This figure is 25 times the volume of the canyon, a similar ratio to that of the ratio of volume of the fans at the mouths of the Californian canyons compared to their volume. The bulk of these very large canyons in the Bering Sea was probably cut during the Quaternary, although initial incision probably started in late Tertiary when the outer part of the continental shelf was exposed. The cutting of the canyons may have been achieved by sliding of masses of sediment deposited on the upper part of the continental slope. There is considerable similarity in form between the Bering Canyons and those off California.

Many of the canyons on the Californian Borderland also spread out as large fans at their lower ends; in many instances the volume of this fan is greater than that of the canyon above it (Emery, 1960) as shown by the following figures in table 3.1. The table shows that

Table 3.1 Volume of submarine canyons and associated fans off California

Canyon	Volume of fan	Volume of canyon 10^9 m^3	Ratio fan to canyon
Hueneme–Mugu	27·6	15·0	1·8
Dume	1·9	0·6	3·3
Redondo	16·7	9·2	1·8
Carlsbad	30·8	1·1	28·0
Coronado	18·4	6·5	2·8

the canyons are used as routes for the transport of sediment from shallower water. In some canyon fans, shallow channels are eroded across the fan. These are mostly only 15–90 m deep and 0·3–3·2 km wide.

Shepard and Buffington (1968) have studied the fan-valley of the La Jolla system. The fan is built into the San Diego trough off La Jolla. The canyon is a steep-walled, rocky submarine gorge in the steep part of the slope, but changes to a fan-valley at its lower end. The channel runs across a fan-shaped prism of sediment, and has well-developed marginal levees, but there are no distributaries, as in many subaerial fans. Below 274 m the valley becomes U-shaped, while at 530 m a terrace about 800 m wide rises above the narrow channel. The terrace extends 10 km to a depth of 660 m. The valley becomes V-shaped below this point and levees are conspicuous below 915 m, gradually broadening and fading out at 1100 m. The canyon is 11·5 km long and the fan-valley beyond it 41·5 km; the respective gradients are 4·2 per cent and 2·3 per cent and are irregular.

The levees are probably formed by the reduction of speed of the turbidity currents carrying sediment down the valley as they overflow. The levees are higher on the outside of bends where the valley curves, but are more symmetrical in the straighter portions. Levees are best developed where the channel is shallower, near the outer limit. Terraces occur mainly on the inside of bends and slope towards the valley side. They are probably formed by slumps of blocks of material from the valley walls. The fan-valleys cut into Pleistocene sediments

(Piper, 1970). Holoçene sands and muds in the fan-valleys are interbedded with sand predominating. Some layers show graded bedding. Sands are thinner and rarer away from the fan-valley. The nature of the sedimentary structures in the sand suggests deposition from powerful currents of declining competence and with a large suspension load. The Holocene sediments are about 2 km thick, showing that 90 per cent of the sediment supplied to the La Jolla Canyon by-passes the fan.

Shelf channels are not very common features on continental shelves. The Hudson channel is one example that is nearly continuous across the shelf but dies out at the shelf margin. The channel heads in a valley 6 m below the surroundings and after 13 km it joins a wider shelf channel at a depth of 27 m. The main channel has basin depressions up to 18 m deep. The channel has an average depth of 30 m below the shelf and a width of 4·8–6·4 km with gently sloping walls. It is connected with the Hudson canyon. The channel sediments are finer than those on the shelf around, and the channel is now being filled with mud. It is thought that the channel was cut by the Hudson River during a glacial low sea level.

Trough-shaped valleys are associated with glaciated areas and run either parallel or normal to the coastline, although the former are more common. One of the best known is the Laurentian trough, which runs for 800 km off southeast Newfoundland. The sides are steep, the floor broad and deep (ranging from 80 km wide to 520 m deep), with elongated basins along its length. Similar features occur off Labrador, south Alaska, and west Norway in association with the fjords. They all have deep basins and are similar to fjords apart from their greater width. Such troughs have also been identified off Antarctica. The troughs were formed by the more extensive glaciers of the Pleistocene, which followed structural weaknesses, whether these were parallel or normal to the shelf margin. The low sea level of glacial maximums would assist in the excavation of the troughs.

The Ganges trough exemplifies the delta-front feature. The Ganges delta has a width of 400 km in the Bay of Bengal and the trough heads 26 km off the crenulate delta margin. The trough has a flat floor 5–6 km across and the sides slope at up to 12° and are straight. The floor slopes outwards continuously with a slope of about 10 m/km, which is less than most submarine canyons. Depth reaches 1080 m, 110 km from its head. The trough may be bounded by faults, and there is a considerable fill of sediment. It becomes more V-shaped down the continental slope and the walls decrease in height, being 300 m at 160 km from the head. Further down, the trough changes into a fan-valley which bifurcates 400 km from the head of the trough into a very complex pattern. The fan-valleys continue to the latitude of Ceylon.

Slope gullies are also associated with deltas at times, and have been described from the Mississippi delta. They are short ravines rarely exceeding 60 m depth. The deepest of the gullies is only 15 m below the ridge, but most are only 9 m deep, varying from 3·2 to 6 km long and about 800 m wide. They are fairly straight and have no tributaries. Gullies seem to change position by the formation of new ones and filling of old ones. They are not associated with strong currents, as fine sediment is found in them. Not all gullies of this type are associated with deltas, although many are. They also occur off San Clemente, California, off a narrow shelf only 3·2 km wide. These are 300–610 m wide and have a maximum relief of 90 m, bedrock outcropping on their sides. Subaerial cutting could account for their heads and they do not seem to be undergoing erosion at present. Another possibility is slumping. Both of these processes would operate more effectively at a lower sea level.

Buffington and Moore (1963) have surveyed gullies on the edge of the continental shelf off San Clemente, California. The sea gullies trend down-slope and have a maximum relief of 110 m, incising the upper slope and shelf break in late Miocene and Pliocene bedrock. They are best accounted for by local tectonic activity, which resulted in terrace formation and buried channels. The gullies were cut in a phase of low glacial sea level (−100 m) and their lower part was drowned as sea level rose, due to both subsidence and then eustatically. The gullies are now partially filled. Turbidity flow probably took place down the gullies when sea level was low.

A fault trough is exemplified by the Tinaja trough of Baja California. The valley runs north–south, parallel to the coast, and is straighter than most other types of canyon or valley. It is partially V-shaped and partially flat-floored with one possible basin depression. It cuts across the Candelaria Canyon at its northern end. Its characteristics support the view that it is faulted in origin, which was confirmed by the location of a fault on its west side by reflection profiling. The fault continues a land fault.

Deep sea channels are the deepest form of submarine valleys and are associated with the abyssal plains. One of the longest is that which starts in Baffin Bay and runs around the Newfoundland Banks. The feature is not a true canyon and has a wide flat floor, 2·5–6·5 km wide and a depth of 90 m below the plain. The channel slopes southwards at a very low gradient of 0·3 m/km between a depth of 4480 m and 5000 m, compared with 57·7 m/km for average canyon gradients. Shepard and Dill (1966) are critical of the Heezen *et al.* turbidity-current explanation of this channel, and they consider that a tectonic rift origin may be more plausible.

A complex series of submarine valleys descend to a depth of 4650 m in the Coral Sea (Winterer, 1970). One of these, the Moresby Valley, is 500 km long, consisting of an integrated drainage system with many tributary canyons and valleys. Sediment from the Papuan coast passes through the system to the basin floor. The Bligh Valley system drains 900 km of the Queensland coast and Great Barrier Reef. The valleys are mainly located in structural depressions but the relief has been intensified by erosion of the valleys into earlier fill (probably pelagic sediment) to depths of up to 750 m. After this phase of entrenchment further deposition has taken place by as much as 500 m of sediment. This sediment is associated with the filling of the Coral Sea basin and is considered to indicate the adjustment of the long profile to this process. The valleys are considered to be graded. The troughs were filled first, but as these overflowed the integrated system was formed when the Papuan Plateau sill was overtopped.

4.3 Origin of submarine valleys

Hypotheses to explain submarine valleys can be broadly divided into two groups—those proposing a subaerial origin, and those proposing submarine processes. There is probably something in favour of both theories. Sea level has oscillated through about 100 m vertically during the Pleistocene period, and tectonic movements can account for further changes of level in many places. On the other hand, now that it is known that submarine valleys can be traced right out into the abyssal plain at very great depths, a subaerial origin for the whole length of the valley is impossible.

The subaerial theories were initiated when the striking similarity in form of sea and land valleys was discovered. Shepard (1948) thought that sea level could be lowered enough to account for the canyons then known. A large mass of ice, with the addition of marginal warping, could result in a fall of about 1800 m in sea level, although the allowance for eustatic fall was ten times more than the normal estimate. Indeed in 1952 Shepard himself abandoned this idea. He then put forward a much more likely theory, in which he suggested that the canyons were formed by a number of composite processes.

Some submarine valleys probably formed subaerially. There are a series of canyons off Corsica which seem to prolong land valleys and which are very similar in form to them. It is probable that these canyons have been cut by normal rivers and have been subsequently down-faulted to reach their present position. The geophysical evidence favours this view (Gaskell, 1960), and the presence of hard rock on the walls of the canyon can most easily be explained in this way.

It has been shown that there is some evidence in favour of continental flexure, which would increase the depth at which canyons are now found. Other more extreme views on the subaerial origin of canyons have been put forward by Landes (1952), but his ideas are very hypothetical. Possible submarine processes must, therefore, be considered. A number of these have been suggested, but the only one for which there is much evidence and which is generally considered applicable over wide areas is the turbidity-current theory. This idea was originally put forward by Daly (1936) and has since been strongly advocated by Kuenen (1950) and many other workers.

4.3a Turbidity currents

The principle on which turbidity currents operate is the fact that water, with a load of sediment in it, is denser than clear water. It will, therefore, flow down-slope. As it flows (providing it is above a critical velocity), it will tend to gather more sediment, the density will increase further, rendering the effect cumulative. In this way such currents, once started, are self-generating. An increasing number of observations show that some such process is active in the ocean. The smoothness of the abyssal plains in those areas to which turbidity currents have access indicates their effectiveness in spreading out over a large area and distributing a large volume of material. The character of the sediments found on the floor and beyond the mouth of many submarine valleys show that material can be carried from shallow water, along the submarine valleys, to accumulate at their mouth in the form of fans. The formation of graded bedding can be explained by the gradual settling out of mixed material. This process can explain the characteristics of rocks now found on land, as discussed by O. T. Jones (1954), but some of these structures may well be due to slumping over a wide area.

Experimental evidence, derived from model experiments carried out by Kuenen (1948), shows that turbidity currents can operate in water mixed with sediment, and that they can attain great velocities. The velocity increases with the effective density of the current. Kuenen estimated that on a slope of 3° a current 30 cm deep should be able to attain a velocity of 0·9 m/sec. By analogy with experimental values and extrapolating further, the conclusion was reached that dense currents of considerable thickness could transport large boulders. A current on a slope of 3° could carry 10-ton blocks with a velocity of 4 m/sec if it were 16 m thick and had a density of 1·5. Although it is generally agreed that turbidity currents can carry much material, it is much less certain whether they can erode the material across which they are moving, particularly if it is solid rock. The fact that the size of the canyons is smaller than the fans at their mouths indicates that turbidity currents are more effective as transporting agents than eroding agents.

Daly originally suggested that during the low sea level of the glacial period much of the continental shelf was laid bare; the sea churned up the sediment to which the lowered base level gave it effective access. Zones of denser water, which flowed over the edge of the shelf as turbidity currents, were created and helped to cut the canyons. In view of the doubtful capacity of the currents to erode the materials of the shelf and slope, this theory is not fully satisfactory.

Shepard (1952) suggests that turbidity currents act with other processes in his composite theory of submarine valley formation. The age of the canyons vary throughout their length; some parts are older than others. The lower parts of the valleys off the eastern USA, at a depth of 1960 m, are cut through Cretaceous rock of shallow water origin. In the head of the Monterey Canyon, wells in the mouth of the Salinas Valley show an old canyon 1525 m below sea level filled with shallow water Miocene sediments. The land must have stood higher in relation to sea level when this canyon was cut. Part of the canyon is thus an old feature. Shepard has suggested that at one time the shelf and slope may have been elevated above sea level and subaerial canyons were cut across them, with fans forming offshore. Subsidence then caused the shelf to sink, and some of the valleys were kept open by turbidity currents, while others were buried. Meanwhile as subsidence continued, the shelf around the canyon was built up by progressive deposition, thus deepening the canyon. After the glacial period the open canyons extended headward across the Pleistocene deposits, cutting the upper and youngest part of the canyon through deposits left on the shelf during periods of low glacial sea level. Although this method of formation may well apply in some areas, it may not account for all canyons, and many problems remain to be solved concerning their formation. The evidence on the whole points to a considerable down-warping or down-faulting of the continental margin in many areas. It is also clear that much material is transported through the canyons to the deep ocean.

Of all the features described, the large submarine canyons and the delta-front troughs are the most difficult to explain. Comments have already been made concerning possible processes, such as turbidity currents. The evidence can now be considered. Submarine valleys occur off all coastal types and normally continue to the base of the slope. Many continue as fan-valleys, but some terminate on the continental rise. Some canyons, especially on narrow shelves, start very close to the shore and most have continuous long profiles, steeper in the upper part than lower down. Few canyons cross shelves wider than 32 km and only the Bering Canyons have heads more than 32 km from the shelf margin, although delta-front troughs occur on shelves up to 130 km wide. Most canyons are located off rivers. The depth changes frequently at the head of the canyon although mud frequently occurs along the canyon. Erosion is, nevertheless, obvious in many canyon heads and many canyons receiving much sediment are not being filled. Much evidence of shallow water material is found in canyons.

The main hypotheses of canyon origin include: 1) erosion by turbidity currents starting in the canyon head, 2) erosion by slow mass movement, slumps, sandfalls, and redistribution of sediment by bottom currents, 3) erosion by bottom currents, and 4) drowning by subsidence of valleys cut subaerially (Shepard and Dill, 1966). The effect of deposition around the canyon should also be borne in mind (Rona, 1970).

Turbidity-current action has been involved by many to account for the formation of submarine valleys. The process has been shown to operate experimentally and is considered further in chapter 5 as a means of sedimentation. The levees alongside fan-valleys could be accounted for by turbidity currents, as could the graded bedding and the distribution of sand in the deep sea fans.

4.3b Other currents

The presence and effectiveness of other types of currents should be considered. Core samples from the floors of several valleys reveal sand with scattered pebbles in the lower portion. If the sediments were turbidites, the pebbles should be at the bottom. The evidence for the operation of turbidity currents is not strong in some instances, and it has not proved possible to start major turbidity currents in apparently suitable canyon heads, although where oversteepening occurs they may form. Evidence of the nature of the sediment in canyon floors does not suggest high-velocity flow. Creep and sand flow, however, is well documented in the canyon heads. The fan-valley levees provide evidence for turbidity currents that fill the channel and can overflow its margins, where erosion is seen. Some mechanism must be able to carry a massive amount of sediment through the canyons.

The presence of relatively strong currents at the deep sea floor must be considered. Such currents could flow down tributary valleys and erode the channels. There are, however, problems in this hypothesis; currents have been recorded as flowing both down and up the canyons, possibly due to tides. Mass movement by creeping and slumping must be considered, the former being important in the steep upper parts of the canyon. Shepard and Dill reach the conclusion that turbidity currents are often rather sluggish, and that creep and other mass movement processes must be considered. Currents of up to 0·5 knots (25 cm/sec) flow in the canyons and must also be taken into account. Much still remains to be learned concerning canyon formation, despite recent advances in technology and observational techniques.

The work of Rona, Schneider, and Heezen (1967) exemplifies recent methods. They studied the continental rise off Cape Hatteras and describe the Hatteras Outer Ridge and the Hatteras Transverse Canyon (which trend parallel to the regional isobaths), and the Hatteras and Pamlico Canyons (which trend perpendicular to the isobaths). Continuous geostrophic contour currents have shaped the Hatteras Outer Ridge, while episodic turbidity currents have contributed to the formation of the submarine canyons. The Hatteras Outer Ridge lies on the ocean side of the transverse canyon and may extend northeast as far as the Hudson Canyon. The ridge crest is about 400 m above the transverse canyon floor and is 100 m higher than the opposite bank. Sediment is about 20 m thick and thins towards the troughs. The Transverse Canyon begins at 4400 m on the continental rise, is 90 km long and 5 km wide, and has a relief of 300 m. Other Outer Ridges and transverse canyons exist, for example on the Blake–Bahama area and the Caicos Outer Ridge. Geostrophic currents act parallel to the isobaths and can cause erosion and deposition. The Outer Ridge may be a constructional feature initiated by gravitational gliding of sedimentary strata on the continental rise, modified by subsequent sediment mantling. It would deflect flow to form the Transverse Canyon landward of it.

4.3c Other processes

Photographic evidence by Laughton (1968) has shown that erosion is active in deep sea channels surveyed in 1958 between the Biscay and Iberian deep sea plains. A further survey was carried out in 1966, and this showed benches on the cliffs in the channels, indicating that the channels have been produced by erosion. The channel has been cut through stratified sediments (revealed by seismic reflection profiles), and photographic evidence indicates erosion of the cliffs on which a 183 m section has been exposed.

One method by which submarine canyons can be eroded has been recorded as a result of underwater observations made by Warme, Scanland, and Marshall (1971). Rock borers were found to be so prolific in the Scripps submarine canyon that the attrition rate through their activities was estimated at between 2 and 10 mm/year on the rock in which their bores were so prolific. The animals included polychaetes, bivalves, and sipunculoids.

A combination of rim up-growth and canyon down-cutting has been suggested for the canyons off southeastern New Guinea by Von der Borch (1969). The canyons were probably initiated soon after the outline of New Guinea was established. The canyons have acted as conduits for sediment movement during the period of shelf up-growth and progradation. In this area the coastal zone is subsiding, causing drowned relief and deep embayments. Meanwhile the adjacent Huon Peninsula is rising and Pliocene and Pleistocene coral reefs have been elevated to 3000 m; no shelf exists. The whole coast is one of low-energy wave action. The canyons head less than 1·5 km seaward of the estuaries with which they are associated. The axial gradients of the canyons are steeper than those of the shelf or rivers. The up-growth of one canyon rim amounts to 100 m in the last 25,000 years, while its total depth is 1500 m at the shelf break. The whole feature could, therefore, have formed during the Pleistocene. Axial deepening may have been limited because the slope seems to continue inland as the pre-erosion surface slope gradient.

The canyons on the continental slope off Cape Hatteras have been described by Rona (1970), who proposes a composite origin. Between the canyons the strata have a convex upwards form, overlying horizontal strata. The geology suggests a composite origin, similar to that proposed by Shepard in 1952. Surfaces of unconformity developed on Eocene to Miocene strata outcropping on the upper continental slope. Canyons were incised into a post-Miocene surface, possibly during low glacial sea level; meanwhile the intercanyon area received sediment and built up. Three cycles of building and partial excavation following fill of the canyons can be identified. The channels in the canyons provide paths along which sediment reaches the abyssal plains from the shelf, while the intercanyon area is building up. In this way the continental slope has prograded seaward. This latter process seems to be a common phenomenon in the world. Canyons are certainly important pathways for sediment to reach the deep ocean, and they should always be considered with their associated fans, as suggested by Gorsline (1970).

Normark (1970) has discussed the development of deep sea fans and considers that over a long period of time they will become uniform. Deposition is localized on supra-fans, which form over the lower sediment at the end of large valleys that occur on the upper part of deep sea fans. The supra-fan has numerous smaller distributary valleys. New supra-fans form on different parts of the main underlying fan following rapid aggradation, which results in channel meandering. Isolated depressions and remnants of channels occur during the build-up of the supra-fans, but in time the whole feature can become symmetrical.

The seaward extension of submarine canyons off the northeast USA have been studied by Pratt (1967). Seven major canyons cross the continental rise off this area. Two types occur: firstly, river extension canyons stretch into the deep sea, and second, slump extension canyons are short and discontinuous. There seem to have been two cycles of canyon formation in the area. All the extensions have high right banks and they bend to the left at the top of the rise. The two longest are the Wilmington and Hudson Canyons, and these may be partially erosional where they deepen and swing right half-way down the rise. The Hudson Canyon is 240 km long beyond the 2000 m depth line and then extends for a further 64 km in an outer gorge over 600 m deep before branching into two distributaries. These long canyons are thought to be partially erosional in their outer parts because they are several hundred metres deep. An oversteepened stretch has cut back into the area of levee formation. The outer gorges cut through an oversteepened portion of the continental rise. The river extension canyons are assumed to be formed by turbidity currents derived from the rivers, They change to depositional canyons on the rise with large levees, the highest being on the right. The extra large canyons are probably associated with flood discharges under glacial conditions with outwash adding to the load. They are probably Pleistocene.

Most workers agree that a number of different processes are involved in the formation of submarine valleys. The most important appear to be turbidity currents, although their erosive role remains to be firmly established. Details of their action will be discussed in the next chapter, as they are very important in sedimentation processes in the deep abyssal plains. Other processes of significance include the upbuilding of the shelf between the canyons and the processes of mass movement by slumping and creep. Changes in elevation of the oceanic margin are also important; tectonic activity plays a part in some areas. Thus a composite origin is probably correct for nearly all canyons; their formation spans a considerable period of time, with different processes—both erosional and depositional—playing a major part at different times.

Further reading

DAVIES, J. L. 1964: A morphogenic approach to world shorelines. *Zeit. für Geomorph.* **8,** Sp. Nu. 27*–42*. (Suggests a dynamic classification of coasts.)

EMERY, K. O. 1960: *The sea off southern California*. New York: Wiley. (All aspects of nearshore oceanography are dealt with in this book.)

EMERY, K. O., UCHUPI, E., PHILLIPS, J. D., BAWM, C. O., BUNCE, E. T., and KNOTT, S. T. 1970: Continental rise off eastern North America. *Bull. Amer. Assn. Petrol. Geol.* **54,** 44–108. (Detailed account of continental marginal structures.)

GUILCHER, A. 1958: *Coastal and submarine morphology*. London: Methuen. Translated by B. W. Sparks and R. H. W. Kneese. (Concentrates on coastal morphology, but also covers the continental margin.)

KING, C. A. M. 1972: *Beaches and coasts*. 2nd edition. London: Arnold. (The points made in the section on the coast are treated in more detail.)

INGLE, J. C. 1966: *The movement of beach sand. Developments in sedimentology* **5**. Amsterdam: Elsevier. (Describes experiments on sand tracing in California.)

RUSSELL, R. J. 1967: *River plains and the sea coast.* Berkeley: University of California Press. (Discusses general points concerning coastal development.)

SHEPARD, F. P. 1973: *Submarine geology.* 3rd edition. New York: Harper and Row. (A comprehensive account of submarine features and processes operating in the deep sea.)

SHEPARD, F. P. and DILL, R. F. 1966: *Submarine canyons and other sea valleys.* Chicago: Rand McNally. (A detailed account of a wide variety of submarine valley features.)

ZENKOVICH, V. P. 1967: *Processes of coastal development.* Edinburgh: Oliver and Boyd. (A translation of the Russian edition, edited by J. A. Steers, giving examples of coastal features from coasts of the USSR, and discussing coastal processes in detail.)

4 The morphology of the open ocean

1 Submarine ridges: 1.1 Types; 1.2 Distribution; 1.3 Description; 1.4 Structure; 1.5 Processes
2 Island arcs and deep sea trenches: 2.1 Distribution; 2.2 Relief and type of trench; 2.3 Structure; 2.4 Processes
3 Abyssal plains, hills, and other sea mounts: 3.1 Abyssal plains; 3.2 Abyssal hills; 3.3 Sea mounts and guyots; 3.4 Coral atolls and other reefs

Away from the continental margins the structure of the oceans becomes much more uniform, but nevertheless, from the morphological point of view a considerable variety of relief forms are hidden beneath the waters of the open sea. Some of the features to be considered in more detail have already been mentioned in connection with the global tectonics and broad elements of oceanic structure discussed in chapter 2.

The pattern of the major feature of the open ocean, the ridge system, has already been described. It is worth, however, pointing out that not all the ridges in the ocean basins are similar in structural type and morphology. There are ridges that are alive and active, there are ridges that were once alive but are now no longer active as sources of the new sea floor; and there are aseismic ridges that were never zones of sea floor spreading. An example of an aseismic ridge is one which runs across the centre of the Arctic Basin close to the north pole. This is the Lomonosov Ridge, a marked feature topographically, separating two deep sea basins on either side of it. Structually it is different from the active ridge that lies between it and the edge of the very wide continental shelf off northern Europe and Asia. The active ridge, called the Arctic Mid-ocean Ridge, has the rift zone and active seismic character of an active ridge along which sea floor is being created, and it is continuous with the mid-Atlantic Ridge. The active ridge bisects the distance between the Lomonosov Ridge and the shelf edge, with the Amundsen deep sea basin on one side and the Nansen basin on the other (shown in figure 4.1). The characterstics of the active ridge are exemplified by the mid-Atlantic Ridge, which has been intensively studied by all the means available and is therefore relatively well known. This ridge seems to be rather more mature than that of the southeast Pacific Ocean, which has less spectacular relief but is more active in its greater spreading rate. This young ridge has yet to develop a central rift-valley. At present it is only a broad gentle rise, lacking the rift and stepped plateaux of the more mature form in the Atlantic and Indian Oceans. On the other hand, the Darwin Rise, running from southeast to northwest across the Pacific Ocean through the Hawaiian Islands, is in a senile stage

Plate 10 Surtsey (*opposite*) : a new island growing off the southwest coast of Iceland illustrates the outpouring of material along the mid-Atlantic Ridge. The upper picture shows the eruption of fragmentary material and the lower the lava flows which consolidate and allow its continued existence. (*Icelandic Photo & Press Service.*)

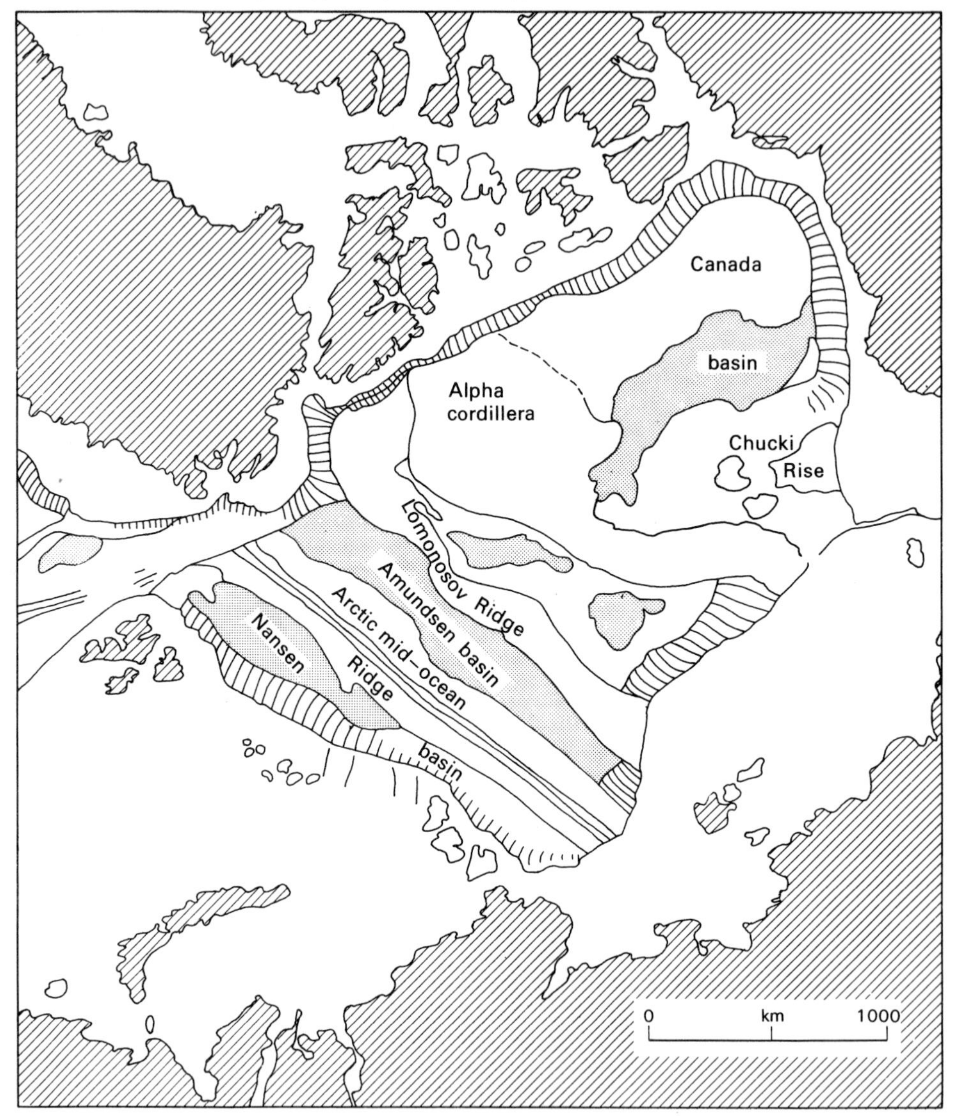

Figure 4.1 Structural features of the Arctic Ocean. The deep basins are dotted. (*After Demenitskaya and Hunkins, 1970.*)

of development, its age apparently increasing towards the northwest. Signs of old age in ocean ridges are found in a steady subsidence of the ridge foundations, connected with the densification of the crustal material as the ridge increases in age. Subsidence also gives rise to other features of interest in the open sea relief. These are the isolated, relatively small positive features of the deep ocean areas—the guyots, sea mounts, and abyssal hills that stud the Pacific in particular in very great numbers, estimated at about 10,000 (Menard, 1959).

They occur in the open ocean far from continental structures. These hills of the deep sea are volcanic in origin. Sea mounts are defined as features having an amplitude of more than

1000 m local relief, and they are often circular in plan. They differ from guyots in usually having a sharp summit, while guyots are characteristically flat-topped. Their flat top is usually associated with subaerial erosion when the guyot was exposed above the sea. If this feature is always formed in this way, then there is a valuable relationship between the depth of water over the flat top of the guyot and its age. This relationship is based on the assumption that guyots sink with age and that, therefore, the deeper ones are the older. If this reasoning is sound, then it is not surprising that the Darwin Ridge is characterized by many guyots and associated features like the coral atolls. A doubt has, however, been cast on the assumption that guyots always derive their flat top from subaerial erosion when they were exposed above sea level. This doubt has come about by the discovery of features that have all the characteristics of a typical guyot in the newly developing rift of the Afar Triangle at the southern end of the Red Sea, where the new rift ocean runs inland on the north coast of Ethiopia. The feature, described by Taziefe (1970), is of volcanic origin and despite the fact that it has formed above sea level it has the flat top typical of a guyot. This example shows that the flat top could be an original feature of a guyot rather than a secondary feature due to erosion. If the same mechanism could occur beneath the sea then flat-topped guyots need not necessarily imply a change of level of their foundation to the extent of the depth of water over their crests. Many guyots, however, are covered by shallow water sediments, corals, or other shallow marine life, in which instances the evidence for their former position at sea level is strong.

The large number of sea mounts and guyots indicate that volcanic activity under the deep ocean is widespread and more or less continuous, since features in all stages of growth and decay are represented. Sometimes the features increase in age systematically, indicating the slow movement of the source of volcanic activity relative to a fixed point. Where the sea mounts indicate a considerable period of formation and where they are riding on a moving plate, then there must be some coupling between the source of volcanic material and the site of the volcano. In fact the two must move together in these circumstances. The indication that the sea mounts and guyots of the northwest Pacific Ocean get older further away from the active volcanoes of the Hawaiian Islands agrees with evidence that the age of the basement of the Pacific deep sea floor increases also in this direction. In fact ages from upper Jurassic to the west of the Emperor Sea mount chain have been suggested, becoming progressively younger towards the eastern side of the ocean. There is thus an interesting relationship between the age of the sea floor crust and the age and character of the sea mounts and guyots, both features being explicable in terms of plate movement and sea floor spreading.

The western Pacific north of the equator represents one of the oldest parts of the oceanic crust, and it is in this zone that the senile Darwin Ridge, with its deep sunk guyots, occurs. The sinking of the foundation of the sea mounts often produces a moat around the foundation of the volcanic cones. This feature is well exemplified around the Hawaiian Islands, where it is probably genuinely the result of subsidence around the volcanic piles. Elsewhere the moat feature may be more the result of non-deposition of sediment around the rising ground where currents are concentrated. Many of the isolated volcanoes that form the steep sea mounts are situated on a general zone of greater elevation, often described as an archipelagic apron. This feature is composed mainly of solid rock, often with only a thin veneer of sediment.

Evidence that many of the sea mounts formerly reached sea level is clearly available in the large number of coral atolls that are typical of the tropical waters of the Pacific Ocean in particular. These coral atolls attracted the attention of Darwin in his cruise across the Pacific in the *Beagle*. He was the first scientist to propose a reasonable theory to account for the ring of coral that forms the typical Pacific atoll. His theory was opposed by that put forward by Murray, based on evidence he obtained while on the cruise of the *Challenger*. Murray's theory was not published until 1880, and in it he suggested that the flat area on which the coral grew was an accumulation of marine sediment that built up until it was sufficiently near the surface for corals to grow. The emerging reef would grow outwards, as it could no longer grow upwards. The central lagoon would form as accumulating sediment killed the living coral, and solution gradually lowered the level to form the lagoon.

Darwin's theory, on the other hand, was based on the idea of subsidence. His view was that a fringing reef grew around a volcanic pile that had emerged above sea level. As volcanic activity died out erosion lowered the volcano, while the fringing reef grew into a barrier reef. The island foundation continued to subside, according to Darwin, and the barrier reef eventually became an atoll when the final remnant of the volcano disappeared below sea level. Darwin based his theory on a five-day visit to the Cocos-Keeling Islands, a shortness of stay that did not impress his critics, among whom Alexander Agassiz was one of the foremost. Darwin's major problem was to account for the subsidence and to find evidence for it. In this matter he was assisted by Dana and W. M. Davis, both of whom were geologists and geomorphologists. Davis put forward the geomorphological evidence that gave considerable weight to Darwin's theory, noting the drowned character of the valleys inside the barrier reefs and the lack of cliffing on the headlands.

It was not until the twentieth century that technology was able to provide some positive evidence in favour of Darwin's theory of subsidence in the formation of atolls. In 1901 a borehole was put down on the island of Funafuti, and it reached a depth of 339·5 m, passing through nothing but coral material the whole distance. This lent strong support to the theory, but opponents suggested that perhaps the coral was not *in situ*. Recent boring has further vindicated Darwin's view, which is now no longer seriously questioned as the correct explanation for many and probably most atolls. The reason for the subsidence has also become clearer, in the light of the processes associated with ridge formation, growth and decay, as exemplified in the Pacific Ocean.

Another important locality in which volcanoes are active in the open sea is in the island arcs. These features, and their associated deep sea trenches have already been mentioned as the zones along which oceanic crust is consumed where plates come into collision. Some detailed studies of the characteristics of these particularly mobile zones of the oceanic crust are further considered in this chapter. These volcanic outbursts originate along the Benioff Zone, the plane that slopes down at about 45 degrees below the deep sea trenches to a depth of about 700 km, where the deep focus earthquakes finally fade out entirely.

The slopes of the deep sea trenches provide some of the most spectacular relief on the earth's surface. Photographs taken during the descent of the bathyscaph to a depth of 6600 m in the Puerto Rico Trench show jagged cliffs and outcrops of rock on the trench walls (Heezen and Hollister, 1971). Talus slopes, masses of rubble, and pitted eroded outcrops as well as fractures have been photographed on the inner walls of the Puerto Rico, Kermadec, Tonga, Peru–Chile, New Hebrides, West Solomon, New Britain, Palau, and South

Sandwich Trenches. This list includes all types of trenches, including the two in the Atlantic Ocean and ones on either side of the Pacific Ocean. Many of these features could not have been formed by deep sea currents. Manganese coating of the outcrops hinders interpretation, although pillow lavas, characteristic of submarine cooling, can often be identified. Slopes can attain gradients of 20–40 degrees, while some walls of rock overhang. The evidence in the Puerto Rico Trench, particularly of the dip of the sediments on its floor, indicates that it is still actively subsiding as sediments continue to collect in it. The South Sandwich Trench on the other hand, is empty, and bare rock outcrops on the trench walls and pumice from recent eruptions litters the trench floor. The floor of this trench has been photographed at a depth of about 6000 m. The emptiness of the floor may be the result of a vigorous current that flows along the length of the trench. That the trenches are recent and active areas of the earth's crust is supported by the general lack of sediment in them. The inner walls exhibit signs of compression and the outer ones signs of tension,

Table 4.1 Ocean basin hypsometry

Depth in m	Cumulative area in percentage	Percentage area
0–200	7·49	7·49
200–1000	11·91	4·42
1000–2000	16·29	4·38
2000–3000	24·79	8·50
3000–4000	45·73	20·94
4000–5000	77·42	31·69
5000–6000	98·62	21·20
6000–7000	99·85	1·23
7000–8000	99·96	0·11
8000–9000	99·99	0·03
9000–10000	100·00	0·01
10000–11000	100·00	0·00

a pattern that agrees with views on the forces that create these features in one of the most mobile and active zones of the earth's crust.

As the ocean floor relief has become better known, more accurate estimates of the hypsometry of the different provinces that can be distinguished have been prepared by Menard and Smith (1966). The most common levels are near sea level and at a depth of about 5000 m. Table 4.1 gives the cumulative areas at different depths. Nine physiographic provinces are differentiated in the ocean basins: 1) continental shelf and slope, 2) continental rise, 3) ocean basins, 4) rise and ridge, 5) ridges not known to be volcanic, 6) volcanoes, 7) island arcs and trenches, 8) composite volcanic ridges, and 9) poorly defined elevations. Each province can be identified in terms of its hypsometric pattern. The first province showed a skewed distribution with a peak in the area–depth curve below 1000 m. The second province is symmetrical, with a low peak about 3500 m, while the third (the ocean basins) shows one steep peak with a flat top between 5000 and 6000 m. The fourth has a rather less steep peak between 3800 and 5000 m. The sum of all the provinces produces a marked sharp peak at 5000 m, with a secondary peak below 1000 m. The similarity in form of the curve for provinces 3 and 4 (the basins and the rise and ridge provinces)

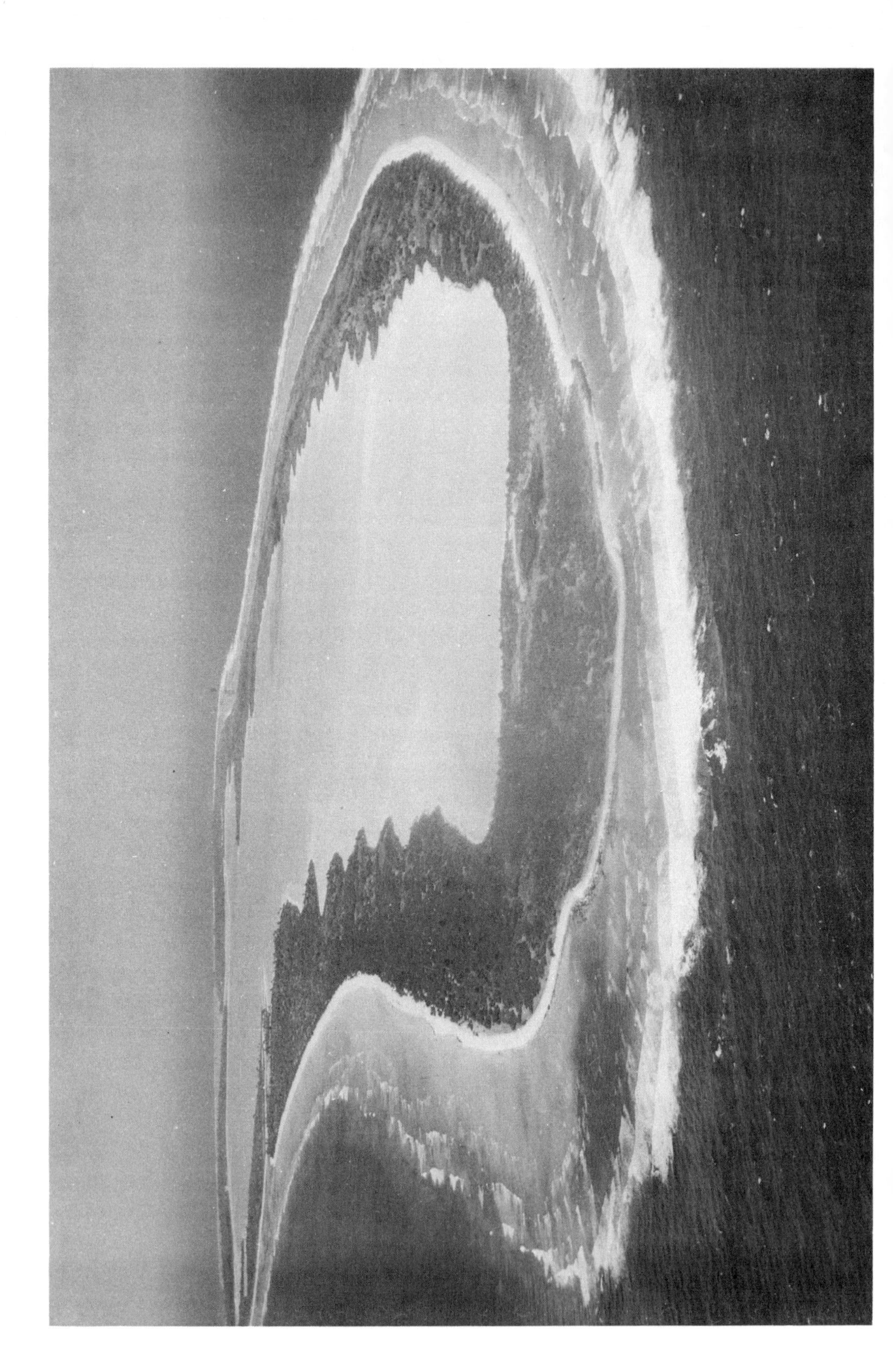

suggests that the latter is an elevated equivalent of the former. The formation and collapse of ridges have been responsible for sea-level fluctuations in some instances. The mean depths of the main ocean basins are: Pacific 3940 m, Atlantic 3575 m, Indian 3840 m, and the Arctic Ocean 3729 m. The total ocean volume is estimated as $1{\cdot}350 \times 10^9$ km^3.

1 Submarine ridges

1.1 Types

Many submarine ridges are major relief features, sometimes rising more than 10 km above the adjacent sea floor and often running more or less continuously for up to 60,000 km. A profile across the mid-Atlantic Ridge is similar to one drawn across the Rocky Mountains (figure 4.2).

The ridges differ in character so that it is unlikely that any one hypothesis will account for all the major ocean ridges. Before the more recent views of the origins of the ridges are considered, different theoretical methods of ridge formation, as proposed by Hess (1954), will be mentioned. The first hypothesis suggests that the ridge is the result of large outpourings of basalt along linear fissures, which build up into piles of basalt lavas and pyroclastics. The peaks of some of the larger volcanoes may emerge to form oceanic islands of basic rock. The pile of material which results from prolonged eruption is lenticular in form, as its base depresses the oceanic crust on which it rests; but owing to its porosity it may be less dense than the original material from which it was derived, and thus it makes a conspicuous relief feature. Once formed, this pile of material would be liable to slow subsidence.

The second method by which submarine ridges could form, according to Hess, is associated with forms producing tension in the crust, possibly due to rising convection currents. This process is associated with breaking up of the substratum, which is probably peridotite, and its mixture with the basaltic crust under the ocean. There is some evidence for this possibility in the blocks of peridotite which have been found on St Paul's Rock. At a time when the hypothetical convection current was active, the ridge would be higher and part of it might well emerge, while a reduction in vigour of the current could well lead to rapid subsidence. At all stages the density of the disturbed part of the crust would be lower than the adjacent areas, enabling it to stand higher than its surroundings and thus to form a positive feature on the ocean floor.

The third possibility is associated with the area in which possible convection currents turn downwards into the deeper regions of the mantle. This process would tend to thicken and buckle the basaltic crust, which could be associated with andesitic volcanism and diorite intrusion. It is a process rather similar to that which was thought to initiate island arcs. These possibilities are illustrated in figure 4.3.

Hess suggests that an example of the first process may be found in some of the Pacific Ocean ridges, many of which are crowned with volcanoes erupting basic rocks. The second type he considers explains some of the features of the mid-Atlantic Ridge, while a possible example of the third process may be found in the Walvis Ridge, which links the mid-

Plate 11 Coral atoll (*opposite*). Gardner Island is a typical coral atoll, showing the ring of coral and the central lagoon. (*US Airforce Photo.*)

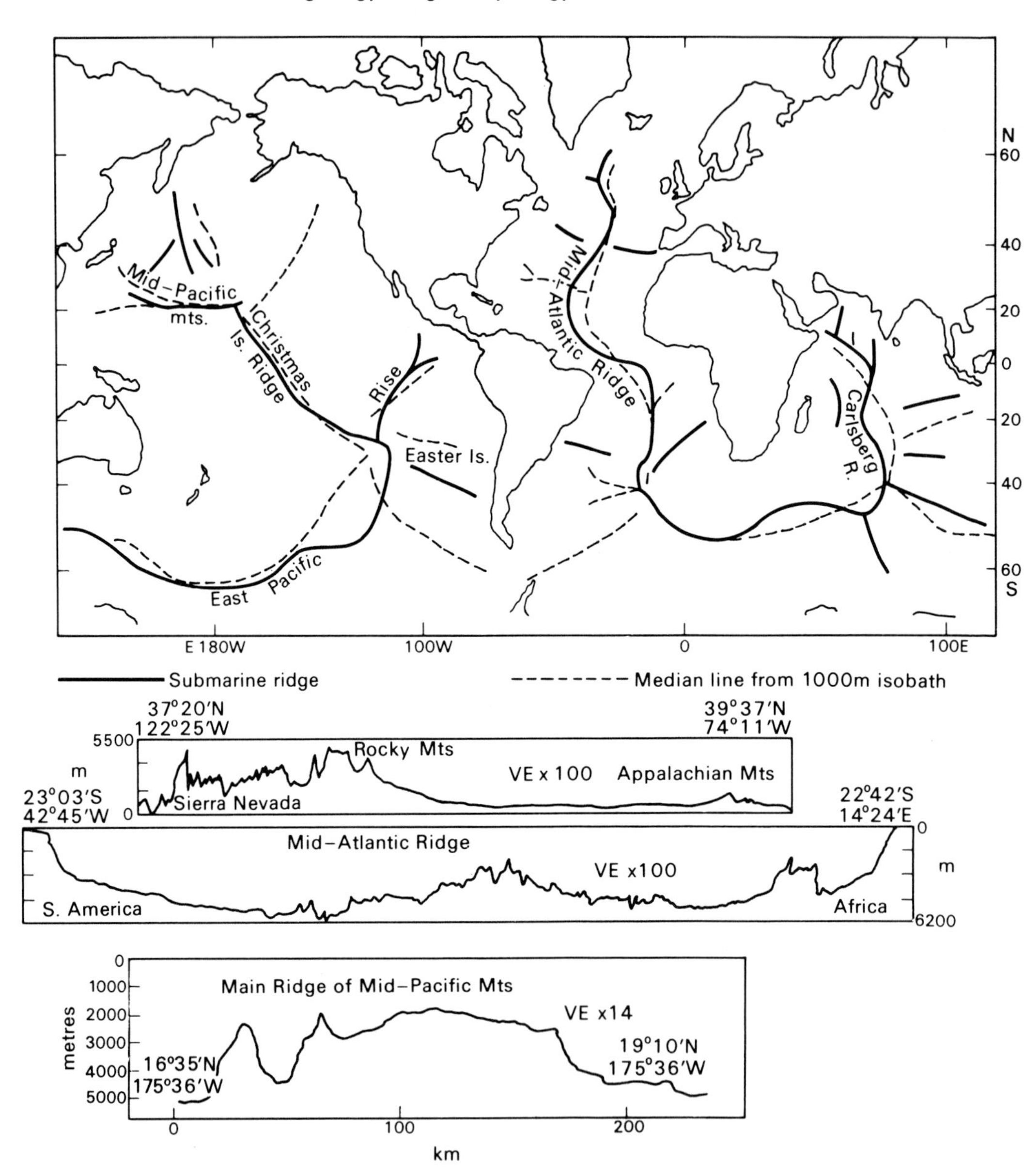

Figure 4.2 Profiles across the mid-Atlantic Ridge and mid-Pacific Mountains, compared with a profile across the Rocky Mountains and Appalachians. (*After Dietz* et al., *1954*.)

Atlantic Ridge to the coast of Africa, where it is continued in an ancient fold of late pre-Cambrian date in southwest Africa.

The median ridges in many parts of the oceans are broad, gently sloping rises, but in the north Pacific there are steep-sided, narrow submarine mountains. The former type are associated with a continuous narrow band of earthquakes and much volcanic activity,

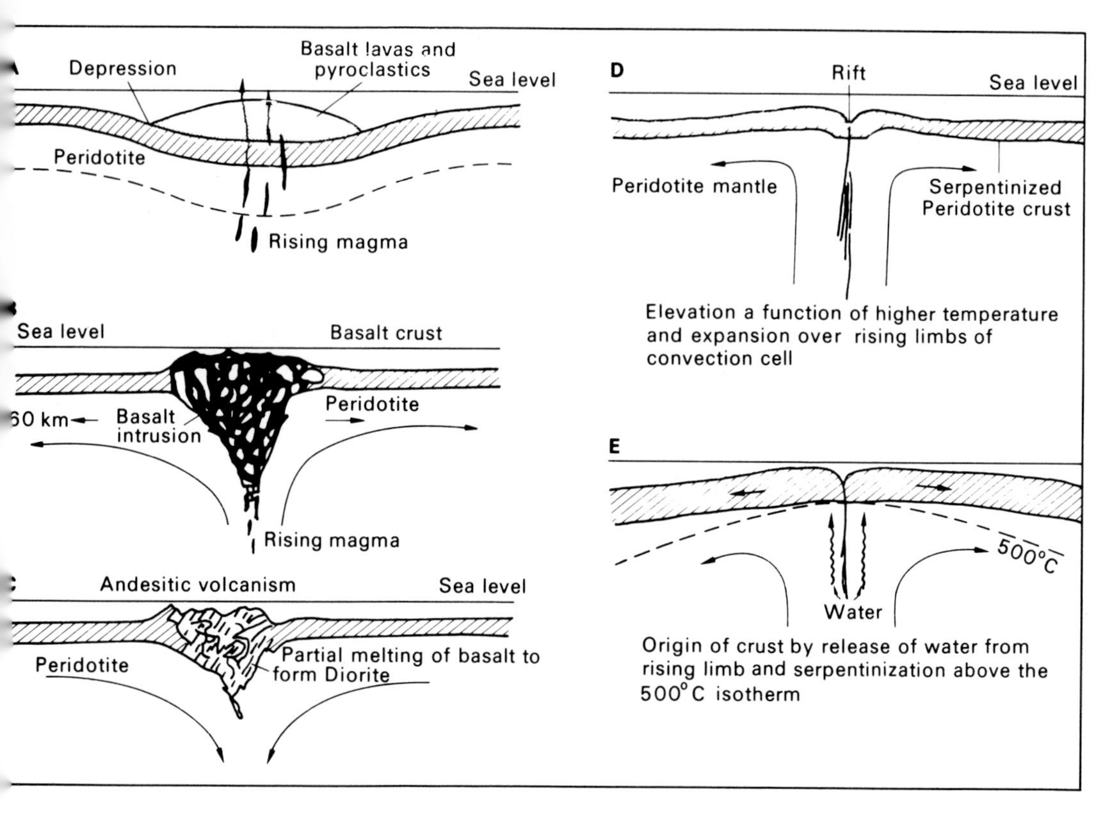

Figure 4.3 Diagrams to illustrate possible methods of formation of oceanic submarine ridges. (*After Hess, 1965.*)

while the latter appear to be in regions of tectonic quiet. Heat flow under the broader ridge is very high. The East Pacific Rise is a good example of the broad type. It is associated with seismic activity, volcanoes, and sea mounts. The mid-Atlantic Ridge is much steeper and narrower; it has volcanic islands and guyots.

Menard (1958) tentatively suggested that the different form of the ridges may indicate a difference of age and stage of development. The gentle rise is the early stage, and the steep type (exemplified by the mid-Pacific Mountains) a later stage. If this is so, submarine ridges may not be permanent features of the ocean floor. They do not appear to differ fundamentally from the material of the surrounding ocean basin. It is possible that broad rises may be temporary features, which are first elevated and then subside to leave a narrow ridge, capped with guyots.

There are certain corollaries to these arguments which are of interest. The hypothesis suggests intermittent movement, with ridges extending at times above sea level. In this state they would form convenient stepping stones by which fauna and flora could be dispersed around the oceans and continents. Thus Easter Island could have been connected across the Pacific with South America or Japan. The problem of the age of the ridges and rates of development is significant in this respect.

1.2 Distribution

The pattern of oceanic ridges presents a number of challenging features. In some oceans the ridges are centrally placed; elsewhere they are not. In three places they run into the land, and at other points they meet or intersect. The pattern is shown in figure 2.11. The Most continuous ridge is that which runs all across the Arctic Ocean from the coast of Siberia, centrally down the Atlantic Ocean and into the Indian Ocean, where it bifurcates. One branch runs up into the Red Sea and the Gulf of Aden (the Carlsberg Ridge), while the other continues unbroken to the south of Australia, where it bisects the distance between Australia and Antarctica. The ridge then runs across the southeast Pacific as the East Pacific Rise, before turning northwards to run into the Gulf of California. A branch of it diverges southeastwards towards the coast of Chile. The east Pacific system also includes the Juan de Fuca and the Gorda Ridges in the north.

Although the ridge is apparently continuous over great distances, and forms one unified system, it is frequently broken by lateral displacement along fractures at right angles to its trend. These fracture lines are mainly in the form of transform faults.

1.3 Description

The mid-Atlantic Ridge is a broad, fractured feature, occupying the central third of the ocean. It has been most studied between 17°N and 54°N, where it is 800–1400 km wide and 2·5–4 km high (Mason, 1960). It rises in the form of a series of steps, becoming more rugged towards the crest. The mountains are bordered by a fractured plateau with a depth of 3200 m. Beyond this high plateau lie three steps at depths of 3660–3840 m, 3930–4200 m and 4200–4400 m respectively, the depths being rather greater on the east than on the west of the ridge. The depth then falls off to the abyssal plains at 4400–4760 m on either side. At the crest there is a deep rift valley, which is 1000–4000 m deep and 25–50 km wide. The floor is rough and contains some mountainous ground. This feature is a very striking aspect of the relief of the whole ridge.

Seismic observations show that near the crest the sediment layer may be missing, and rock of low density is exposed at the surface in places. This rock is 2·8 km thick and rests on a denser material beneath, in which waves travel faster than in the normal ocean crust. This denser material could originate in several ways. It could represent an abnormally low mantle velocity, or a mixture of mantle and crustal rocks due to intensive intrusion of igneous material, or it could be due to volcanic activity or serpentine formation, as suggested by Hess (1954). This process involves an increase of volume of 25 per cent, which would account for the density intermediate between that of the crust and mantle. This process could explain the elevation of the ridge, but it fails to explain the high heat flow, which is more than 6 times the average value.

Another point about the unusual character of the central part of the ridge is its high magnetic anomaly, which suggests that it is underlain by highly magnetic intrusive rocks. It has also a small negative isostatic anomaly. The significance of the serpentinization idea is that the process operates mainly at relatively low temperatures below 500°C. But with the high heat flow, temperature could be raised above this and the process would then reverse, leading to shrinking and hence subsidence of the ridge, a process aided by the liberation of water in deserpentinization.

In many ways the mid-Atlantic Ridge resembles the African rift-valley, which appears to be a feature resulting from tension in the crust. If this analogy is sound, it suggests that tension is also a factor to be considered in the central parts of the oceans. The idea of local tension as a result of rising convection currents fits in with the position of the zones of excess heat flow and with other factors. On the other hand the results could be due to local uplift, as a result of crustal differentiation.

The central rift-valley is a feature typical also of the Indian Ocean ridges. It continues up into the Carlsberg Ridge and those in the Gulf of Aden and the Red Sea. The northwards continuation of the mid-Atlantic Ridge into the Norwegian Sea is narrower and complex. There may be a fracture zone near Jan Mayen, possibly associated with the recent eruption of Beerenberg. One section of the system forms the Reykanes Ridge running southwest from Iceland. The rift-valley in this part appears to be absent north of 57·5°N, but was found between 60 and 62·5°N. The features that link the Atlantic and Arctic Oceans are associated with the Nansen sill. The report of a deep trench through the sill may represent the continuation of the central rift-valley into the Arctic Ocean, where the sea is at its narrowest, thus showing the ridge to be continuous. The continuation of the ridge across the Arctic Ocean is not symmetrical as in the Atlantic (see figure 4.1). The seismic belt associated with the ridge runs parallel to the continental slope from Svalbard to Severnaya at a constant distance of 465 km from the shelf break. This seismic belt is also parallel to the Lomonosov Ridge and the same distance away from it. It thus lies midway between the shelf and the Lomonosov Ridge. The Lomonosov Ridge, which is the main relief feature, is not, therefore, part of the main active ridge system. A reinterpretation of the soundings of the region of the seismically active zones does indicate the presence of a ridge, which is a true part of the main oceanic ridge system. The seismic belt continues into the interior of Siberia via the Lena delta, following the Verkhyansk trough. This then is the third place where an oceanic ridge runs inland. It may link with the Baikal rift-valley.

One part of the main oceanic ridge system that differs in topographic form is the East Pacific Rise (Hess, 1965). The ridge does not have a median rift-valley, and the heat flow pattern differs, in that the maxima form two belts on either side of the crest, which is smoother and broader than the mid-Atlantic Ridge. Hess suggests that the East Pacific Rise still consists of the original ocean floor and is at an earlier stage of development than the rifted ridges.

Another ridge feature in the Pacific, by contrast, is thought by Hess to be much older than the rest of the oceanic ridge system. This is the Darwin Rise that trends southeast to northwest in the central and western part of the Pacific. It is suggested that once the Darwin Rise extended northwestwards to the present coast of Asia. Its southeastward extension may be present in the south Chile Rise. The Darwin Rise is thought to be a dead ridge, which is slowly subsiding. The evidence for its subsidence will be mentioned in more detail in a later section dealing with coral atolls, as they provide evidence of sinking. Menard has suggested the following history of the Darwin Rise. Before 100 million years ago a rise 10,000 km long and 4000 km wide, and 2–3 km high, formed in the southwestern Pacific in water more than 5 km deep. The rise extended from the Tuamotu Islands to the Mariana Islands. Early in its development faulting formed long ridges and troughs near its crest and transverse fractures formed in the central region. There are many volcanoes, which became truncated as they became inactive. The load of the growing archipelagos cracked the crust and basalts flooded out, partially burying the faulted ridges. The rise then started to subside and has completely disappeared during the ensuing 100 million years. All that can now be seen of the rise are some faults, much volcanic rock, and abnormal velocities of 8·4 km/sec in the upper mantle. Exceptional heat must have allowed expansion and extremely active volcanism. The faulting seems to have taken place before the volcanic activity reached its main phase, in which linear archipelagoes were created. The basalt flooding followed. The linear archipelagoes are unique to the Darwin Rise, and each ridge seems to have had a different history in detail.

Menard (1964) has suggested that volcanism reached a peak between 60 and 100 million years ago, during which period huge volcanoes and aprons of volcanic rock formed, and many island stepping stones were formed. Shortly after this phase the rise started to sink by 2 km, and the volcanoes became atolls or guyots. Simultaneously the East Pacific Rise started to grow. The mantle beneath the Darwin Rise is abnormal, having a high velocity of 8·36 km/sec, and is therefore probably denser, but the heat flow appears to be normal. The minimum life of a rise, including development and decline, is probably about 2×10^8 years.

The Darwin Rise was probably once continuous from Chile to northeast Asia. It was initiated in an oceanic area, by contrast to the mid-Atlantic Ridge, which was probably initiated under a land area. When the Darwin Rise was initiated the Pacific Ocean was larger in early Mesozoic times. Since the inception of the rise the Pacific has been contracting. Continents have migrated towards it from the east and island arcs from the north and west, as they over-rode a downwards sinking convection cell. This movement still operates all around the Pacific. It is possible, but speculative, that the whole Pacific was formed around the Darwin Rise as it split. The system is now thought to be dead by Hess.

1.4 Structure

Evidence of the structure of the oceanic ridges can be derived from the geophysical techniques mentioned in chapter 2. The study of earthquakes along the ridges has also revealed much of their character and structure.

Sykes (1967) has studied the pattern of earthquakes on the mid-oceanic ridges and shown that two distinct types of dislocation give rise to shallow earthquakes on the ridges. These are associated with the ridges themselves and with the fracture zones that offset the ridge crest along its length. The mid-Atlantic Ridge has been offset by a fracture zone at 35°N, 35°W by 128 km, resulting in a steep-walled fracture zone, with a maximum depth of 4400 m, striking N 75°W (Fox, Lowrie, and Heezen, 1969). The magnetometer data indicate an offset axial growth rate of 1 cm/year. Serpentine and a little gabbro was collected from the southern wall of the fracture, which offsets the central rift-valley. The fracture zone and offset are similar to those in the equatorial zone, although the dimensions are less. There is also a systematic change in strike of the fracture zone. The measured strike does not agree exactly with that which should occur theoretically if the pole of spreading is situated at 58°N, 37°W, as given by Morgan (1968). Sykes (1967) has shown that the earthquakes that originated in the fracture zones are characterized by strike-slip movement on a steeply dipping plane. The pattern of earthquake occurrence agrees with the idea of transform faulting. It is, however, the reverse expected if the faulting were simple offset faulting, as illustrated in figure 2.9. The spatial distribution of earthquakes also rules out simple offset movement. Earthquakes are mainly restricted to the zone between the ridge crests where transform faulting causes movement in opposite directions. They do not extend beyond this zone, where the movement would also be in opposite directions if simple transcurrent offsetting were taking place.

The faulting that occurs on the ridge itself also gives rise to earthquakes, but in this zone the faulting is normal. It is responsible for the formation of the steps noted in the relief of the mid-Atlantic Ridge, and is an indication of tension. Similar normal faulting has been found on the ridge near Siberia and in east Africa. The data were obtained mainly from the Atlantic, although the Rivera Fracture on the East Pacific Rise and others in this area were also considered. A similar pattern was also observed on the Macquarie Ridge south of New Zealand. The movement recorded was the reverse of that required to offset the ridges.

In transform faulting the opposite movement should not continue beyond the ridge crests as it is due to the lateral spreading of the zone on either side of the crest. In the equatorial Atlantic the crest of the mid-Atlantic Ridge has been displaced by 35° between 15°N and 5°S, to enable the ridge to maintain its central position. Nearly all the seismic activity occurs between the offset ridge crest and is limited to a belt less than 20 km wide. The normal faulting that occurs around the ridge crest has a much wider spatial scatter. The conclusion is reached that transform faulting can only exist with crustal displacement and the evidence cited supports the oceanic spreading hypothesis.

The mid-Atlantic Ridge is perhaps the best known of the ridges. The geophysical data provide evidence of its character in places where the relief does not. The strong magnetic anomalies along the crest can be identified where the central rift is lacking. High heat flow along the ridge crest has been measured both in the north and south Atlantic, the thermal anomaly being from 2 to 5 times the normal heat flow in a belt about 200 km wide or less. On either side of this belt the heat flow seems to be less than normal. The structure has been worked out from seismic observations. On the flanks the ridge has normal oceanic structure, consisting of several hundreds of metres of sediment. Layer 2 below is mainly made up of volcanic rocks about 1 km thick, with a velocity of 4·0–5·8 km/sec. Beneath is the main oceanic crustal layer 4·5–5 km thick, with a velocity of 6·2–6·8 km/sec. Below the Moho the velocity is 8·0–8·5 km/sec. The velocity is lower under the ridge crest, the mantle velocity here being 7·3 km/sec over a belt 400–800 km wide, except in one or two sections where it is nearly 8·0 km/sec or more. Layers 2 and 3 cannot be differentiated under the crest, and layer 3 appears to be missing, although layer 2 may be thicker. The cover over the mantle material is, however, in general much less, being 2·5–3 km instead of the normal 6 km.

The gravity anomalies show a mass deficit up to 30 km deep about 0·25 g/cm^3, dying out to zero about 1000 km from the ridge axis. This could be accomplished by a phase change from garnet–peridotite to feldspathic-peridotite. It could also be due, according to Hess, to higher than normal temperature with fracturing and dilation as well as the formation of serpentine in the zone above 500°C. Thermal expansion of a rising column could explain the gravity anomaly. A rising column 750 km high and 100°C warmer than a normal column would explain both the relief of the ridge and the gravity anomaly. The expansion would be about 2·8 km. Vulcanicity along the ridge is limited to Jan Mayen, Iceland, Azores, Ascension, Gough, and Tristan da Cunha, all of which are active at present, especially Iceland, which is the location of the most voluminous outpourings. A decrease of pressure could account for the eruption of basalt.

Tobin and Sykes (1968) have collected similar data from the northeast Pacific seismic zones. In this area also, offset and active oceanic ridges are linked by transform faults, sometimes called great faults. The ridges in this area are the Juan de Fuca Ridge and the Gorda Ridge, both short ridges trending roughly northwards, and offset by the linking Blanco Fracture zone. Nearly all the earthquakes on the Blanco Fracture zone occur between the two ridge crests. The Mendocino Fracture zone is also seismically active where it links the Gorda Ridge to the north Californian coast. This fracture zone appears to be an east–west continuation of the San Andreas Rift. The seismic activity indicates deformation of the oceanic block north of the Mendocino Fracture zone and east of the Gorda Ridge. The Gorda Ridge has a prominent rift zone at its centre, but this feature is lacking in the Juan de Fuca Ridge. Most of the seismic activity seems to take place where there is a median ridge. The Juan de Fuca Ridge is linked by the Queen Charlotte Island Fault zone to the northeast Aleutian arc. The feature is a dextral transform fault. Winterer *et al.* (1968), however, show that sediment is not displaced by the Pioneer Fracture zone, although layer 2 is, thus giving some indication of the age of the movement. The complexity of the structural relationships in this zone suggests a change in the spreading pattern in the northeast Pacific in the last 10 million years, and this may account for the non-parallelism of the fracture zones and the complexity of the magnetic pattern. Spreading is at present southeast to northwest, whereas it was previously east to west, as indicated by the great fracture zones of this area, which were formed more than 10 million years ago. These include the Mendocino, Murray, Clipperton, and others. The area to the east of the ridges seems to be a small separate plate that is underthrusting beneath the Washington and Oregon coasts, causing volcanic activity in the Cascade Range.

1.5 Processes

In discussing the nature of the evidence some of the processes have been mentioned already. The ridges mark the location of areas of rising material that then spreads out to move as a plate across the surface of the earth, over oceans and continents alike. The process that leads to ridge formation is related to the changes undergone in the rising material. Layer 3 is uniform in thickness in the ocean, being $4{\cdot}7 \pm 0{\cdot}5$ km in 80 per cent of the measurements of refraction profiles. The Moho cannot be a phase change to denser material and Hess suggests that it is determined by the depth of the shallowest level that the 500°C isotherm can attain on the ridge crest. As the crust moves away from the crest it cools and the discontinuity is permanently frozen into the crust. At 500°C or below, the periodotite + water becomes serpentine + heat, thus rising material becomes serpentinized and new oceanic crust is formed. The reaction can be reversed as the crust sinks again and so the crust is disposable. The reaction goes the other way at temperatures above 500°C. The high heat flow could be explained in part by the reaction, and basalt in eruption can also contribute to the high heat flow. The reverse process at the sinking point would increase the density and thus enhance the sinking and keep the circulation going. Dredging of serpentine from the northern wall of the Puerto Rico Trench and its location on the flank of the ridge south of St Paul's Rock, and ultramafic rock on the Carlsberg Ridge all support this view.

Menard (1965) has pointed out that 50–56 per cent of the whole rise-ridge system is central in an ocean. The pattern can be made more coherent if it is related to the shield areas. A line of ridges almost encircles South America, Antarctica, and Africa; roughly circular ridges surround North America and Europe. Quantitatively the ridges are considered to form a circle round the shields if they lie no more than 1000 km away from the shield centre. By this criterion 87 per cent of the ridges lie on circles around the continental shields, the radii being between 1600 and 2200 km. Thus it is suggested that the pattern of shields controls the pattern of ridges, which in turn controls the pattern of oceans in many areas. If the continents provide the primary control then some process connected with them must cause the ridges. Menard suggests that separation of light material to form the continental crust could leave a denser residue, which would sink. This would produce flow around it, stretching the crust in a ring around the sinking mass. If this process occurred under Africa and South America the rings would meet at the site of the mid-Atlantic Ridge in the central part of the Atlantic Ocean. The type of movement envisaged could result in the stretching characteristic of the mid-Atlantic Ridge. Movements of the blocks away from the East Pacific Rise in the northeast Pacific Ocean cannot be explained by this hypothesis. Movement in the Pacific could be related to sinking beneath the North American shield and the associated surface stretching. The theory can also account for the offsetting of the ridge system by the existence of convection acting from each side of the rise. The convection cells would be roughly circular in plan and shallow in cross section.

The general consensus of opinion on the nature and origin of the oceanic ridges considers that they represent zones of linear tension in the earth's crust, through which new ocean crust is formed as the marginal zones move away. The pattern of magnetic reversals, the heat flow, the distribution of seismic and volcanic activity, and the structure all support this main conclusion. Many details remain to be filled in, but the broad pattern is remarkably consistent. If the crust is being formed at the sites of the ridges, then it must be consumed elsewhere unless the earth is expanding. The opinion is generally held that expansion is not taking place so that the absorption of oceanic crust must be taking place. The zones in which this is occurring are thought to be the island arc and deep sea trench systems. These features are thus genetically related to the central ridges and to each other.

2 Island arcs and deep sea trenches

2.1 Distribution

Island arcs and deep sea trenches are mainly features of the Pacific Ocean. Their only occurrence in the Atlantic is in the Antilles, which appear to have pushed through the narrow linking zone between North and South America, and in the south Atlantic where the South Sandwich Trench has pushed through the narrow gap between Tierra Del Fuego and Graham Land, an appendage of Antarctica.

The trenches form an almost continuous ring around the Pacific Ocean, while the island arcs are most typically developed in the western part of the Pacific Ocean (figure 4.4) There is a contrast between the trench features of the eastern and western Pacific Ocean. In the west the island arc-trench unit is separated by a small ocean from the continental margin. In the east the Andean type of trench occurs where the down-warp is immediately adjacent to the land. The sediments associated with these different types of trench differ in

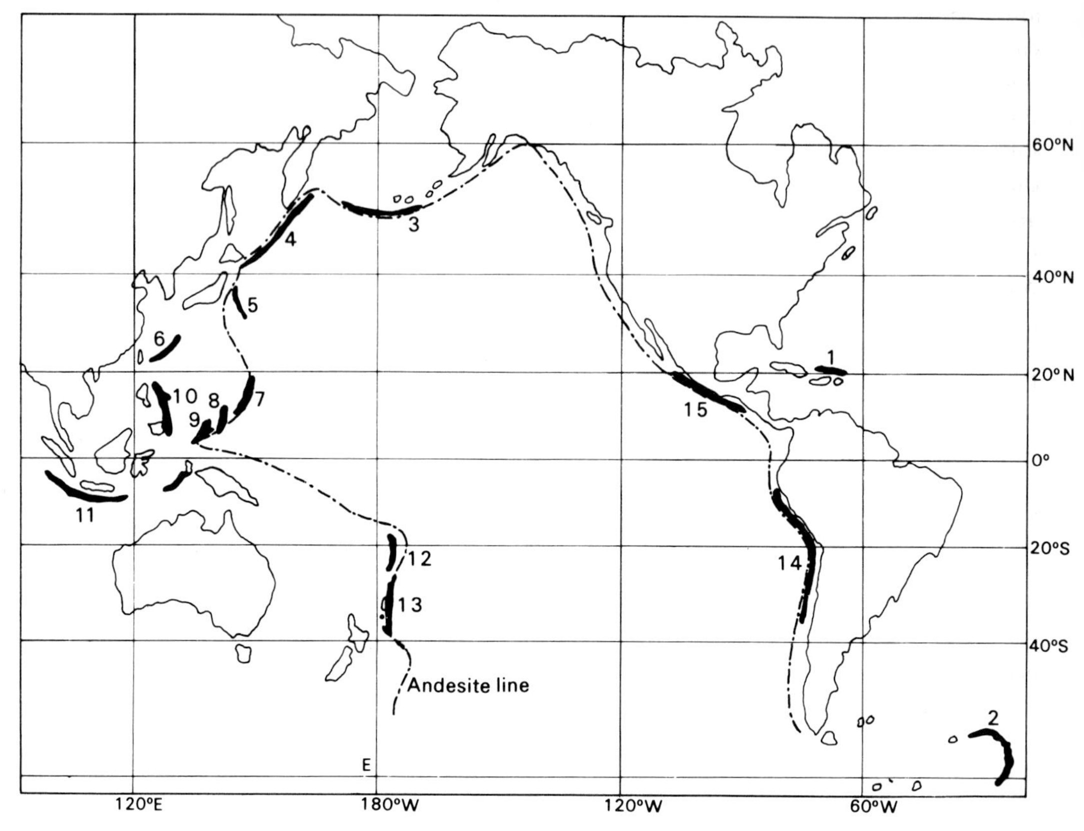

Figure 4.4 Map to show the deep sea trenches and Andesite line. The numbers refer to the following trenches:

1 Puerto Rico
2 South Sandwich
3 Aleutian
4 Kurile
5 Japanese
6 Riu Kiu
7 Nero
8 Yap
9 Pelew
10 Philippine
11 Sunda
12 Tonga
13 Kermadec
14 Peru and Chile
15 Middle American

origin and character. Thick turbidites are found in the Andean type, while volcanic sediments are typical of the island arc type of trench.

Some trenches occur in the Indian Ocean, and others in the so-called mediterranean or marginal seas, including the Caribbean, already mentioned, the Mediterranean, Indonesian, and Melanesian basins. Some trenches are not associated with island arcs, including the Romanche Trench in mid-Atlantic, and the Vema and Diamantina Trenches in the Indian Ocean.

2.2 Relief and type of trench

Fairbridge (1968) lists four types of trench. First, there are the peripheral type on the margins of the main oceans. These are normally associated with island arcs or the young mountain ranges. They are asymmetrical in relief with typical deep ocean floor outside the arc and may involve a total relief of 10,000 m depth plus 7000 m height in the adjacent mountains, making a total relief of 17 km. The second type is the marginal sea reversed trench. These trenches occur in the seas marginal to the Pacific and are parallel to island arcs; they are also asymmetrical and face away from the Pacific. The third type is the transverse or oblique trench which crosses oceanic ridges or continental structures. Some are transverse oblique, or *en echelon*. They tend to be symmetrical and are associated with the fracture zones that cut across the ridges. The final type lie parallel to the first or second, forming the double island arc system in which an intermediate trough lies between the inner volcanic arc and the outer non-volcanic arc. The deeper trenches exceed 10,000 m, the deepest being the Challenger Deep in the Mariana Trench at 11°20′N, 142°–143°E; it is 11,022 m deep. The strip over 10,000 m deep is 32 km long and 0·8 km wide. Several trenches are over 2000 km long, including the Kurile–Kamchatka Trench of the first type, and the longest of all the Peru–Chile Trench, is 5900 km long. The Java Trench is 4500 km and the Aleutian Trench 3700 km. The Ryukyu Trench, at 2250 km, is the longest in the second type, and the Chagos, at 2450 km, is the third type. The Bali Trench in Indonesia is the longest in the fourth group (2300 km).

The maximum width only just exceeds 100 km, a width attained by the Kurile–Kamchatka and Japan Trenches, the Peru–Chile Trench, the Makran and Puerto Rico Trenches, the last two being in the Atlantic. None of the last three types attain this width. The limited width and great depth mean that the trenches have fairly steep sides, but the common vertical exaggeration of profiles is misleading. Common slopes are 4–8°, steepening downwards to 10–16°, although parts of the Tonga Trench slopes at 45°. These gradients are steeper than all other oceanic relief, with the exception of the slopes of atolls and barrier reefs, and some fault scarps. Most trenches are V-shaped, although the floor is often flat, varying in width from a few hundred metres to several kilometres. Some trenches have marginal benches or steps, some of which are small basins filled with sediment.

Ludwig, Hayes, and Ewing (1967) describe the Manila Trench, which is 4940 m deep and is a shallow arcuate depression extending as a narrow continuous furrow sub-parallel to the west coast of Luzon from 16°40′N to 13°N. Seismic reflection profiles illustrate the variation in the sea floor relief and sub-bottom structure. Except for the Luzon Trough, no detectable amount of sediment is present east of the trench floor. A fairly uniform layer of sediment overlies the oceanic basement in the adjacent South China Sea basin. The sediments are continuous along the west wall of the trench and end abruptly as a perched ledge just west of the trench floor. The steep slopes of the ledges suggest that the trench has subsided or the sea floor has extended near the axis of the trench, as step-faulting also occurs. The Luzon Trough forms a sediment-filled feature midway between the trench and the continental shelf.

2.3 Structure

The arcs and trenches are active tectonically and show various geophysical characteristics that provide a clue to their origin. Isacks, Oliver, and Sykes (1968) have discussed the seismic character of the island arcs and trenches. Seismic evidence indicates transform faults and other seismic activity beneath the island arcs. The intensity of earthquakes is greater in the active compressional zones. The feature most characteristic of some of the island arcs and trenches is the pattern of intermediate and deep focus earthquakes. The epicentres of the shocks occur to a depth of 600 km, which is below the normal level of

strong material. The strength of the crust must be maintained to this depth to account for the depth pattern of earthquakes. On the ridges and elsewhere in the ocean and on land, all earthquakes are shallow.

A term that is used in connection with the evolution of island arcs is the 'Benioff' zone, which refers to the zone of earthquake activity that takes place along the plane where the oceanic crust disappears downwards into the mantle (Mitchell and Reading, 1971). The activity in this zone is associated with its two distinct units, which are the volcanic ridge or arc, and the deep sea trench in a modern active system. Mitchell and Reading suggest that the rocks of the island arc system fall into three groups. First, there are the volcanic rocks of the arc which are undeformed and subaerially erupted volcanic rocks, block-faulted volcanic rocks, and granites. There are also metamorphic rocks formed under conditions of low pressure and high temperature. Second are the deposits of the deep sea trench and metamorphic rocks formed under conditions of high pressure and low temperature. These are characteristic of the Benioff zone, where the colder crust is subjected to high pressures as it moves downward along the Benioff plane. The third group of rocks consists of ocean floor rocks, which are gabbros, ultrabasics, and pelagic sediments.

The pattern is clearly shown in the Tonga Trench, where the zone in which the earthquakes increase in focal depth slopes down into the mantle at 45°. Within this zone the earthquakes are limited to a narrow plane less than 20 km wide. The belt of deep focus earthquakes is limited to the active trenches, where movement downwards of the crust, as a strong material, is taking place. Where no active movement is taking place or where erosion exceeds the rate of sinking, the trench may become filled with sediment.

The seismic character of the Tonga–Fiji, Kermadec, and Kurile–Kamchatka area has been considered by Sykes (1966). He identified a zone of intense seismic activity beneath the inner or island margin of the oceanic trench. The Tonga–Kermadec Trench runs in an almost straight line northwards from New Zealand to near Samoa, where it turns abruptly west towards Fiji. Figure 4.5 shows the associated tectonic activity of deep and shallow earthquakes and active vulcanicity. The feature links with the New Hebrides and the Tonga–Fiji arcs. The New Hebrides is a reversed feature in relation to the others because the depth of focus of earthquakes increases to the east and not to the west as is normal at the oceanic margin (figure 4.9). In some areas the seismic zone extends continuously to a depth of 650 km.

The Tonga–Kermadec Trench sequence is one of the most active in the world, containing about a half of all deep focus earthquakes, which occur mainly in the Tonga arc. Gravity anomalies as well as parallel chains of islands and volcanoes indicate the tectonic activity. The two arcs together run almost along a great circle for 2000 km. The northern bend of the system outlines the change in trend of the Pacific basin as a whole through about 90° from north-northeast to west-northwest to include Melanesia. Near Samoa, at the bend in the feature, uplift of some of the islands has taken place and there is evidence of tensile stress in the crust over a shallow depth range. Deep focus earthquakes also follow the trend of the system towards Fiji.

Deep shocks are restricted to the Tonga–Kermadec system, Indonesia, and the Philippine

Plate 12 Two aerial views (*above, opposite*) of Hull, Phoenix Islands, show breaches through the reef formation from the ocean to the lagoon. (*David R. Stoddart.*) The exposed coral reef (*below*) is at the Madrepore Lagoon, Port Denison, on the Great Barrier Reef. (*Paul Popper Ltd.*)

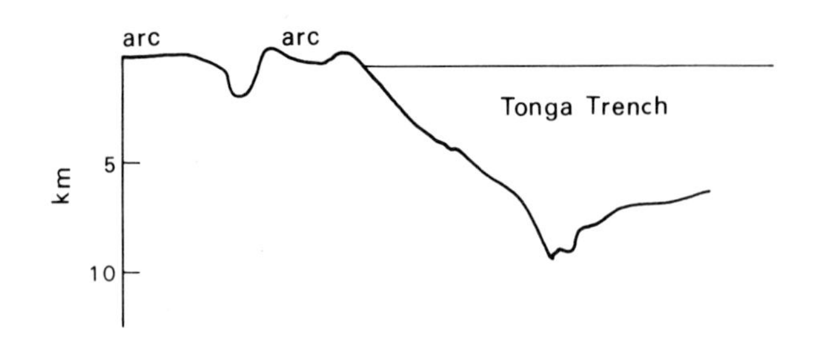

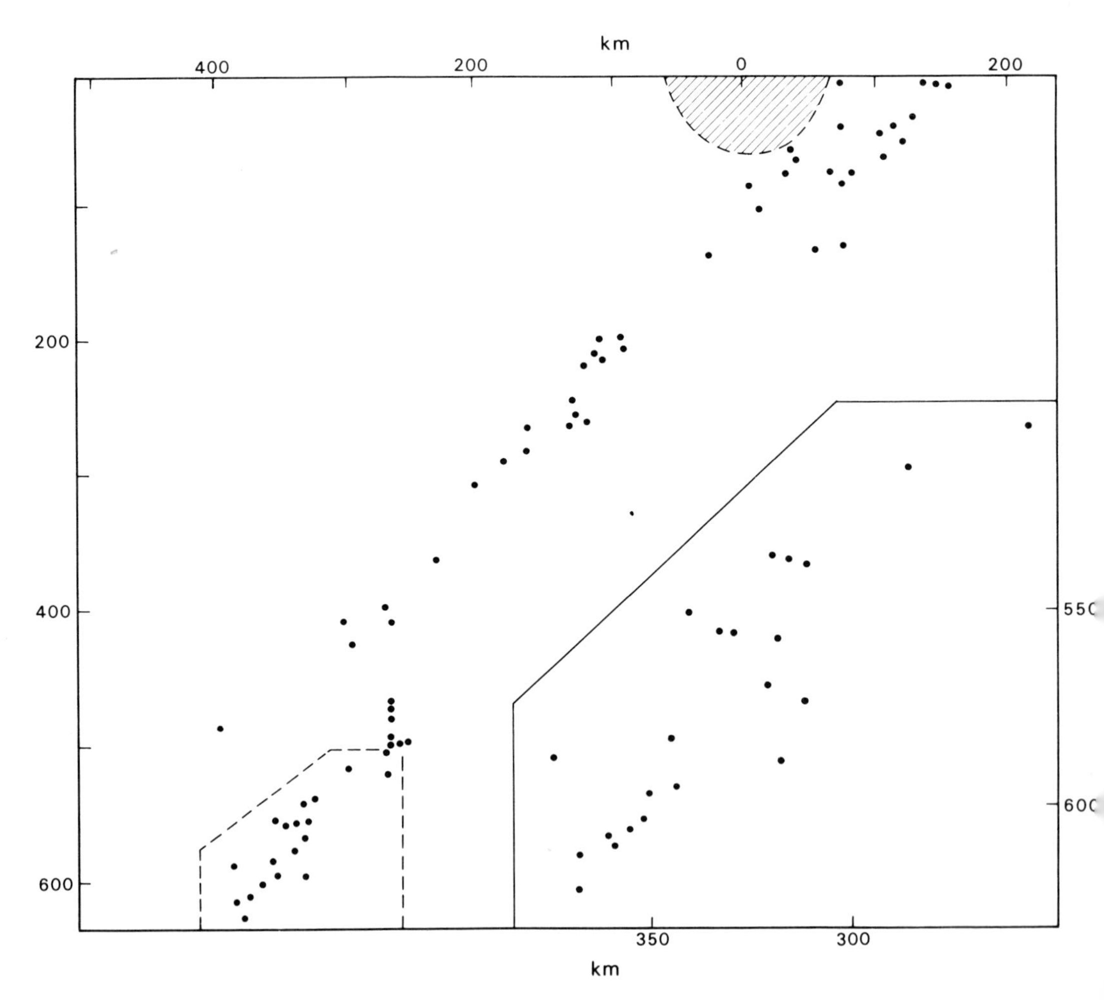

Figure 4.5 Section perpendicular to the Tonga Trench to show pattern of earthquake centres in 1965 A profile of the trench is shown above. (*After Isacks, Sykes and Oliver, 1969.*)

arcs from the Marianas to Kamchatka and South America. The continuity of the earthquakes on the dipping planes varies; most of the planes dip at 45°. The number of earthquakes varies with depth. At the surface there are about 100, falling off to 2/year at 300 km while in the Tonga–Kermadec region the value rises again to 12/year at a depth of 550 km, falling to only 0·3/year at 675 km.

The island arcs in the Japan region show a somewhat similar pattern. The distribution of volcanoes, deep sea trenches, and deep earthquakes suggests three arcs, shown in figure 4.6. The northern arc is the Kurile arc, the central one is the north Honshu arc, and the

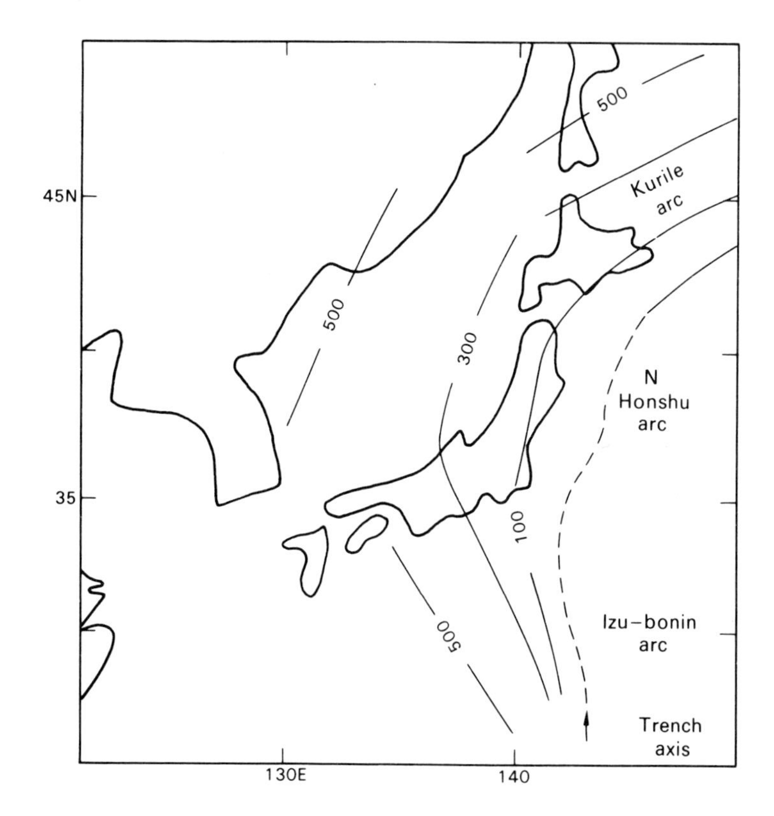

Figure 4.6 The Japanese Island arc and deep focus earthquakes isolines of shock depths.

southern is the Izu–Bonin arc. The dip of the deep seismic plane is at 60° in the Izu–Bonin arc, but only 30° in the North Honshu arc, and intermediate in the northern arc. The motion on the plane is predominantly dip-slip. The axes of compression are nearly parallel to the dip of the seismic zone,with tensional axes perpendicular to the zone.

In the Tonga arc the seismic plane is under compression below a depth of 80 km. Where the orientation of the deep mechanism is variable there may be contortions in the deep seismic zone. The shallow earthquakes are caused by thrust-faulting under the inner islandward margin of the Tonga–Kermadec Trench. Transform-faulting occurs at the north end of the Tonga arc, while hinge-faulting occurs where the trust system meets the transform system. These trenches exemplify the island arc–trench system of the western Pacific, where structural activity is complex and active.

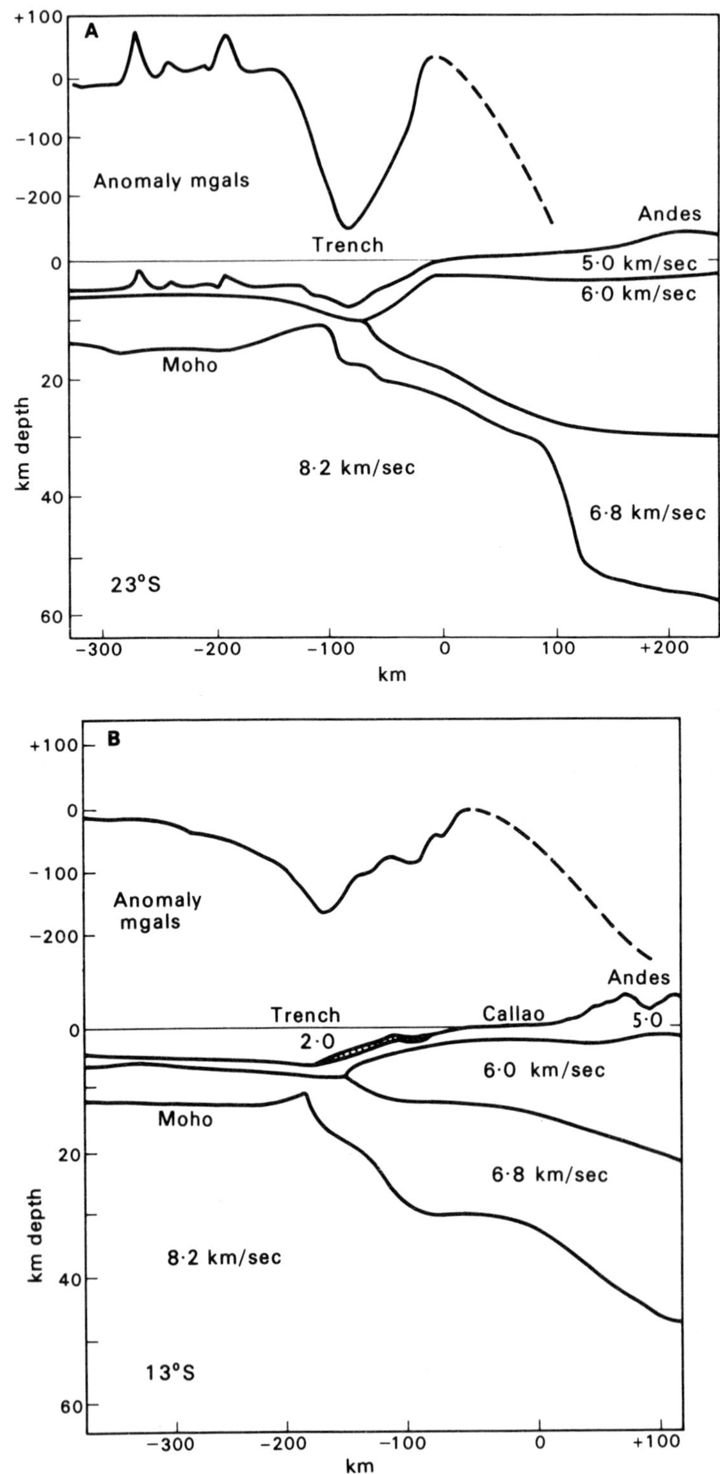

Figure 4.7 Profiles across the Peru–Chile Trench. **A**: near 23°s; **B**: near 13°s. (*After Hayes, 1966.*)

The Peru–Chile Trench forms a contrast, and has several different characteristics, being much more closely related to processes operating in the Andes, with which the trench is associated. The geophysics of the Peru–Chile Trench have been discussed by D. E. Hayes (1966). The results are based on 26 profiles, showing the relief and free-air gravity anomalies and the total intensity of magnetic anomaly. The trench is not a relief feature south of 40°S, but can be traced by a strong negative gravity belt. Seismic reflection profiles indicate a continuation of the structure to Drake Passage. North of 32°S lies a sediment-free section extending to 8°S, while the sediment-filled trench runs to 57°S. The main line of gravity negative anomaly runs along the maximum depth of the trench, where it is a relief feature. A second line of negative anomaly runs parallel to the trench, roughly along the coastline of Colombia, Ecuador, and Peru. The Bolivar geosyncline is associated with this feature. The crust appears to thin rapidly on the ocean flank of the trench. Crustal thickness is normal for the transitional zone between continent and ocean beneath the axis of the trench. The seismic and gravity results agree well. There seems to be no major changes in crustal structure along the length of the trench. The magnetic data suggest a difference in the magnetic state of the materials on either side of the trench.

The trench varies considerably along its length. The main trench province extends from 8°S to 32°S and here the trench is V-shaped and narrow, with maximum depths of 6000 m, very rugged relief to the west, and little sediment in the trench axis. In the sedimentary province the sediment thickness is at least 2 km and the floor is flat, the steep rugged ground to the west being blanketed in sediment. The terminal zone of the trench extends to 1°N to 4°N, where maximum depths are about 4000 m and the profile is poorly defined. The Carnegie Ridge, with a very narrow V-shaped trench in it, runs off at 1°N to 4°S to link with the main east–west ridge in this area. The crustal pattern is shown in the sections in figure 4.7. The profiles illustrate the type of ocean–continent transition characteristic of this type of trench, with gradually increasing depth of the Moho continentwards, as layers 2 and 3 increase in thickness.

The observations of Scholl *et al.* (1970) are of interest in view of the theories of trench formation. Sediment amounts in the trench depend on the erosion of material in the Andes, which lie parallel to the tectonic trench throughout its length of 4800 km. Sediments from the Andes have access to the topographic trench from 18°S to 30°S, a distance of 2000 km, where the Andes and trench are only 250 km apart. The maximum relief between the crest of the Andes and the trench trough is 15 km. The trench is filled with late Tertiary turbidites south of 41°S and these also fill the central valley. At its maximum the trench is 8000 m deep, varying from V-shaped to U-shaped; south of 33°S it is flat-floored and asymmetrical. The sedimentary sequence is undeformed and thus cannot be subjected to underthrusting, actually along the trench.

Sediment available to fill the trench increases rapidly southwards between 32 and 41°S, and is low between 23° and 32°S, owing to the desert climate. Table 4.2 indicates the estimated volume of sediment removal by tectonic activity compared with the estimates of denudation. Most of Chile has been emergent since the mid-Cretaceous and the continental margin is overlain by 100×10^3 km^3 between 23° and 42°S, which denudation could have provided in the time available. The trench between 27° and 44°S contains 70×10^3 km^3 of turbidites, which probably accumulated during the low sea level of the late Tertiary period. This amount is also of the right order. The evidence does not favour rapid loss of

Table 4.2 Sediment infill in the trench off South America

Region	Estimated volume of sediment deposited during late Tertiary glacial ages—10^3 km^3				Estimated volume of sediment removal via spreading—10^3 km^3			Denudation in contributary drainage basins during glacial ages—cm/1000 years	
	Trench turbidites	Shelf and slope	Onshore	Total	Rate, cm/year	Volume loss in 3 million years	Total sediment accounted for	Required by sedimentary volume	Highest estimate
27–31°S	1	1	0·5	2·5	0	0	2·5	3	5
					5	18	20·5	21	5
					10	35	37·5	38	5
36–42°S	43	2	9	54	0	0	54	46	50
					5	66	120	102	50
					10	132	186	159	50

material by convergence, as no evidence of crumpling of the sediments has been found and their volume does not allow for any removal.

2.4 Processes

The gravity surveys of Vening-Meinesz (1934) have shown that the deep trenches are associated with strong belts of isostatic anomaly. Some of these negative anomalies lie symmetrically over the topographic deeps; others are asymmetrical, with the maximum negative anomaly on the continent side of the deep trench. The Yap Deep, shown in figure 4.4, is an example of the former, and the Nero Deep (off Guam) of the latter. The anomalies indicate a deficiency of density, maintained by some process that prevents the establishment of isostatic equilibrium. It seems likely that the arcs are associated with the junction zone between the continents and oceans.

According to Vening-Meinesz (1934) and Umbgrove (1947), there are two types of island arcs. The single island arcs and their associated deep trenches lie nearer to the margin of the true ocean, for example the Mariana Islands. The more complex double island arcs, such as the deep south of Java in the East Indies, are found in the continental section of the ocean margin.

The theory put forward by Vening-Meinesz and Umbgrove is still of interest. They suggested that a down-warp of the crust takes place at the edge of the deep ocean structure, possibly along the deep-seated shear zones, associated with the deep focus earthquakes. The down-warp could be initiated by the underthrusting of the oceanic basement beneath the continental crust. This process could account for the great depth of water, forming the deep trench, while the buckling down of the crust could account for the isostatic anomaly. On the continental side of the trench the crust is up-warped, but owing to its thinness, it does not emerge. It does, however, tend to crack, and volcanic activity produces volcanoes which may extend above sea level to form the island festoons along the curved crest of the up-warp.

The double island arcs are thought by Vening-Meinesz and Umbgrove to occur where compression causes buckling in areas where the continental crust is rather thicker. The outer arc is the non-volcanic one, while the inner one is essentially volcanic in type. The Aleutian Islands illustrate both types—at their western end is a typical single arc, with the deep trench lying close to the island chain, but as the arc is followed eastwards it becomes double.

Stille (1955) has shown that some of the trenches are fairly recent. The New Britain–New Hebrides Deep in Melanesia is post-Tertiary. He points out that there are four features associated only with the Pacific border: 1) deep trenches, 2) Meinesz negative anomaly zones, 3) active volcanoes, and 4) deep focus earthquakes. All of these could be accounted for by thrusting of the continents over the ocean floor.

Seismic surveys have shown that there is little evidence for a thickening of the continental crust beneath the deep trench. The Puerto Rico Trench has been studied in detail. North of this West Indian trench the structure is typically oceanic with no continental crustal material. The gravity anomaly of −226 mgals over the trench is explained by the very thick sediment, 4–8 km thick, in it. Ewing and Heezen (1955) have suggested that the negative anomaly could be explained by a thin crust and a thick layer of sediments. This could be the result of crustal tension, an idea supported by Worzel and Shurbet (1955), but not generally accepted and possibly not the best explanation of the data.

The trenches and arcs are now generally thought to be the places where the oceanic crust is moving under the continental crust and being consumed. Thus in the Peru–Chile Trench the sediments accumulating should be swept under the continental margin. The evidence presented above and in table 4.2 does not however, support this view. Scholl, Christenson, Huene, and Marlow (1970) suggest either 1) underthrusting has not occurred in the late Tertiary, or 2) has been slower than the estimated rate of 5–10 cm/year that has been suggested, or 3) underthrusting has not involved the sediments in the trench. They reach the conclusion that the third possibility is the most likely in view of the strong evidence elsewhere for the sinking of plates at these tectonically active trench zones. Hayes also suggests that high-angle normal faulting near the base of the continental slope could account for the trench structure, with a downward flexure of the area further offshore. The structural thinning under the outer flank of the trench could have provided the zone of weakness for flexuring.

The crustal thinning on the flank of the trench is found in other trenches, such as the Puerto Rico Trench and the middle America Trench. The local tension is included in the general model given by Isacks, Oliver, and Sykes (1968). They suggest (figure 4.8) that a slab of hard lithosphere sinks under the adjacent slab into the asthenosphere. The model shows the recorded deepening of the mantle below the island arc.

It appears that the lithosphere moves down into the mantle as a coherent body. This is

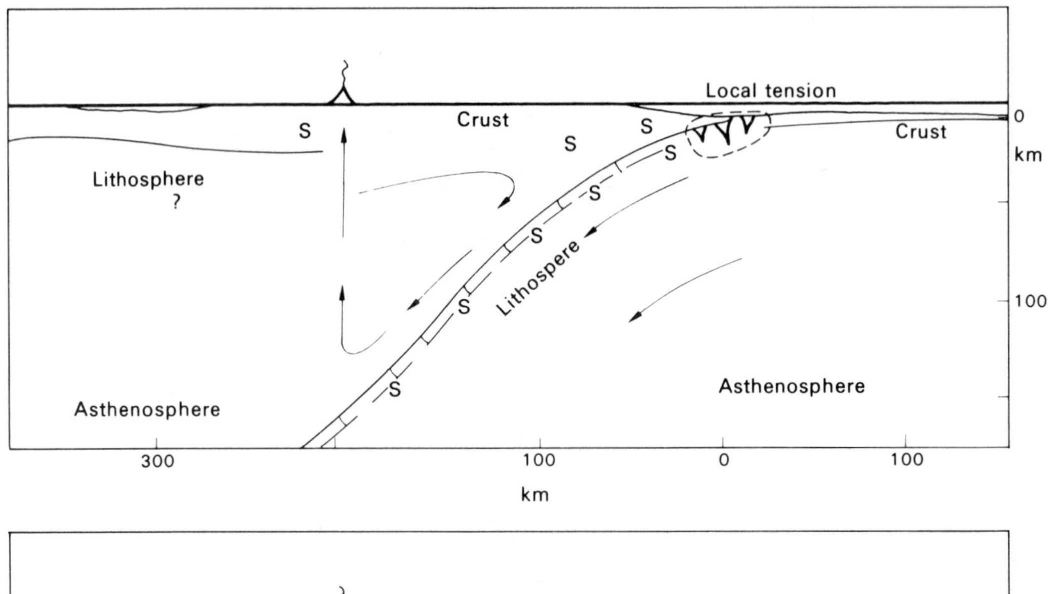

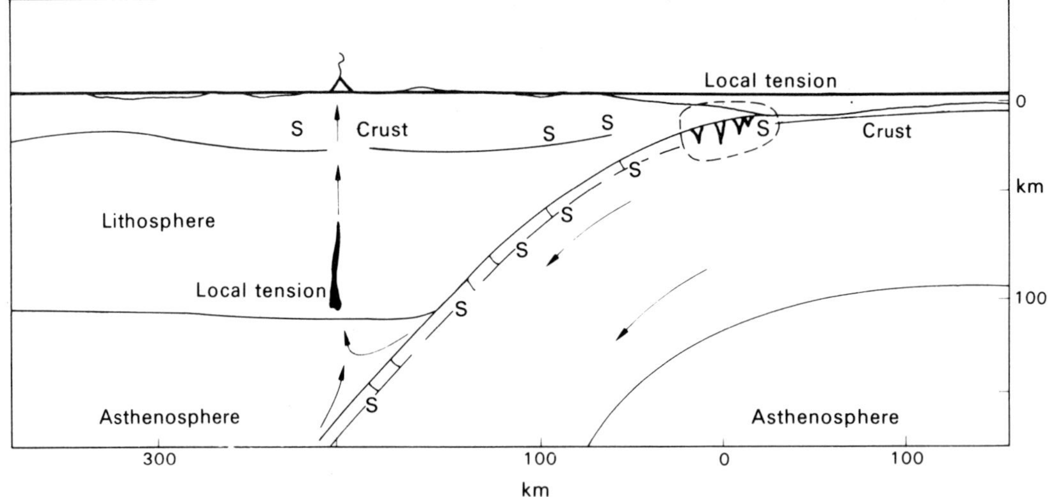

Figure 4.8 Diagram of possible structural patterns in the zone of deep sea trenches and island arcs. (*After Isacks, Oliver, and Sykes, 1968.*)

important because it provides the necessary strength at depth to account for the pattern of deep focus earthquakes. It is only under these conditions that strong rocks extend blow the normal level of about 100 km. The lithosphere normally moves down away from the ocean, but in some areas it moves down towards the ocean. This accounts for the occasional reversed arc. Figure 4.9 illustrates the arrangement as a block diagram, showing the lithosphere reaction to the compressional forces that are driving the two sections of crust towards one another. This pattern is found in the junction zone between the Tonga Trench and the New Hebrides Trench in the area of the Fiji Islands. The force creating the Tonga Trench is directed westwards, while that causing the New Hebrides Trench is directed eastwards.

An interesting relationship has been demonstrated between the length of the seismic zone in the island arcs and the calculated slip rate of undercutting. The length of the seismic zone delimits the distance along the sinking plate on which deep focus earthquakes occur, and it is measured perpendicular to the elongation of the arcs. The length is greatest where the deepest earthquakes occur. The average rate of oceanic spreading from the world rifted ridge crests is 1·3 cm/year over a 10 million year period. This rate must also apply to the rate of underthrusting. The relationship between the calculated slip rate and the length of the seismic zone is shown in figure 4.10. The correlation between the two variables gives a correlation coefficient of $r = +0{\cdot}8743$, with a regression equation of Y (length of arc in km) $= -0{\cdot}11 + 1{\cdot}21 \times$ (slip rate in cm/year). The more rapid rate of spreading is associated with, and perhaps caused by, the more rapid slip rate. The period of 10 million years

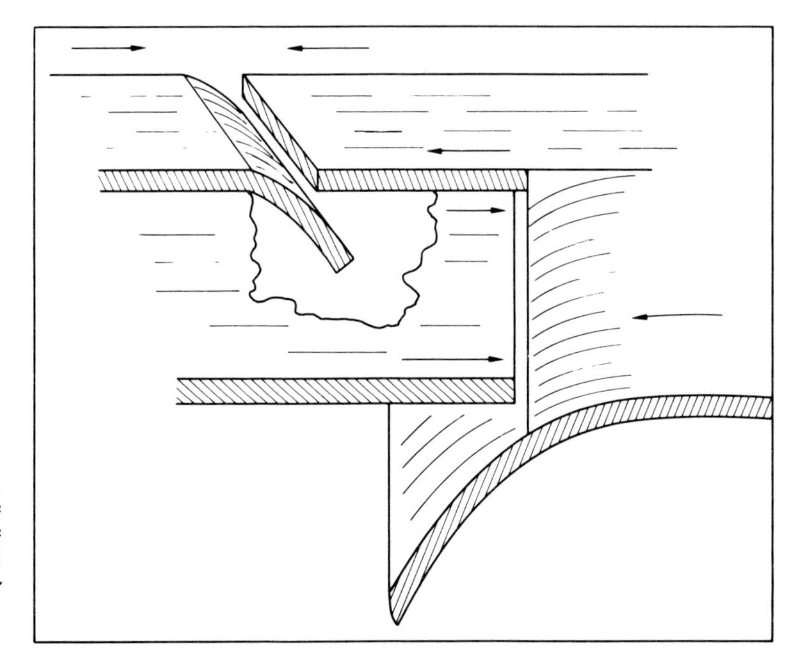

Figure 4.9 Block diagram of the junction of plates at the sites of deep sea trenches and island arcs. *(After Isacks, Oliver, and Sykes, 1968.)*

indicates the time during which the present arc system has probably developed. It may also represent the time required for the assimilation of the downward-trending lithosphere. Thus where underthrusting is most rapid the lithosphere penetrates further before it is re-absorbed and weakened so much that it cannot respond to fracture by seismic activity.

The method of assimilation of the down-warped lithosphere may vary. In places there is evidence that portions may become detached and sink deeper. The isolated deep focus earthquakes below the North Island of New Zealand could result from this occurrence. In other areas the down-warped lithosphere may come into contact with a harder layer and become deformed along its lower edge. The contortions in the plane of deep focus earthquakes in the Tonga Trench may exemplify this. Such a state could halt down-warping. The depth to which the lithosphere descends depends on the process of absorption.

Orowan (1965) has studied the possible causes of crustal spreading in terms of the physical nature of the response of the rocks to stresses of various types. Convection in the mantle appears probable from evidence of deep focus earthquakes. The form is likely to be relatively narrow, rising hot dykes. These dykes could be reflected on the surface by the ridges in the oceans. Their curvature and the fracture zones across them can be explained by this mechanism. The non-Newtonian dyke convection would involve a mechanism that drives the hot ascending flow to the zones midway between the continents. The velocity of the westward drift of the Americas is calculated from gravity spreading pressure under the mid-Atlantic Ridge and from the energy released in circum-Pacific earthquakes. Both methods yield a result of the order of 1 cm/year. The plasticization and fluxing of

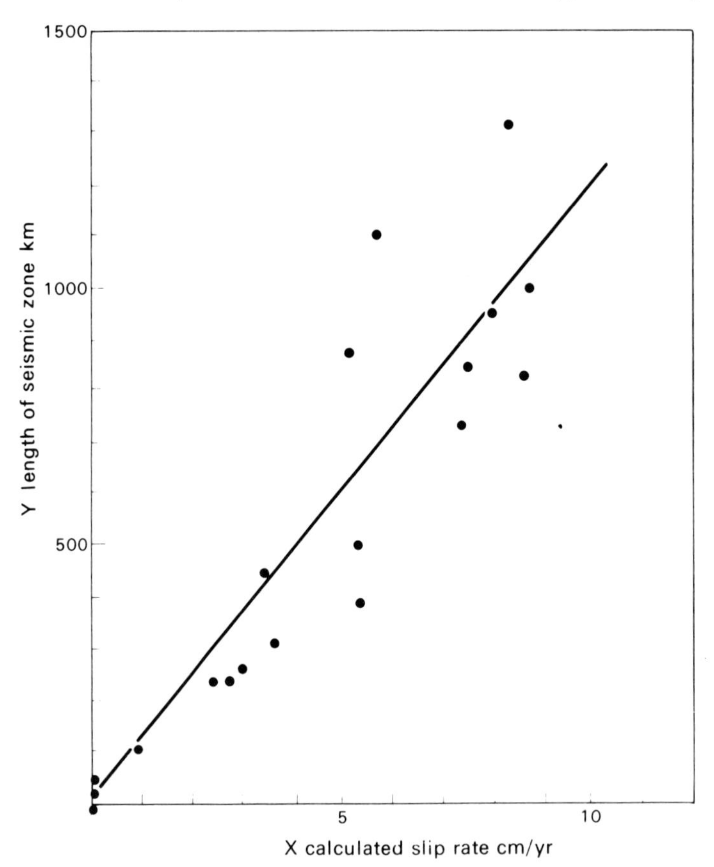

Figure 4.10 Graph relating the calculated slip rate against the length of the seismic zone in the marginal trench systems. The equation for the line is $Y = -0{\cdot}11 + 1{\cdot}21X$, $r = 0{\cdot}874$. (*After Isacks, Oliver, and Sykes, 1968.*)

the crust causing the orogenic processes to operate may be based on the formation of water by serpentinization. The serpentine would flow continentward from the ocean ridges and water would be precipitated where the flow turns downwards at the continental marginal shear planes. The accumulation of water and low-density volatiles in this zone would lead to the reduction of density and its inversion. The result would be an upward discharge of lava, causing geosynclinal subsidence. It is estimated that this occurs at a rate of between 0·1 and 0·2 mm/year. The reduction in density takes place beneath the shear plane, thus causing the reversal of density.

One reason for the concentration of island arcs in the Pacific has been suggested by

Orowan (1965). He suggests that they occur where an oceanic plate is meeting a continental one. It seems that many of the trenches are fairly old features. The Aleutian Trench probably dates from late Cretaceous or early Tertiary, while geological evidence indicates a similar date for the Puerto Rico Trench. Deep-seated granite was intruded in the island at this time. The great Java Trench in the Indian Ocean can be traced into the Ganges delta area and the Indo-Gangetic plain, which is a filled geosyncline. The system links with other filled geosynclines that date back at least to the Mesozoic. The extent of filling seems to be the major difference between these structures. On the other hand, some of the marginal trenches of the western Pacific may have formed during the last 10 million years.

There is now little support for the tectogen theory of trench formation. The strong negative gravity anomalies imply current activity. The evidence concerning the nature of the trenches and arcs on the whole favours the new global tectonic hypothesis. In this theory trenches and arcs represent the zones along which the oceanic plates are being consumed at the margins of the continental plates. They occur especially where a continental crustal section comes into contact with an oceanic one. Thus they mark the zone against which the westerly-moving Pacific plate meets the Eurasian and Australian plates, and where the easterly-moving Pacific plate, east of the East Pacific Rise, meets the American plate both in the northern and southern hemispheres. The pattern agrees with the slip vectors calculated for earthquake displacement in the circum-Pacific zone. Thus many lines of evidence link in explaining the pattern of trenches and arcs and their association with the zones of spreading along the oceanic ridges. Both features represent the most active tectonic zones in the ocean basins.

3 Abyssal plains, hills, and other sea mounts

3.1 Abyssal plains

Abyssal plains have a flatness unique in the ocean floor, but they are often interrupted by abyssal hills and areas of gently undulating relief. An abyssal plain is defined by Heezen and Laughton (1963) as an area having a gradient less than 1 : 1000. The slope of abyssal plains is determined by the critical bottom slope across which sediment can be transported by suspension flows. Bye (1971) has calculated from momentum and conservation equations that 1 : 1000 is the critical bottom slope. Gradients flatter than this must give rise to sedimentation, and the area of deposition may extend over a few hundred kilometres. He concludes that suspension flows provide a suitable mechanism for the very flat surfaces of abyssal plains, and accounts for their typical gradients.

Abyssal hills are defined as having a height of up to a few hundred metres and a width of a few kilometres by Heezen and Laughton. The abyssal plains usually lie at the base of the continental rise, with the hills on their seaward side, at the lowest point in a basin; the gradient of the rise is usually between 1 : 100 and 1 : 700, while that of the plains varies between 1 : 1000 and 1 : 10,000. Abyssal plains occur in small as well as large oceans, and also in all latitudes, including the Arctic ocean. The Sohm abyssal plain exemplifies their characteristics. It lies in the western Atlantic off Newfoundland and is T-shaped with a maximum depth of 5830 m. Hills break the smooth floor towards the south; the area covered is 906,000 km^2. There are smaller and more isolated abyssal plains in the eastern

Atlantic and similar ones occur in the northeast and extreme south Pacific. Another type occurs in the western Pacific, and they are called archipelagic plains. The percentage area covered by plains in the Pacific is less than that in the Atlantic owing to the smaller influx of sediment in proportion to the area. This is partly due to the marginal trenches, as the plains are associated with gaps in the trench system.

Off the western USA, in the Pacific, extensive areas of smooth flat plain exist, for example in the Gulf of Alaska (Emery, 1960). These areas can be directly correlated with the rate of deposition of sediment. Where a deep trench intercepts the sediment, the sea floor beyond is very much more irregular. Most of the sediment in the area off Alaska and northern California has been supplied from the land. Where the supply of sediment is plentiful, the irregularities of the sea floor have been buried to produce the flat, even abyssal plain. Further offshore the abyssal plain is largely covered by organic sediment or volcanic material.

The largest Pacific abyssal plain occurs south of Alaska in the northeast Pacific north of the Mendocino Scarp. The plain lies at the foot of the continental rise which is scored by submarine valleys and surmounted by sea mounts. Abyssal plains are found in all the other oceans, including an extensive one in the Canada Basin of the Arctic Ocean, which may have an area of 388,000 km^2, more than twice that of the plain in the Australian Bight. Even larger ones occur in the Ceylon Basin of the Indian Ocean, with an area of 518,000 km^2. Smaller ones occur between the Lomonosov Ridge and the Alpha Rise and the Asian Arctic coast. Abyssal plains also occur in the Gulf of Mexico and in the deep sea trenches, such as the Puerto Rico Plain, which is the widest in a trench. In the western Pacific the flat areas are in the form of archipelagic plains. They lie beyond the base of the pedestals on which the island groups stand. In character they are similar to many other abyssal plains and are probably formed in the same way. They should be differentiated from the archipelagic aprons that occur closer to the island group pedestals. The latter need not have a smooth surface, while the former are smooth. Most of these features occur in the central Pacific. The aprons are probably the result of partial burial of a hilly landscape by sediment.

The abyssal plains have been shown to be covered by sediment as a result of seismic surveys. Their relief confirms this, because their gradient decreases away from the source of sediment. The evidence for turbidity current action includes: 1) the presence of coarse sand and shallow water fossils in the sediments, 2) the thickness of the sediments suggests rapid deposition, which is greater than the rate of sedimentation in the normal deep sea environment, and 3) abyssal plains only occur where turbidity currents can reach from the land. They do not occur where deep sea trenches can intercept the sediment, so that they are more common in the Atlantic than the Pacific. Channels and associated fans extend down to depths exceeding 5000 m. Abyssal plains are particularly well developed in the Atlantic, where they occupy 15 per cent of the ocean floor. There are two large ones in the western Atlantic and more numerous, smaller ones in the east, where basin-type structure is better developed.

The amount of sediment on the plains is similar to that on the rises. In general persistent reflection horizons cannot be traced in the abyssal plains, although they can be on the rises and at the seaward margins of the plains. This is due to the nature of the sedimentation, which is probably the result of turbidity currents on the margins, where coarse material can reach. In the central area pelagic sedimentation is more important, and the material derived from turbidity currents (which is mixed with pelagic sediment) is of similar calibre, both being fine clay. The sharp seaward margin of the reflecting horizons could be the effect of sudden change in sedimentation over a wide area, such as an ash fall or change of material due to change in environmental conditions (for example, the change from calcareous to inorganic sediment resulting from climatic effects). Heat flow measurements on the abyssal plains indicate a value similar to the mean oceanic value of 1·2 m-cal-/cm^2/sec, or $\frac{1}{10}$ to $\frac{1}{2}$ that on the ridges.

Heezen and Laughton (1963) suggest seven possible theories for the distinctive characteristics of abyssal plains. These include their flatness, continuity of gradient over long distances, exponential shape, and the presence of terrigenous and shallow water sediments on them. The hypotheses are: 1) atectonic areas, 2) mylonitic plains or areas of intense tectonic activity, 3) subaerial erosion surface, 4) lava plains, 5) areas of long-continued sedimentation, 6) uniquely fine sediments distributed by bottom currents, and 7) burial of original relief by turbidity currents. Several of the hypotheses can be ruled out at once. The first four need not be considered in detail as they all suffer from obvious difficulties. The fifth does not account for the exponential form of the plains. Bottom currents are also unlikely to have produced this shape. The seventh possibility accounts for all the known facts.

3.2 Abyssal hills

The deep sea floor is diversified by a large number of small positive features including abyssal hills, sea mounts, guyots, and coral atolls, most of which are genetically related. An attempt has been made by Krause and Menard (1965) to analyse the geometrical character of some of these features statistically. They studied 15 east Pacific profiles 111 km long to ascertain the relationship between the different dimensions of the hills found on them. The aim of the analysis was partly to classify the submarine hills and to assess their distribution in terms of their dimensions. The relationship between the number of hills, N_0 and the height of the hills, H, in a 111 km profile was given by $N_0 = -\frac{9}{15} \log_{10} {}^{H+3}$, a relationship was also found between the number of hills on all 15 profiles and the width of the hills, W, in the form $N_{15} = 120W^{-3/2}$. The profiles were located in the region between San Francisco and the Hawaiian Islands and were selected as being typical of specific processes. The five processes included were pelagic sedimentation, turbidity current sedimentation, slumping of sediments, tectonic deformation, and volcanism. Profile 15 was typical of turbidity sedimentation, while volcanoes were associated with profiles 8, 10, 11, and 12. All the profiles sampled the abyssal hill province. The frequency distribution of depths at 36 m intervals summarize the data and show that some of the profiles have a normal depth distribution. Some are skewed to the shallower depths. The abyssal hill profiles show a random distribution of height, suggesting a uniformly operating process or processes. The hills were classified in terms of their height and breadth into a number of categories, such as pinnacles, plateaux, and broad low swells. The analysis suggested that there was some relation between the frequency of abyssal hills and their height, but no horizontal periodicity was found (with one exception). There was a stronger relationship between the frequency of abyssal hills and their widths. This relationship provides a basis for classification, providing classes that have geological meaning. A certain predictability of dimension is indicated as well as a consistant mechanism, which can be related to the energy needed to form the hills, given as $W_f = \frac{1}{12}\pi\rho g h^2 w_a^2$, where W_f is the energy required, ρ is the difference in density between rock and sea water, h is the height of a conical hill and w_a is its basal width. Thus the energy required is proportional to the square of the height.

The abyssal hills of the central equatorial Pacific have been described by Moore and Heath (1967) in the area around 8°20'N, 153°W, where the hill relief is from 50 to 270 m. The hills are elongated, trending north to south, the largest lying on a ridge trending parallel to the former Darwin Rise. The base of the first layer has about twice as much relief as the hill surfaces, so that the first layer ranges between 100 and 440 m in thickness, being greatest in the basins and least on the hill tops. The rate of sediment accumulation must have varied by a factor of 4 over 10 km distance if the sediment base is synchronous. As a result the sedimentation has made the relief smoother. The study area was chosen as being typical of the Pacific and as undisturbed by tectonic activity at present as possible, and also by the effect of turbidity currents, at depths between 4900 and 5180 m. The hills, classified as low to high abyssal hills in Krauss and Menard's classification, slope generally between 2 and 3°. Two fairly continuous reflectors were located within the sediment cover, with an overlying discontinuous one. The first two may be synchronous stratification planes. The higher of the two could also be associated with a change in the physical or chemical nature of the sediment, and this seems more likely. The hills could result from igneous intrusions beneath the lower reflecting layer, which would agree with their dome-shaped form. Such a mechanism would also account for outcropping and reworking of Tertiary sediment. The effect of deep sea currents on sedimentation also requires further elucidation.

3.3 Sea mounts and guyots

Sea mounts and guyots are particularly numerous in the Pacific Ocean, although they also occur in the Atlantic. A sea mount, which has more than 1 km local relief, and which may have a sharp top, is differentiated from a guyot by the flat top of the latter. There are about 10,000 sea mounts and guyots in the Pacific alone; some of them rise 3000 m above the adjacent ocean floor (Menard, 1959). They are also associated with coral atolls.

Distribution of the three features is of interest. Menard has shown that the coral atolls are concentrated in the southwestern part of the Pacific, in a line extending west-northwest from Hawaii, with only a few exceptions. Distribution of sea mounts on the other hand is much more extensive, although they are found in greater number in the centre and particularly in the northeastern part of the ocean. Guyots are concentrated in the west and in the area south of Alaska, as shown in figure 4.11. When their distribution is considered in detail, they show linear patterns in many areas. The distribution of the three features indicates that different parts of the Pacific basin have been volcanically active at different periods.

There seems to be little doubt that both sea mounts and guyots originated as volcanoes, erupting beneath the ocean. They are often circular in plan; their slopes tend to be steeper than land volcanoes, owing to the more rapid solidification of the lava in sea water. The difference between the pointed sea mounts and the flat-topped guyots is due to the elevation of the guyot above sea level at some stage in its development. The top has been truncated by subaerial and marine erosion. Its present depth below sea level is the result of subsequent sinking. The tops of many guyots are now about 1500 m below sea level. The variation in depth, which is considerable, probably reflects a variation in age, the deeper ones being the older, as subsidence has continued longer.

Guyots are grouped in three main areas. There is a north–south line running from Kamchatka to the latitude of Hawaii, a group south of Alaska, and a large group west of Hawaii, running from the Mariana Islands to the Marshall Islands. Those in the Gulf of Alaska average 915 m depth.

The depth over the guyots can be explained either by a rise in sea level, which has

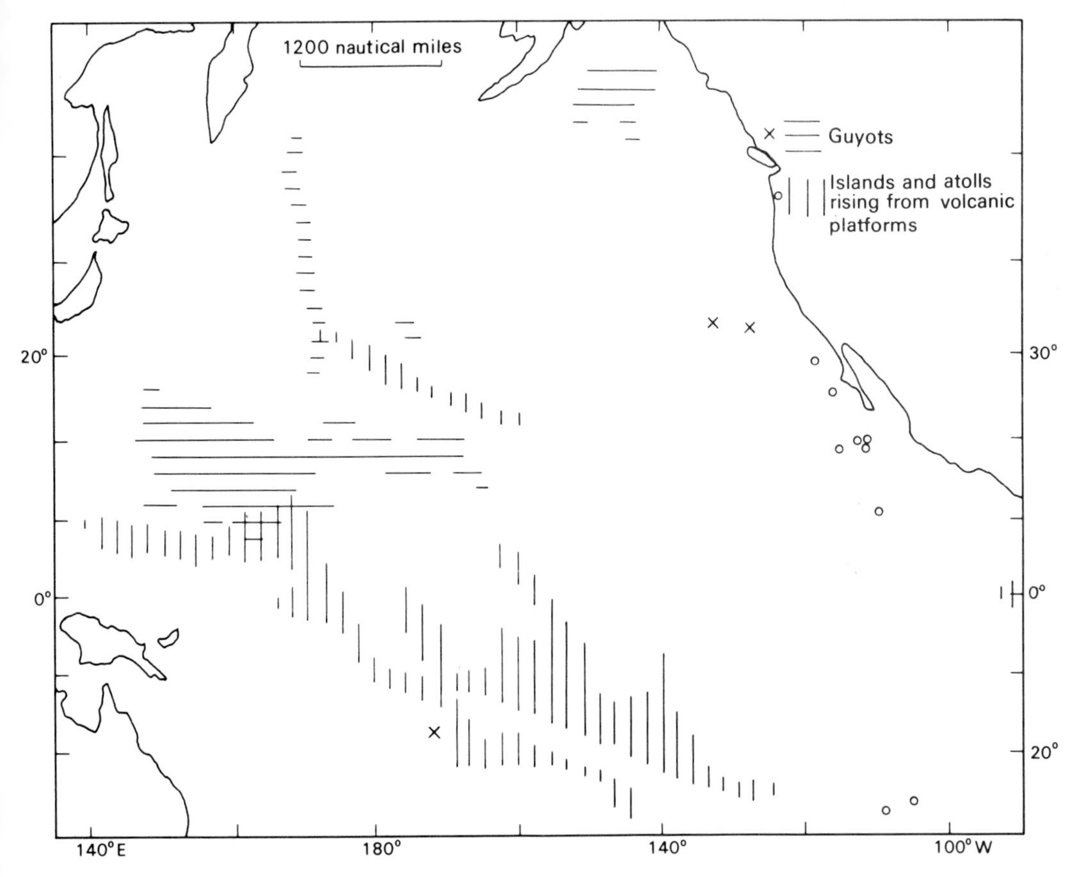

Figure 4.11 Map to show the distribution of guyots and islands, including atolls, rising from a volcanic platform. (*Adapted from Menard, 1959.*)

drowned the guyots, as suggested by Revelle, or by their foundation having subsided. This is the most likely explanation, and is favoured by Menard, as well as being supported by geophysical evidence. The denser basic substratum of the ocean is bent down under the weight of the overlying erupted material, which is of relatively lower density of 2·4, and can therefore, form a positive feature. Once erupted the volcano tends to sink as the weight of the lava depresses the crust into the void left by the erupted material.

The Hawaiian chain illustrates the stages well. Volcanic activity has shifted southeastwards along the linear group in this area. At the northwest end the volcanoes are oldest and these now form atolls; the central section consists of volcanoes, now no longer active; while the southeastern end has the active volcanoes of the Hawaiian Islands. Further evidence in favour of the sinking of the volcanoes to form guyots is the frequent occurrence of a trough of deeper water around the foot of the extruded material, where the crust has bent down.

The linear arrangement of sea mounts and guyots may be related to elongated zones of up-warping in the ocean floor. This process generates heat, which may help to account for the subsequent volcanic activity. Some of the many sea mounts of the eastern Pacific are related to the four major east–west fracture zones in this area. These zones are 100 km wide and 2000 km long. The fracture zones, along which there has been lateral displacement of at least 160 km, form steep asymmetrical ridges, with a total relief of 1600 m.

The Cobb sea mount, lying 500 km west of Gray's Harbour, Washington, provides an example of these features. It is an ancient volcano rising 2743 m to within 34 m of the surface, and has an area of 824 km^2. Slopes average 12° and there are four terraces at 915, 183, 146, and 82 m. A basalt pinnacle rises from the highest terrace with 45° slopes; it has a flat fissured top 200 by 400 m. The feature has been dated at 27 $\pm$ 6 million years and its morphology suggests an Oligocene date for the volcano, which has subsequently been terraced by wave abrasion as it subsided to 914 m. The terracing is related to Pleistocene glacio-eustatic still-stands. The feature is associated with the East Pacific Rise, but its date makes the relationship obscure, and until more data are available it is impossible to resolve the problem.

Many of the sea mounts and guyots are fairly old. Shallow water Cretaceous material has been dredged from the crests of some of them, including corals, thus indicating that the guyot summit was near sea level at this time. Miocene material has been obtained from others. At the present time the central Pacific is on the whole quiet seismically. There are only a few active volcanoes within the Andesite line. However, considering that all sea mounts and guyots must have been active volcanoes at one time, the Pacific cannot always have been as quiet as it is now, unless the period over which the 10,000 volcanoes formed was a very long one. At the present time there are four active clusters of volcanoes within the Pacific. Assuming that there are 20 volcanoes to each cluster, there must have been 500 active clusters to form all the submarine hills. It is suggested by Menard that all the large volcanoes of the Pacific basin could have been produced since the Cretaceous period at the present rate of volcanic activity.

Menard (1969) has shown that volcanic sea mounts increase in size with distance from the mid-ocean ridge crest. They must, therefore, remain active as they drift away. Very few rise above the ocean crust formed in the last 10 million years by sea floor spreading. They must normally develop after this length of time. Guyots and atolls normally form on old crust and they must remain active for tens of million years. The palaeobathymetry of the Darwin Rise is thus compatible with sea floor spreading. It remains difficult to see how the volcanoes can remain in contact with their magma source as they drift. The sea mounts and guyots provide useful evidence of sea floor spreading. Wilson (1965b) has considered this aspect of the distribution and character of these features. Fossils have been dated on nine islands and three sea mounts in the Atlantic, and five more islands have been dated by other means. A plot of the results show that the ages of the features increases away from the mid-Atlantic Ridge. Values of 15–18 million years occur fairly near the ridge, while those off Africa are between 25 and 35 million years, increasing to 135 million years near the coast. In the western Atlantic ages of 90, 110, and 135 million years have been recorded (figure 4.12). Similar evidence is found in the Pacific, where 10 out of 11 chains of islands get older away from the East Pacific Rise. On some of the Pacific Islands there is evidence of uplift, one group lying in the central Pacific and the others about 200–750 km from the

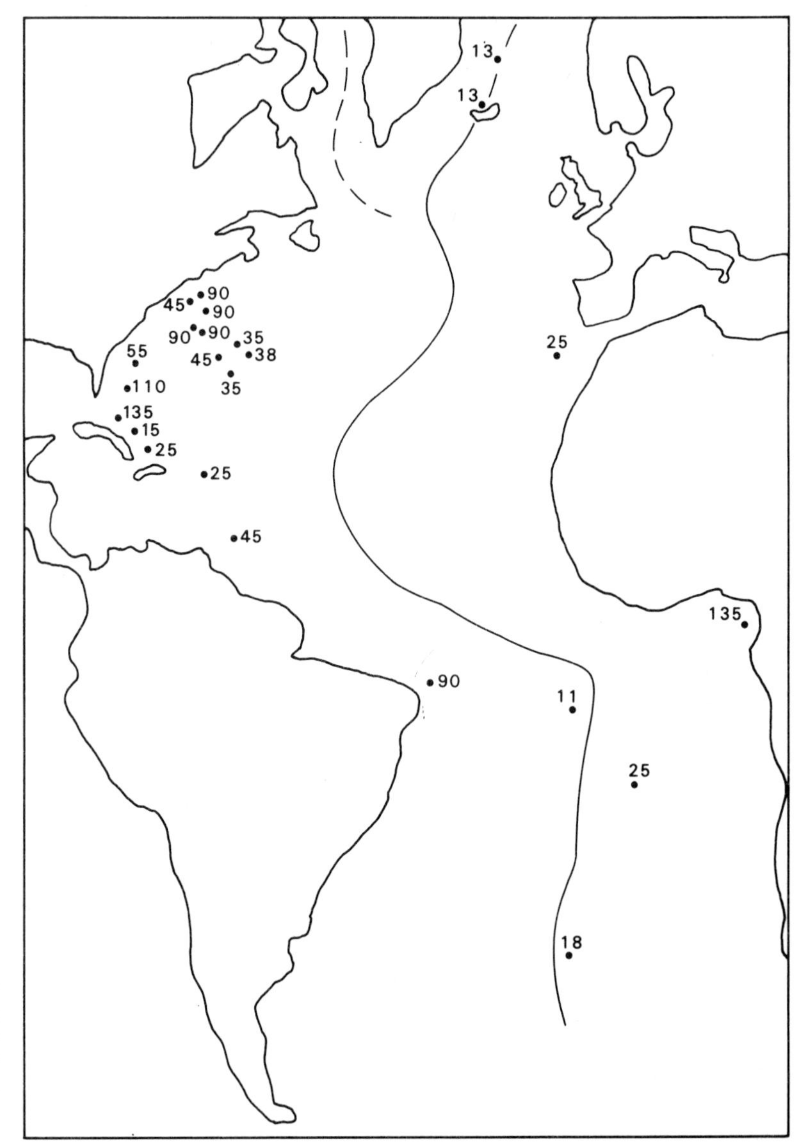

Figure 4.12 Ages of volcanic rock in the Atlantic relative to distance from the mid-Atlantic Ridge. (*After Bullard, 1968.*)

island arcs. A third group lies at a similar distance to the trenches off central and south America. The explanation for the upbending in those positions has been mentioned in connection with tensional up-warps where the crust is meeting another plate at the edge of the Pacific plate. In these areas the natural tendency of oceanic islands to subside has been counteracted by a tendency for uplift. It is this general tendency of oceanic islands to subside that is responsible for the great number of sea mounts, guyots, and coral atolls in the oceans.

A gravity survey of a sea mount in the Atlantic between the mid-Atlantic rift-valley and the Sohm abyssal plain by Le Pichon and Talwani (1964) reveals a double peaked feature at 35°N, 46°W. The sea mount rises to 2000 m above the sea floor from a depth of more than 4500 m. There is a complex of sea mounts in this area on the edge of the abyssal plain. The crust in the area is of oceanic type with thinner layers than usual. The gravity survey revealed a single gravity low, which was thought to be due to a secondary magma chamber at the base of the crust.

The geomorphology of the Erben Guyot has been described by Boyce and Smith (1968). In size it is large, being 4000 m in height, with the flattish top typical of guyots. Its summit lies at 550 to 915 m and it is situated 48 km south of the Murray Fracture zone scarp; its slopes are up to 45°. It is probably formed of bedded lava and volcanic debris, and its two peaks may be volcanic vents. Subsidence of the guyot is probably associated with movement along the fracture zone. The top of the feature is an elliptical platform 6·4–10 km across, with a small summit peak rising to a depth of 412 m. The guyot was probably formed by complex volcanic activity.

An example of a guyot that need never have been at sea level is described by Malfait and Dinkelman (1972). This is the Horizon Guyot, which is situated south of the Hawaiian Ridge and west of the Hawaiian Islands. Its summit is at a depth of 2000 m and its surface is capped by Tertiary nannoplankton, which has been truncated to form layers. Chert also outcrops and there are hard rock terraces formed of fluid lava flows. The material includes some of Eocene date. The guyot is a ridge 300 km long, and the sediments on it record a period of 100 million years, although the strata are not continuous in time, and interruptions occur. The form of the guyot can be explained by submarine erosion and deposition, which has been complex. The changing pattern of productivity at the surface is one factor that could help to account for the complex sedimentary record. Currents reaching more than 15 cm/sec were recorded 12 m above its surface, and ripples and dunes were observed on the surface, indicating movement of material under present-day processes. There was evidence that the sediment was moving up-slope.

All the islands, sea mounts, and guyots of the Pacific are associated with vulcanism, although in the atolls the volcanic material is buried beneath thick growth of coral. During the last 100 million years, sea floor volcanoes have provided as much material as land volcanoes in 3000 million years; the former have thus been much more active than the latter. Owing to high oceanic pressures in deep water, submarine volcanoes probably erupt lava rather than ash where the depth is about 5000 m. Ash is more likely where depths are less than 1500 m. Volcanoes in the Pacific range from 1 to 10 km high.

Guyots provide good evidence for subsidence of the volcanoes as they acquired their flat top when exposed at the surface. The flatness could be due to either erosional truncation by waves at sea level, or the formation of an atoll with a flat lagoon resulting from sediment infill. Another alternative is the filling of a caldera by pelagic sediment. Large calderas, however, only develop in shallow water. A final possibility is deposition of almost fluid lava on top of a cone of fragmental material. Samples dredged from guyot tops support the wave erosion hypothesis most strongly.

Submarine volcanoes are characteristically grouped. Circular groups occur in the Galapagos, elliptical ones in the Marquesas, and linear ones are exemplified by the Austral Islands and the Emperor sea mounts. In many cases the activity moves along the group, giving active volcanoes at one end and atolls or guyots at the other. The groups are often surrounded by moats with arches beyond, as around the Hawaiian group, where the arch lies 330–370 km from the volcanoes. The spread of volcanoes along a major lineation could be due to tension where the arch cuts the lineation. The bulk of the volcanic material in the Pacific is in the form of aprons, which may have a thickness of 2 km, giving a total volume of 20×10^6 km^3, or more than 80 per cent of Pacific volcanic material.

3.4 Coral atolls and other reefs

The formation of coral atolls was at one time a problem that aroused much interest; rival theories were hotly debated. These features, much more numerous in the Pacific than elsewhere, are closely associated with sea mounts and guyots, which form their foundations. The distribution of coral reefs and atolls is determined by the conditions that coral polyps, and associated reef-building organisms, require for successful growth. They need a water temperature averaging 21°C or over. Clear water less than 46 m deep is also essential. They are, therefore, restricted to the shallow water of the tropical zone, and only rarely extend beyond 20°N and S. There are some coral formations up to 30°N near Hawaii, in the West Indies in the Atlantic, and around the Indian Ocean, although they do not form typical atolls in these areas. True atolls are nearly all restricted to the tropical Pacific. The fact that corals will only thrive in shallow water is important from the point of view of the rival theories, put forward to account for the atolls. These features consist of a ring of reef coral, growing actively on the seaward margin and enclosing a lagoon in the centre, no solid rock being visible in the true atoll.

Two major theories were proposed during the nineteenth century. Darwin (1842), who studied coral formations during the voyage of the *Beagle*, suggested that the coral atolls were formed from the simple fringing reef, which develops round an emergent island. The fringing reef widens to a barrier reef as the island round which it is forming slowly subsides, until the island completely disappears below the surface and an atoll surrounds the lagoon over the former position of the island (figure 4.13). This theory was supported by Dana (1885) and Davis (1928), the latter adding the geomorphological arguments in favour of subsidence. He drew attention to the lack of cliffing in the barrier reef stage and the drowned nature of the valleys, which often were found to have a great thickness of alluvium. The opposing theory was put forward by Murray (1880), supported by Agassiz (1906) and Semper (1881). Their idea was that any submarine platform could be built up by deposition, or lowered by erosion, until it was at a suitable elevation for corals to grow. The corals then built up an atoll, which developed a lagoon as the reef-building corals on the edge grew more vigorously than those in the centre. Their observations were mainly made in the East Indies, and their arguments may well apply correctly to some of the reefs in this area and in the Indian Ocean, where there is some evidence of emergence.

An early attempt to show which of these theories accounted best for the Pacific atolls was made in 1901, when a borehole was put down to a depth of 339·5 m on the atoll of Funafuti. Throughout this depth coral material was passed through, lending strong support to the subsidence theory. The opponents, however, argued that the coral was in the form of

talus, which had slipped off the edge of the submarine platform, rather than being in the position of growth, but for this view there was little or no evidence.

More recently further experiments (including boreholes and seismic studies) have shown that in the deep Pacific Ocean, atolls are almost certainly the result of subsidence of a volcanic cone, at a rate at which the reef-building organisms could maintain their hold. However, rising sea level, due to subsidence, has not been the only process operating. As was pointed out by Daly long ago, sea level was lowered during glacial times and the temperature fell. These changes must have had some effect on the development of the

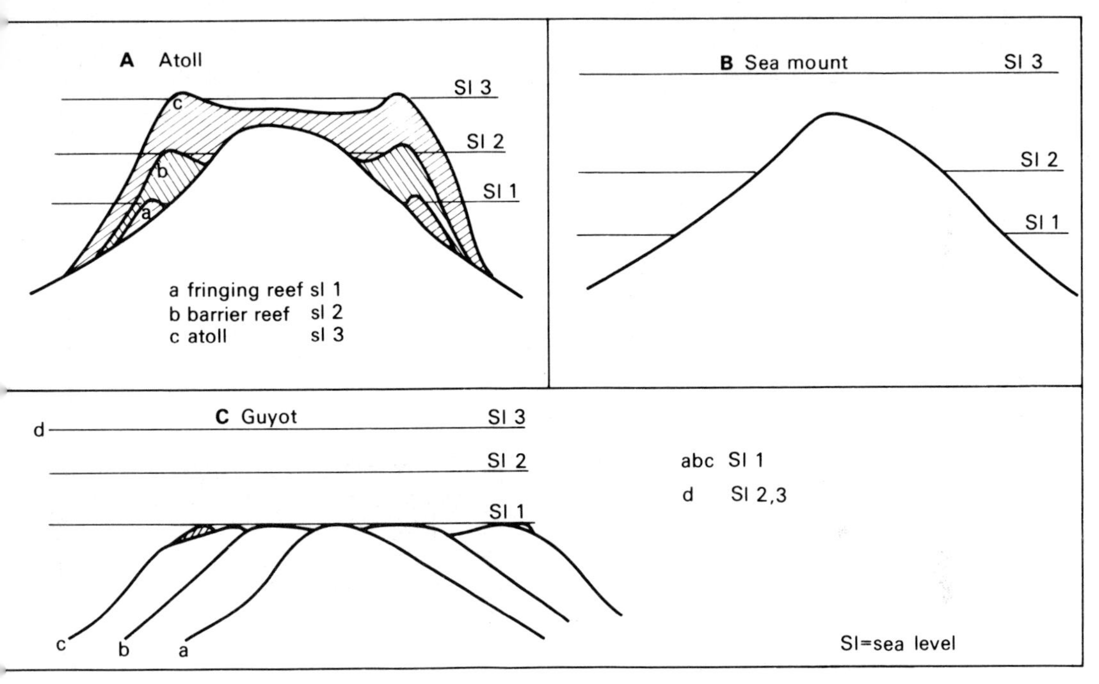

Figure 4.13 Schematic diagram to illustrate the formation of **A,** an atoll on a slowly subsiding volcanic foundation. **B** shows the formation of a peaked seamount, resulting from rapid subsidence, which was too fast for coral growth to keep pace, or where conditions are not suitable for coral growth. Such features may also form if the peak never reaches shallow water. **C** shows possible formation of a flat-topped guyot; stage a, volcanic island growing up to sea level; b, erosion by waves keep pace with the growing volcano; c, flat-topped area near sea level is formed as volcanic activity ceases (coral may grow on this platform); d, sea level rises to drown the guyot, as its foundation subsides. If coral growth is active atolls may form in this way.

atolls. In areas near the limiting temperature control, the sea probably became too cold for corals to grow. The coral in these areas may have died, leaving the island foundation to be trimmed by the waves. When the sea rose again as deglaciation took place and the water became warmer, the corals returned in some instances and built up a new atoll on the planed-off island. In this type the coral need not be very thick to form an atoll.

In other areas reef corals perhaps never returned, and the island, now sunk beneath the

Plate 13 Different types of coral formation under water: *top,* Grand Cayman Islands; *centre,* off British Honduras. (*David R. Stoddart.*) *Foot,* the coral reef at Korolevu, Fiji. (*J. Allan Cash.*)

surface, due to the post-glacial rise of sea level, will form a guyot. Drowned atolls have been described by Fairbridge and Stewart (1960) on the Alexa Bank, which is on the Melanesian Border Plateau. This submerged bank has a raised rim, as do true coral atolls, and it lies at a depth of 21 m. while the central depression is at 30 m depth. The outer slopes of the bank drop away steeply at 30–40° for thousands of feet. The reason why corals deserted this bank is not easy to find. It is in equatorial waters, so a lowering of temperature is not likely to be the cause. On the other hand, corals can be killed either by emergence for a short period, or too rapid submergence, so that their upward growth cannot keep pace. Other possibilities suggested are the deposition of too much volcanic ash, or the formation of foul water in enclosed basins, which are found in this vicinity. Much coral escaped destruction, however, with the changing sea level and other conditions, as shown by the many atolls now found in the Pacific.

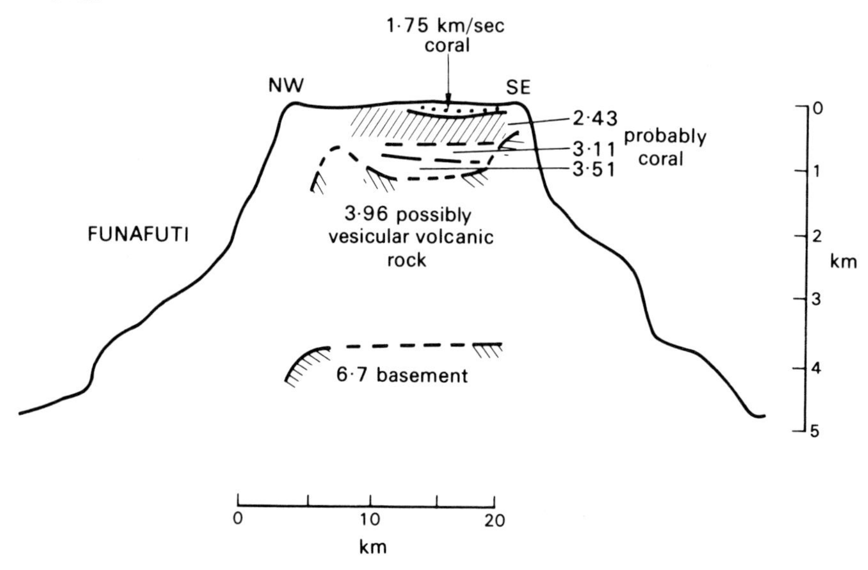

Figure 4.14 Section through the atoll of Funafuti. (*After Gaskell, Hill, and Swallow, 1958.*)

The island of Funafuti has now been explored more thoroughly by seismic survey which showed that there were several thousand feet of material having a low velocity of wave travel; this is probably coral rock. This material extended all across the lagoon of Funafuti, while similar results were obtained from the island of Bikini and Eniwetok (figure 4.14). At this last island the results were also confirmed by a borehole, which penetrated to the base of the coral (Kuenen, 1954). This core gives some evidence of the rate of sinking of the island. Boreholes were put down on both sides of the atoll, reaching volcanic rock beneath coral formation at depths of 1280 m and 1400 m. The coral appeared to be in the position of growth. The top of the Miocene was found in this atoll at a depth of 183 m, indicating subsidence at 15–18 m/million years. The top of the Eocene was found at 854 m, corresponding with a subsidence of 18–21 m/million years. The rate of up-growth during the Quaternary must have been much more rapid, as the glacial changes of sea level were much more rapid than the Tertiary rate of subsidence, assuming that coral formed at the high glacial sea levels was destroyed as sea level fell to its low glacial level. Thus the corals building up the atoll during the post-glacial rise of sea level must have been working 200 times faster than those in the Tertiary. Although this is quite possible the results remain tentative. It is clear, however, that glacial changes of sea level must be taken into account in the consideration of atoll formation.

Shepard (1973) has suggested that the foundations of many of the atolls were formed during volcanic activity in the Cretaceous period. The volcanoes that emerged were attacked by erosion and truncated. Where the conditions were favourable, corals grew on the banks around the islands, which have slowly subsided since. In some instances the reef-building organisms have kept pace to form atolls; in others they did not gain a foothold before the water was too deep and the truncated islands have now become guyots. Subsidence is indicated by the lowering of the Moho by 2·5–3·0 km beneath them.

It has been suggested that coral atolls develop from the preliminary forms of fringing and barrier reefs (the atoll forming the final stage), when all the original foundation has disappeared below sea level. Fringing reefs may either form facing the open sea or be protected by a barrier to seaward. On the exposed side of the reef, as with other reef types, the outermost feature may be an algal ridge, formed predominantly of calcareous algae, and not corals, of which *Porolithon* is common. Corals flourish on the inner slope of the reef flat, which is built up by dead coral, sand, and other detritus, interspersed with some living coral. The lagoon, separating the reef from the shore, is often flat-floored, its depth varying between 30 and 100 m, although in places coral may still grow within it.

The largest barrier reef is that off the Queensland coast of Australia, extending from 9°s to 22°s. The feature is not a continuous barrier reef, but in the south especially consists of many reefs spread over an irregular platform. To the north, however, the reefs become more linear in pattern, facing the ocean and separated by narrow channels. The width of the shelf is 130 km near the Torres Strait in the north, narrowing southwards to 30–50 km width near Cairns, increasing again southwards to about 160 km wide.

The origin of this great barrier reef zone is related to the structure of the Queensland coast. The offshore zone has been faulted down in the recent past, probably at the end of the Tertiary period. It has been suggested that a fault system let down a peneplained surface to a suitable depth below sea level for reef-building organisms to become established, in an area in which other conditions necessary for their growth were favourable. The flat and shallow floor between the reefs can probably be explained by sedimentation in the lower areas and coral growth elsewhere.

There are about 425 atolls in the world (Stoddart, 1965) and these show considerable variation in shape. They often vary from the theoretical circular shape that they would be expected to have if they were growing on drowned volcanoes, to square, triangular, or linear, with many irregular shapes. Stoddart assessed 99 atolls, mainly from the Pacific, where they are most abundant (some from the Caribbean were also included). Three shape measures were used: $F = A_1/L_1^2$, A_1 is atoll area and L_1 is maximum length. $R_c = A_1/A_2$, A_1 is the area of the atoll and A_2 is the area of the circle with the same perimeter. $R_e = d/L_1$, where d is the diameter of a circle with the same area as the atoll and L_1 is the longest atoll axis. Two other indices were used: I_e, the ellipticity ratio, $L_1/2b$, where $b = A_1/\pi(L_1/2)$, and radial line index $I_r = \sum_1^n \left| \left[100 \cdot \frac{r_1}{\sum_1^n} - \frac{100}{n} \right] \right|$ where r is the length of the radial lines 1 to n = 16 from the centre of gravity. A linear relationship was found between area and perimeter and between length and breadth. The ellipticity index was found to be the best of the indices and the F value for the shape measures. The mean value of I_e for all the atolls was 2·867. The main conclusions from the study were that atoll shapes are relatively homogeneous, although a few have very unusual shapes. A significant difference between different island groups was suggested. Some results are given in table 4.3.

Table 4.3 Atoll shape measures

Atoll group	Number in sample	F	R_c	R_e	I_e	I_r mean values
Caroline	27	0·348	0·554	0·616	2·750	30·38
Marshall	16	0·395	0·524	0·593	3·656	33·56
Maldive	17	0·375	0·644	0·677	2·429	22·13
Gilbert and Ellice	9	0·324	0·478	0·641	2·284	30·34
Total sample mean	99	0·349	0·569	0·638	2·867	29·78
Standard deviation		0·133	0·186	0·142	—	14·73

Most atolls provide evidence of subsidence that is often more complete than that provided by guyots. The whole sequence of deposits that rest on the volcanic core can be sampled and dated where boreholes have penetrated to the igneous material that forms the foundation. Geophysical work has given the depth of the volcanic material at 1006–1980 m on the world's largest atoll at Kwajalein (Stoddart, 1969). Drilling on Mururoa Atoll in the Tuamotus reached basalt at depths of 438 and 415 m in two bores. The dating agreed with that carried out in the Marshall Islands 7000 km away. On Midway Island basalt was reached at 157 and 384 m after penetrating 120 m of post-Miocene limestone. A well at Andros Island in the Bahamas went through 161 m of surface limestone and then dolomite was penetrated to a depth of 4488 m, where shallow water facies of lower Cretaceous were found. The recent work corroborates Darwin's hypothesis. In some areas, however, reefs rest on shallow foundations, but these are not true atolls. Some atolls have been elevated, while others have been drowned.

Coral atolls must be considered as ecological systems that are basically biological in origin. The dominant corals that affect the reefs are those of tropical seas in which scleractinian corals are dominant. Faunas are larger and more diverse in the Indian and Pacific Oceans than in the Atlantic Ocean. Temperature exerts a major control on reef distribution. The hermatypic corals must live in shallow water, because of light needs. Most coral reefs fix 1500–3500 g carbon/m^2/year; their main predators are fish and the echinoderm, *Acanthaster*. The latter may be aggressive predators with high consumption. Dark water induced by intense rain or flood water may kill corals, and hurricanes often cause damage that requires 10–20 years recovery.

The reef forms a morphological feature as well as being an active biological system of high productivity. The main features of a reef are shown in figure 4.15. Patterns of coral formation vary considerably; there are typical atolls, fringing and barrier reefs, as well as linear reefs or faros. Reefs are zoned laterally and transversely as a result of changing environ-

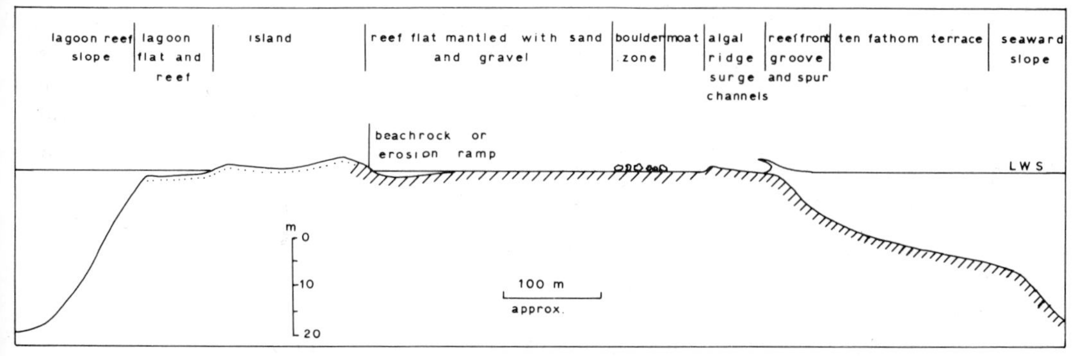

Figure 4.15 Zones of coral reef. (*After Stoddart, 1969.*)

ment due to exposure, giving windward and leeward features. Different features form at different elevations and variety of forms and species are characteristic of different zones.

Growth rates of massive corals may be 1 cm/year, while branching corals may increase at 10 cm/year. Rates of reef growth can reach 2·5 cm/year extrapolated from colony measurement, while values of only 0·02 mm/year over 70 million years has been obtained from borehole data. There are probably great variations in both rates. There is only a thin veneer of modern coral on most reefs, despite long-continued subsidence, owing to the effects of recent sea level changes. Slightly raised reefs date from 90–160 thousand years B.P. and there is evidence for a sea level at present height at 30–35,000 B.P. The reef forms a rigid framework, and sediments associated with it are organic in origin. Owing to recent changes of conditions and sea level, many modern reefs are not in phase with their inherited geological characteristics.

Further reading

HAYES, D. E. 1966: A geophysical investigation of the Peru-Chile trench. *Mar. Geol.* **4,** 309–52. (A detailed account of the zone along which the east Pacific floor is meeting the South American continent.)

HEEZEN, B. C. and HOLLISTER, C. D. 1971: *The face of the deep*. New York and London: Oxford University Press. (Photographs taken in the deep sea are discussed and illustrated. Many include deep sea sediments and structures.)

HESS, H. H. 1965: Mid-ocean ridges and tectonics of the sea floor. In W. F. Whittard and R. Bradshaw (editors), *Submarine Geology and Geophysics, Proc. 17th Colston Res. Soc. Bristol.* London: Butterworth, 317–34. (The tectonics of the ridge systems are discussed.)

ISACKS, B. L., SYKES, L. R., and OLIVER, J. 1969: Focal mechanism of deep and shallow earthquakes in the Tonga–Kermadec region and the tectonics of Island Arcs. *Bull. Geol. Soc. Amer.* **80,** 1443–70. (An account of the most active island arc-trench system.)

MENARD, H. W. 1964: *The marine geology of the Pacific*. New York: McGraw-Hill. (An account of many aspects of the Pacific ocean structure and morphology, which refers to the sea mounts and guyots.)

MENARD, H. W. and LADD, H. S. 1963: Oceanic islands, seamounts, guyots and atolls. In

M. N. Hill (editor), *The sea*. Vol. **III.** New York: Wiley, pp. 365–87. (A description of the positive small features of the sea floor.)

MOORE, T. C. and HEATH, G. R. 1967: Abyssal hills in the central equatorial Pacific: detailed structure of the sea floor and sub-bottom reflectors. *Mar. Geol.* **5,** 161–80. (Describes abyssal hills in an area where they are plentiful.)

MORGAN, W. J. 1968: Rises, trenches, great faults, and crustal blocks. *J. Geophys. Res.* **73,** 1959–82. (Draws together ideas concerning a number of different features of the deep sea morphology.)

STODDART, D. R. 1969: Ecology and morphology of recent coral reefs. *Biol. Rev.* **44,** 433–98. (Describes the biology and morphology of coral atolls and includes an extensive bibliography.)

SYKES, L. R. 1966: The seismicity and deep structure of island arcs. *J. Geophys. Res.* **71,** 2981–3006. (Describes the significant features of island arcs.)

5 Ocean sediments

The systematic and scientific study of deep ocean sediments dates from the *Challenger* cruise of 1872–76. The results of this work were published mainly by Murray (1891 and 1912), during the later years of the nineteenth century. He was able to provide a reasonably accurate map of the distribution of the major types of deep sea sediment. The sediment of the shallower water is also of interest and important from many points of view, both scientific and economic.

Some brief mention has already been made of the type of resource supplied by the sediments of the shallow water. The highest sediments are those of the foreshore and backshore zone of a beach. These sediments are extremely valuable in that they provide the best natural defence against erosion for a coastline where they are plentiful and thick. They also provide the necessary facilities for many seaside activities, such as swimming and surfing. The search for suitable beach fill material from rather further offshore indicates the value of the natural beach sand, in that its artificial replenishment is often thought to be the best form of coast defence. Sand and gravel deposits offshore are also used for construction purposes where inland sources of supply are not enough. The shelves also supply smaller quantities of more valuable materials, including sedimentary gold and diamond deposits, where there are drowned beaches offshore. Sedimentary phosphates from the edge of the continental shelf are of great potential value as a component of fertilizers. From the deep oceans the manganese nodules may eventually prove of value if they can be collected economically. These are, however, hardly sediments, in that they form by crystallization from sea water where sedimentation is slow enough not to bury them. The main value of the deep sea deposits is from the scientific data that can be obtained from their study.

The study of deep sea cores has gradually extended back in time as longer and longer cores have been extracted with steadily improving coring techniques. The recent work in this field of the deep-drilling ship *Glomar Challenger* has enabled cores to be drilled that pass right through the whole sedimentary sequence to the basement rock beneath. The cores so obtained have provided valuable confirmation of the global tectonics along several lines of evidence. The type of evidence that can be obtained from a study of cores is related to the nature of the sediment, whether it is organic or inorganic. If it is organic, the nature of the fauna it contains is of great interest in providing evidence for the conditions under which the sediment accumulated. In using this type of evidence it is necessary to take several aspects of the environment into account. For example fauna may change for several reasons. There may be evolutionary changes; sudden extinctions have sometimes been related to changes of polarity allowing harmful cosmic radiation access to the atmosphere at low level, and hence killing off some species. The relative abundance of species will differ with variations both in temperature and salinity in the upper layers of the sea, with variations of circulation, and also with variations in latitude. The latter two changes could be associated with movements of plates, and would have a much slower tempo than most climatic changes, at least those that are recorded during the Pliocene and Pleistocene periods, during which ice-masses have waxed and waned repeatedly. The variations in circulation and latitude will cause variations in fertility, which in turn reflect changes in species and their abundance. Not all cores are suitable for the analysis of changes in fauna and sediment type. It is thus necessary to distinguish between the sedimentary environments within the ocean basins. These environments are related to the two major types of deep sea sedimentation, and while the study of the slow particle by particle sedimentation provides valuable evidence of changing surface conditions in terms of climatic variation, the study of the turbidity current sediments provides valuable evidence on the processes by which turbidity currents operate and the part they play in the formation of deep sea morphology and sediment dissemination.

In order to obtain the optimum amount of information from deep sea cores it is essential that an accurate time scale be fixed. A number of methods have been used for this purpose, including Carbon 14 analysis of organic matter, and the use of magnetic reversals. The latter method has proved of great value, providing an accurate time scale covering about 5 million years, very much longer than the Carbon 14 method, which can only go back reliably for about 50,000 years. One method can take over where the other ends. Uranium isotopes can also be used for intermediate time ranges.

The study of deep sea cores has revealed a complex sequence of temperature changes covering the glacial period. In suitable cores the record is much more complete than any that can be found on land, and this is one of the great advantages of the study of deep sea cores for this purpose. The best sites are those on slight rises far from the continental margin, because it is in these situations that the slow particle by particle sedimentation is most likely to be undisturbed by turbidity current activity. On the other hand, the sites must not be too exposed and elevated, such as the Blake Plateau for example, because such sites are frequently influenced by the deep sea currents in such a way that sediment cannot collect in the exposed environment.

The most useful evidence on climatic variations obtained from deep sea cores is based on an analysis of the microfossils in the cores. These fossils are the remains of marine plankton, coming from the diatoms, radiolaria, ostracodes, and foraminifera, of which the diatoms are the smallest, being members of the phytoplankton. Diatoms and radiolaria are siliceous, while some of the most important foraminifera from the point of view of micro-palaeontology consist of calcareous tests. Most foraminifera are benthic in habit, but the 1 per cent of known species which are planktonic are so numerous that they make up 99 per cent of the fossils found in the sediments of the ocean floor. The Chalk, for example, consists largely of these minute remains. Many species of foraminifera exist and these have distinct climatic zonal extent, so that they provide valuable evidence of past climatic change when their variation in deep sea cores is studied. One of the important members of this group of foraminifera is *Globorotalia menardii*. This animal apparently spends part of its mature life below a depth of 500 m, a fact that must be taken into account in a consideration of the distribution of the species in deep cores. This particular species is representative of middle latitudes and tropical waters, but it can be used for climatic change assessment because of its habit of changing coiling direction with climate. All the Pleistocene species coiled to the left, while under warmer conditions right coiling predominated. *Globorotalia truncatulinoides* also vary their coiling direction under different conditions. A study of the pattern revealed three zones in the north Atlantic, a left-coiling zone stretching northwestwards from the west coast of Africa (10–30°N) to Nova Scotia; Newfoundland separated two right-coiling zones to the northeast and southwest. A northern species that also changes coiling direction is *Globigerina pachyderma*. The changes are closely determined by temperature changes in the north Atlantic and could be due to fluctuations of sea level in part, or to changes in the strength of the north Atlantic circulation, so that care is necessary in interpreting the results of micro-palaeontological studies. They have, nevertheless, provided a consistent picture of the climatic oscillations associated with the glacial period, showing that the post-glacial amelioration of climate took place about 11,000 B.P.

The original classification of the deep sea deposits was that of Murray, based on his *Challenger* samples. He was able to show that calcareous foraminifera, especially *Globigerina*, were characteristic of large areas of deep sea sediment, particularly in the Atlantic ocean, so that his classification includes this type of ooze. Basically he used as his major criterion for classification whether the sediment was predominantly organic or inorganic. The most important inorganic sediment that he recognized he called 'Red Clay', from the colour of the early specimens of this type dredged from the central Atlantic. Later specimens did not always have this particular colour and darker description has been given of some samples. The name 'Red Clay', however, was retained as a general term for this very fine inorganic deposit. More recently the term 'Red Clay' has been used less frequently, and modern classifications do not all incorporate it. The term 'lutite' is now sometimes used instead of clay, as it is not connected with particular problems of clay mineralogy. Nevertheless the term 'Red Clay' is a useful one to describe the wide spread, mainly inorganic deposits that are so common, particularly in the deeper waters of the Pacific. The organic deposits of the deep ocean are basically divided between the calcareous and the siliceous types of organisms that make up the most important proportion of the deposit. The two major calcareous types are the *Globigerina* ooze and Pteropod ooze, the former being much more widespread, while the latter is restricted to the shallow water of a north–south strip along the crest of the mid-Atlantic Ridge. The two main siliceous oozes are the diatom ooze, forming a belt around the southern ocean, and the radiolarian ooze that forms an east–west belt in the equatorial Pacific. The general pattern of organic oozes can be explained in terms of the zones of oceanic fertility where planktonic creatures develop in sufficient numbers to influence the nature of the organic bottom sediments and to dominate over inorganic materials. The distribution of fertility in the ocean is in turn dependent on the circulation of the ocean. The deep sediment pattern, therefore, provides a useful link between marine geomorphology and geology, on the one hand, and physical and biological oceanography, on the other. The changing pattern of ocean sediment distribution, explored in the analysis of deep sea cores, adds a time dimension to the pattern, thus enhancing its value, and linking it with the knowledge of the growth of the ocean basins.

One of the most important discoveries concerning ocean sediment derived from the study of long cores, particularly those obtained by the *Glomar Challenger*, is that the ocean sediments on the whole are very thin. They are also geologically young. Few sediments older than the Mesozoic period (which started about 230 million years ago) have been located anywhere in the oceans. These findings confirm the youth of the ocean floors and support the ideas of sea floor spreading and plate tectonics. The oldest parts of the oceans appear to be in the northwest Pacific and the western Arctic Ocean, while wide strips of the central Atlantic, Indian Ocean, and eastern Arctic and northeast Pacific Oceans are very recent in age from the geological point of view. Thus the water in the ocean basins is probably older, at least in part, than the floors of the oceans on which it rests.

The oceans are a vast receptacle into which much of the waste product of subaerial erosion eventually finds its way, via the rivers, glaciers, and wind. Innumerable remains of oceanic organisms also accumulate in them, as well as material brought into the water more directly from the interior of the earth via the volcanoes that stud large areas of the ocean floor, particularly in the Pacific Ocean. The sediments also include a small amount of extra-terrestrial material in the form of meteoric dust.

The classification and character of the materials provides useful information concerning the nature of the ocean and the processes at work within it. In the depths of the oceans some of the most valuable data concerning climatic changes during the glacial period have been found. Sampling of sediment in the sea, and particularly in the deep ocean, is a matter that requires optimum efficiency, as it is a difficult and expensive operation. Sampling must be such that enough samples are obtained to provide a reliable estimate of important variables. Kelley and McManus (1969) have studied 450 samples from the continental shelf off Washington State to test the reliablility of sampling plans for 9 textural variables. The problem is made more difficult in that variability cannot be assessed visually and local effects may be of major importance. The 9 variables used were 1) per cent sand, 2) per cent silt, 3) per cent clay, 4) sand–mud ratio, 5) median particle size in ϕ units, 6) mean particle size in ϕ units, 7) standard deviation as a measure of sorting, 8) skewness, and 9) kurtosis. The Wentworth limits were used to differentiate sand, silt, and clay. The 450 stations were located on a series of east–west traverses. Duplicate samples were obtained from each site by grab sampler, then the uppermost 1 cm was twice subsampled. Only 130 samples had the complete set of values calculated. Analysis of variance was used and all the F values for among stations to within stations mean squares were significant at the 95 per cent level. For the 9 variables the F ratios were:

Per cent sand	27·65	Skewness	9·93
Per cent silt	22·50	ϕ mean size	7·86
ϕ median size	17·74	Standard deviation	5·85
Sand–mud ratio	15·54	Inman Kurtosis	4·70
		Per cent clay	1·71

The use of pilot studies to estimate the variability by means of replicated samples to provide variance estimates is recommended, while stratified sampling with depth is often valuable.

1 Classification of oceanic sediment

The deposits in the oceans can be subdivided into two main categories—shallow water sediments and deep water ones. The character of some deep sea sediments, however, suggests that terrigenous sediments find their way into the deep oceans very far from land. The classification is based mainly on the origin and character of the sediments, and not on the position in which they are found.

Thus major divisions are made between terrigenous material derived from the land, volcanic material, and organic material, which includes skeletons of minute oceanic organisms. By far the greater area of the ocean floor is covered by deep sea pelagic sediments, which are easier to classify than the much more variable sediments of the nearshore zone. The pelagic sediments may be broadly subdivided into inorganic and organic deposits as follows:

Table 5.1 Classification of pelagic sediments

I Inorganic deposits	these contain less than 30 per cent organic material, and are often called Red Clay.
II Organic deposits	the common term for the material containing more than 30 per cent organic matter is 'ooze'. These deposits are subdivided according to their main chemical basis and by the predominant organism.
A Calcareous oozes	these contain more than 30 per cent calcium carbonate in the form of minute skeletons of different planktonic animals and plants.
i *Globigerina* ooze	formed of the tests of pelagic foraminifera.
ii Pteropod ooze	contains shells of pelagic molluscs.
iii Coccolith ooze	contains many coccoliths, which are the protective structure of minute Coccolithophoridae.
B Siliceous oozes	these contain a large amount of silica, derived from the siliceous skeletons of small planktonic creatures.
i Diatom ooze	containing remains of planktonic plants in the form of frustules.
ii Radiolaria ooze	containing the skeletons of radiolaria, planktonic animals.

The inorganic deposits forming the bulk of the shallow water sediments can best be classified according to their size. Revelle has suggested that the terrigenous deposits should be classified by size (see Table 5.2). Silt ranges between 62 and 4 microns, while clay particles are less than 4 microns diameter. Coarse clay lies in the range from 2 to 4 microns, medium clay from 1 to 2 microns, and the finest clays go down to 0·12 microns. In order to avoid mineralogical implications the clay grade sediments are often referred to as 'lutites'.

Table 5.2 Classification of sediment by size

Sand: more than 80 per cent coarser than 62 microns		
	Very coarse sand	2000–1000 microns
	Coarse sand	1000–500 microns
	Medium sand	500–250 microns
	Fine sand	250–125 microns
	Very fine sand	125–62 microns
Silty sand:	between 50 and 80 per cent coarser than 62 microns	
Sandy silt:	more than 50 per cent coarser than 5 microns and more than 20 per cent coarser than 62 microns	
Silty mud:	more than 50 per cent coarser than 5 microns and less than 20 per cent coarser than 62 microns	
Clayey mud:	less than 50 per cent coarser than 5 microns.	

At the other end of the scale the very coarse deposits are defined as boulders above 256 mm, cobbles from 64 to 256 mm, pebbles from 4 to 64 mm, and granules from 2 to 4 mm. This is according to the Wentworth classification of particle size, which can readily be converted to ϕ units, where the ϕ scale is logarithmic to base 2 with the sign reversed. Table 5.3 gives the settling velocity of different sized particles in distilled water at 20°C. The figures indicate that the finest particles can be carried very far from land before they eventually come to rest on the sea floor.

Table 5.3 Settling velocities

		m/day
Fine sand	250–125 microns	1040
Very fine sand	125–62 microns	301
Silt	31·2 microns	75·2
Silt	15·6 microns	18·8
Silt	7·8 microns	4·7
Silt	3·9 microns	1·2
Clay	1·95 microns	0·3
Clay	0·98 microns	0·074
Clay	0·49 microns	0·018
Clay	0·25 microns	0·004
Clay	0·12 microns	0·001

Turekian (1968) has proposed a classification of oceanic sediments in which the components of the sediment are considered in terms of their source and method of transport.

1 Pelagic biogenic, remains of organisms: calcareous and siliceous tests and organic and phosphatic material.

2 Non-biogenic: sedimentary components not originating from life processes in the ocean.
   a Pelagic detrital: particles originating near the surface of the sea and deposited by settling.
   b Bottom-transported detrital: material carried mainly by turbidity currents and bottom currents, containing shallow water and land-derived material.
   c Indigenous material: derived within the ocean basin by a variety of processes such as volcanic material and its reaction with sea water, migration, and reconstitution of materials in sediments, weathering of volcanic material exposed above sea level.

2 Character of pelagic sediment

Pelagic sediments are those deposited in the deep ocean. They cannot be defined by depth, as shallow water and hemipelagic material can reach down to the ocean's deepest basins. Pelagic sediments are mainly clays, oozes of calcareous or siliceous type, and some authigenic minerals. The material includes aeolian dust, volcanic ash, suspended river sediment, meteoric dust, skeletons of pelagic organisms, and metallic oxide precipitates. Different elements precipitate at different speeds and each material accumulates at a different tempo, although they have in common a very slow rate of accumulation, varying from 1 to 10 mm/1000 years. This allows plenty of time for solution and oxidation, and benthic animals digest and stir up the material so that borings and faecal pellets are common.

The actual volume of material is large despite the slow rate of accumulation. Menard (1965) points out that to form a layer of steel on the ocean floor at the rate of 1 mm/1000 years would need an annual deposit of $3 \cdot 6 \times 10^8$ tons, a value much in excess of world production. Deep sea sedimentation is not continuous and the bottom of the sea is now known to be less static than was formerly thought. The nature of the submarine relief is important. Sediment on abyssal hills is usually stable and slumping is uncommon. The rate of sedimentation affects the stability. Discontinuity of sedimentation, noted in the Pacific, may be related to variations in the benthic fauna.

2.1 Organic oozes

2.1a Calcareous ooze

Globigerina ooze is one of the most widespread deep sea deposits. It is formed by the accumulation of the calcareous skeletons of foraminifera, mainly the planktonic form, *Globigerina*, which consists of rounded calcareous tests. The average calcium carbonate content of five samples of this ooze from the Pacific was 82 per cent, varying from 75 to 89 per cent. Compared with many other deep sea deposits this ooze is badly sorted, having a sorting coefficient of 6·4, while the median diameter of the samples was 6·4 microns. Quite a large proportion of the material was in the sand size grade, over 100 microns.

The lower limit for the definition of *Globigerina* ooze is 30 per cent foraminiferal tests, although some authorities suggest 30 per cent calcium carbonate of which more than half must be foraminifera. Some of the calcium carbonate in the deposit may be precipitated chemically. In some areas the calcium carbonate may be as low as 30 per cent, but in others it is over 90 per cent, with the mean about 65 per cent. The concentration of calcium carbonate tends to be greater in the shallower water. A higher proportion of calcium carbonate is dissolved in the water as the small tests sink to deeper levels. Calcareous ooze is rarely found in water deeper than 5000 m. The deposit is often nearly white in colour. The number of species in the sediment decreases from the tropics towards higher latitudes. In the Atlantic the amount of lime in the deposit also decreases in the same direction (Sverdrup, Johnson, and Fleming, 1946).

The other calcareous oozes contain remains of pteropods and other organisms. In the pteropod ooze the actual percentage of pteropods seldom exceeds 30 per cent, but they are much more conspicuous because of their greater size. Nevertheless the lime content of the ooze remains high, varying between 50 and 90 per cent, the average being 80 per cent. The depth at which this ooze is found is limited to between 1500 and 3000 m, with the lime content highest around 2000 m. The coccolith ooze is rare, although coccoliths occur in most calcareous oozes. They rarely, however, form the bulk of the deposit.

The source and variability of the percentage of calcium carbonate is an important aspect of this group of deposits. It is found that the percentage of lime in the bottom deposits reflects that in the surface water, according to Trask (1955). The amount of lime in the surface waters increases with salinity and temperature, which helps to account for the higher lime content in the tropical regions. Lime is built into the organisms in the zones where upwelling waters bring nutrients to the surface where photosynthesis allows the organisms to convert the lime into parts of their skeletons. An extra thickness of bottom sediment may accumulate in those areas where warm and cold currents meet. Mixing causes the destruction of many individual creatures, whose remains sink down to the bottom. The calcium carbonate content also depends on the dilution with terrigenous matter, which tends to increase towards the coast, while volcanic material will have a similar effect.

2.1b Calcium carbonate

A very important aspect of the lime content is solution. The process takes place as the tests sink to the bottom and when they reach it, the latter being the more important according to Kuenen (1950). Solution is probably more effective at depth, because of the lower temperature and salinity, and possibly also the greater pressure. Solution cannot take place in stagnant conditions, as there is no means whereby the saturated lowest layer can be moved to allow more solution to take place. On the other hand, it is likely to take place effectively in those areas where the water has come from high latitudes, and in which the carbon dioxide content is higher as a result.

The north-moving bottom water in the south Atlantic Ocean gradually loses its effectiveness to dissolve calcium carbonate, as its degree of saturation increases as it moves north. It is low in the southern part of the ocean. Most of the calcium carbonate in the sea comes from solution of limestone on land, most of which itself was originally deposited in the sea. The original source must have been the early igneous rocks, some of which contain a proportion of lime. There is, therefore, a cycle of calcium carbonate formation, deposition, and solution which goes on in the ocean, and which involves the ancient marine sediments, now in the form of limestone and chalk outcrops on the land.

Calcium carbonate is one of the main biogenic materials. It is found in the form of foraminiferal tests, coccoliths and pteropod tests. The first are formed of calcite. The second form the finest grained material, less than 30 microns, while the third are molluscs with aragonite shells, of relatively large dimensions of 1–2 mm. They occur in depths of less than 3500 m. The percentage of calcium carbonate reflects the relative accumulation of this material and clay. Generally calcium carbonate is highest on the upstanding parts of the sea floor. In some areas, however, where the surface waters are very productive of calcium carbonate, even deep areas have high values, such as along the path of the Gulf Stream at about 40°N and below the upwelling off the coast of southwest Africa. In the Argentine

Plate 14 Detail of diagram of the continental shelf off western Europe (*opposite*). (Depth indicated in fathoms.) The abrupt continental slope is shown at the seaward margin of the very flat continental shelf off south-west England. The sea mounts in the Biscay abyssal plain can be seen, as can the Cap Breton submarine canyon. (*Copyright © 1968 Bruce C. Heezen and Marie Tharp; published by the Geological Society of America.*)

Cap Breton Can.
2650
2640
2115
2700
2790
680

basin, on the other hand, high fertility is not accompanied by calcium carbonate at hardly any depth.

The ocean ridges stand out as areas high in calcium carbonate. The amount decreases to depths of 4500 m, virtually ceasing in deeper water. The depth of rapid decrease of calcium carbonate is called the compensation depth. Below 500 m sea water is undersaturated so solution should occur. Coccoliths, that can only live in the uppermost 100 m (owing to their requirement of light for photosynthesis), occur, however, in much deeper water than 500 m, and foraminifera grow most of their shells below 1000 m. Organic covering on the shells may help to preserve them from solution as they fall to the bottom. Solution when the shells reach the sea floor may be assisted by the production of carbon dioxide by benthic life, which decreases the saturation level, thus enabling solution to proceed more rapidly. The removal of the organic coating by bottom fauna will also help solution. Experiments in the sea suggests that solution is more rapid below 3800 m.

The compensation depth at 4500 m has been considered by Hudson (1967), who suggests that it must be due to an increase in the rate of calcite dissolution, and not to a boundary of equilibrium solubility. He shows that aragonite becomes undersaturated and actually dissolves at shallower depths than calcite. The compensation depth is generally rather deeper in the Atlantic than the Pacific. The main form in which aragonite is found is in pteropod shells, and these appear to dissolve around 3500 m depth. Temperature exerts an effect on the solubility of calcium carbonate, and solution appears to be a phenomenon of deep, cold water. It is important to distinguish between theoretical solubility and actual dissolution. The bulk of the ocean is undersaturated with calcite and all forms of calcium carbonate, thus the compensation level is more likely to be related to the rate of solution, rather than solubility. There may be a balance between the increasing undersaturation with depth and the decrease of temperature with depth, which should result in slower solution. Thus there may be a retardation of solution in deep cold water, even though the saturation is decreasing with depth.

Chave (1967) has discussed modern carbonate sedimentation, and has shown that the generally accepted views may need modification. It is generally held that carbonate sediments are rare beyond 30°N and S, and that in the tropics sea water is saturated with calcium carbonate, leading to the more rapid calcification in the tropics so that there is more calcium carbonate production than in higher latitudes on the sea bottom. Pure carbonate sediments are now found off Florida, the Bahama Banks, northwest Australia, and the Persian Gulf. If, however, carbonate sediments are defined as those with more than 50 per cent calcium carbonate their distribution is much more widespread. Coral banks occur off the Norwegian coast in 50–300 m depth up to 69°N, consisting of corals, echinoderms, molluscs, brachiopods, and bryozoans. Carbonate material is accumulating widely around western Europe, North America, and southern Australia in depths ranging from the shore to 200 m. There are calcareous sands in the Falkland Islands, southwest Chile, and Japan.

Carbonate sediments are likely to accumulate everywhere in those areas where they are not diluted to below 50 per cent by deposits from rivers. For example off Norway the fjords trap the land-derived sediment, and in Australia the desert prevents land material from reaching the coast in large amounts. High latitude waters seem to be saturated with calcium carbonate as well as those of lower latitudes. The rate of carbonate growth with variation in latitude cannot yet be assessed accurately, but such evidence as there is suggests that latitudinal differences may not be great. The abundance of bryozoans in non-tropical sediments may be significant.

2.1c Siliceous oozes

The siliceous oozes become more important in areas where the organic production of calcium carbonate falls off, and where dissolution of this material exceeds its production. One of the factors favouring the increase in number of diatoms (which forms the base of diatom ooze) is a reduction in salinity, for example off large rivers.

Diatoms are phytoplanktonic siliceous algae. They make up a relatively large proportion of the total deposit at higher latitudes and in positions where the sinking currents can dissolve the calcium carbonate, and thus reduce the proportion of this mineral. The tiny plants grow near the surface where light allows photosynthesis, particularly in high latitudes. The area where the Antarctic circumpolar water rises to the surface to the south of the Antarctic convergence, is a zone where much nutrient material is carried up from the depths to be used by diatoms in photosynthesis in the upper layers of the sea. The remains of the organisms then sink to accumulate as diatom ooze.

This deposit is much better sorted than calcareous ooze. Its sorting coefficient is 1·85 compared with 6·4. The bulk of the material falls into the silt grade. There may be quite a high calcium carbonate content in the deposit at times, varying between 2 and 40 per cent. It may also contain from 3 to 25 per cent mineral grains, probably transported by floating ice in the higher latitudes. The line separating the diatom and *Globigerina* ooze follows closely the zone of the Antarctic convergence, where the Antarctic intermediate water sinks down and flows to the north. Diatoms thrive to the south of this line and foraminifera to the north. Some of the smaller and thinner frustules in the diatom ooze appear to dissolve before they reach the bottom, as only the larger ones are conspicuous in the deposit.

The other type of siliceous deposit is found in low latitudes. It is formed where the solution of lime on the bottom is particularly active. Silica is secreted by a group of protozoa called 'radiolarians'. These are planktonic animals with highly complex and ornate skeletons, and although they are greatly outnumbered by the foraminifera in the surface fauna, special conditions of sedimentation allow them to become predominant in the deep sea deposit over a fairly restricted area. The deposit is termed radiolarian ooze when the tests exceed 20 per cent and they can reach a maximum of 60–70 per cent. This type of ooze contains less than 20 per cent calcium carbonate; the average is 4 per cent and sometimes only amounts to a trace. Inorganic material less than 50 microns in size is often fairly conspicuous in this type of ooze, and reaches a maximum of 67 per cent and a mean of nearly 40 per cent. In colour the deposit often resembles the inorganic Red Clay.

Radiolarians have been found in large numbers in some fossil sediments, now outcropping on land—for example the Devonian and Carboniferous rocks of Australia, where deposits are up to 3000 m thick. It seems, however, that these deposits were shallow water sediments, forming in a geosynclinal environment close to the shore. Silica is also deposited in the sea in shallow water sponges, which grow on the bottom.

Siliceous oozes are more common in the Pacific Ocean, partly because the amount of calcium carbonate declines more rapidly with depth in the Pacific. The deep and bottom water of this ocean contains four times as much silicate as the Atlantic and more phosphate and nitrate (Revelle *et al.*, 1955).

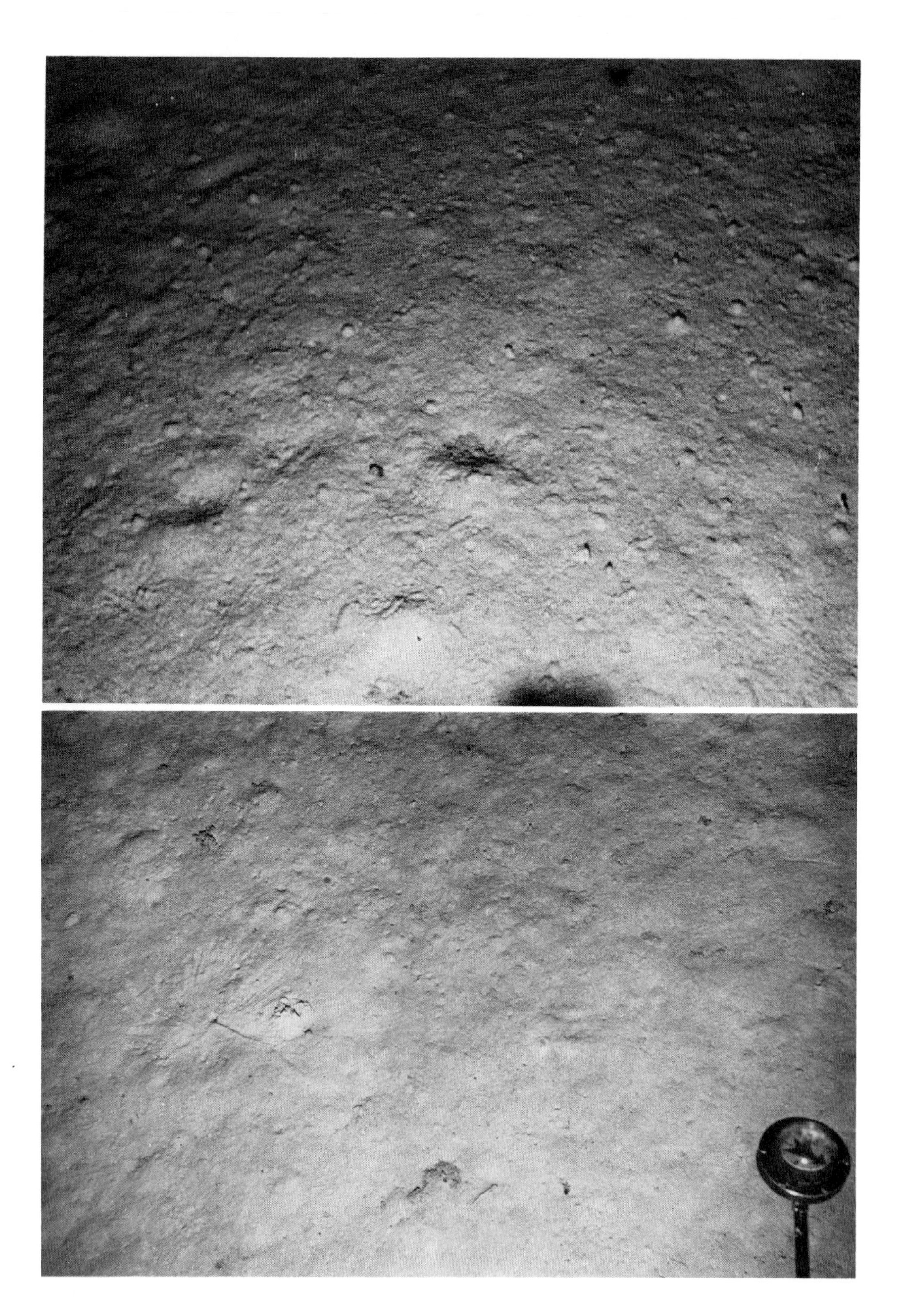

There is a complex relationship between the organic processes, whereby lime or silica is secreted to form the organic portion of deep sea deposits, and those processes that dissolve the remains as they sink to the floor. The relative proportion of the deposit made of organic calcium carbonate or silica also depends on how much it is diluted by inorganic mineral remains. Chemical precipitation of calcium carbonate adds another factor.

The organic silica deposited is highly soluble, so that sediment is only formed in restricted areas. Large diatom tests can accumulate in high latitudes in the summer in zones of high productivity. Rapid growth takes place in the equatorial Pacific because the nutrient supply is good, and the availability of silica in upwelling waters allows radiolaria and diatoms to flourish.

Some organic material can accumulate more effectively in areas where the water is deficient in oxygen; where it is abundant the remains of bones and other organic parts are utilized by other forms of life before they can accumulate as deposits. Thus phosphates and apatite are relatively rare in deep sea sediments. Some local basins become stagnant and can accumulate organic material, such as the Black Sea and Cariaco Trench off Venezuela, where the concentration of organic material reaches 3–5 per cent, the normal concentration being less than 1 per cent.

In summary, sediments rich in organic matter will tend to occur where 1) there is an abundant supply of organic matter, 2) a fairly rapid rate of accumulation of inorganic matter, particularly if it is fine-grained, 3) little oxygen in contact with the sediment. The third factor reaches its most extreme form in seas and basins in which the water is stagnant in the lower levels, as in the Black Sea. On the other hand, deposits will be low in organic content 1) if there is a small supply of organic material, 2) if there is a relatively low rate of accumulation of non-decomposable material, and 3) if there is an abundant supply of oxygen. The reason for the second condition is at first sight not obvious, as it might be supposed that if the deposition of inorganic material were fast, it would dilute the organic portion of the sediment. It does, however, protect the organic material from solution, especially when the inorganic material is fine-grained so that water cannot readily penetrate through it. It also prevents the destruction of the organic matter by benthic animals. Richards and Redfield (1954) have shown that there is an inverse relationship between the organic content of the sediments and the oxygen content in the overlying water in the northwest of the Gulf of Mexico.

Lime is the most important single element of organic deposits of the deep sea environment. The average lime content of all the pelagic sediments has been calculated to be about 37 per cent, while it forms only 25 per cent of all terrigenous deposits. The calcium carbonate is unevenly deposited in the major oceans. There is an average of 41 per cent lime in the Atlantic sediments, while those of the Pacific have only 19 per cent. According to Revelle *et al.* (1955) present pelagic sediments are high in calcium carbonate.

2.2 Inorganic clay deposits

A large proportion of the ocean floor, including about half the Pacific, is covered by inorganic material, often called 'Red Clay'. It forms the deposit where organic remains are

Plate 15 Iberian abyssal plain (*above opposite*). Bottom of ooze typical of area of slow deposition extensively reworked. In the right foreground are the radial feeding tracks of a buried mud feeder. The tubes of several tubeworms can also be seen. Cliff south of Freen Deep, northeast Atlantic (*below opposite*). The sediment slope is marked by the radial feeding tracks of the worm buried in the hole on the left. (*A. S. Laughton, National Institute of Oceanographic Science, Wormley.*) *Above,* area of picture 1½ × 1½ m, depth 5340 m; *below,* area 3 × 4 m, depth 5009 m. (*See Laughton, 1968.*)

dissolved before they reach the bottom. Red Clay may be of extremely fine grade, 83 per cent being in the clay grade and only 17 per cent in the silt grade. One sample had a median diameter of 1·1 microns; the upper quartile was as low as 3·5 microns; it was fairly well sorted, with a sorting coefficient of 2·86. This value is an extreme example of the deposit rather than the average. Some Red Clays contain no calcium carbonate; others have up to 29 per cent, the average being about 7–10 per cent. Siliceous remains are also few, amounting to a maximum of 5 per cent, a minimum of zero, and an average of 0·7–2·4 per cent. The average of 126 samples showed that the maximum proportion of particles under 50 microns was 100 per cent and the minimum 31 per cent, with an average of 86·5 per cent.

Red Clay, now sometimes referred to as brown clay, is brownish red or chocolate brown and is very fine-grained. Its colour is due to the presence of ferric hydroxide or oxide, with a little manganese oxide. It contains many clay minerals, as well as decomposed volcanic dust, which at one time was thought to constitute the greater part of the deposit. The extremely finely divided clay minerals resemble those found on land, and have probably come from the land. Their fine size allows them to drift for great distances, owing to their very low settling velocities.

Inorganic clay accumulates very slowly, so that it takes a long time for objects dropped on the sea bed to be buried. Thus sometimes remains of creatures, such as shark's teeth, ear bones of whales, or inorganic matter, such as pebbles dropped from floating icebergs, lumps of pumice, and other debris are found amongst the Red Clay. Some of these objects can be dated, for example the shark's teeth were Tertiary in date, showing that the clay has accumulated very slowly.

Inorganic clay is often found in the deeper parts of the oceans where the calcareous oozes do not collect. The mean depth for 126 samples of clay was found to be 5407 m, while the minimum depth was 4060 m, and the maximum depth was 8282 m. The mean depth at which the other deposits occur are: *Globigerina* ooze, 3612 m; pteropod ooze, 2072 m; diatom ooze, 3900 m; and radiolarian ooze, 5292 m. The siliceous oozes occur deeper than the calcareous ones because silica is less soluble than calcium carbonate, while Red Clay is hardly soluble at all.

Clay minerals are derived from pelagic, bottom-transported, and indigenous sources. In the Atlantic Ocean most clay is derived from the continent around. The ratio of kaolinite to chlorite depends on the climate, as kaolinite is characteristic of tropical climates, which destroys chlorite. In temperate and arctic climates chlorite is preserved or formed. In the Pacific, indigenous components are important because of extensive volcanic activity, but throughout the Pacific detrital sedimentation also occurs, indicated by the presence of illite, or detrital mica, which is derived from the land. The most important clay mineral associated with indigenous volcanic activity is montmorillonite, which is closely linked with zeolite phillipsite, a mineral supplied by submarine volcanoes in association with sea water. This latter mineral may make up more than half the clay accumulation where other clay sedimentation is slow. It is virtually restricted to a belt extending northwest off South America towards Japan in the Pacific, but is absent in the other oceans.

Quartz occurs in the Atlantic mainly in high latitudes as a result of glacial activity and off the Sahara, where it is blown out to sea by the wind. In the Pacific the greatest concentration is off Washington and Oregon, and off New Zealand.

A relationship between the clay minerals in the bottom deposits and those collected at sea as wind-borne dust in the eastern Atlantic is discussed by Chester (1971). The average amount of dust collected at sea in the eastern Atlantic was 57 micrograms/m³ in the northeast trade winds. These winds brought Saharan dust over the sea. The amount of dust in the variable wind zone of the inter-tropical convergence region averaged 0·23 mg/m³; the southeast trade winds, bringing dust from the Namib and Kalahari deserts of South Africa, averaged 1·14 mg/m³; and the lowest values were found in the southerly winds of the south Atlantic, which had an average of only 0·07 mg/m³. The character of the dust reflected the local soil characteristics and was also related to the minerals in the deep sea sediments. Chlorite had only a small variation with latitude, but there was a tendency for greater abundance in polar regions. In the latitudes sampled between 40°N and 50°S chlorite made up less than 10 per cent. Kaolinite, on the other hand, was higher in amount in the low latitudes, as it is preferentially formed under tropical conditions. Amounts exceeded 50 per cent in the equatorial parts of the Atlantic. Montmorillonite is probably mainly derived from volcanic material *in situ* in the deep sea deposits. Its pattern in the aeolian dust is complicated, and it averaged about 20 per cent and showed little relation to the deep sea sediment pattern. Illite forms under a wide range of conditions and its percentage presence was the result of dilution by other types of clay minerals. The pattern of aeolian dusts confirmed that deep sea clays are mainly detrital in origin, apart from montmorillonite, which is mainly authigenic in origin. Quartz was important in the northeast trade-wind samples, but not in the southeast trade-wind area. Again the detrital origin of quartz in the deep sea sediments was confirmed. Aeolian dust probably makes a considerable contribution to the inorganic material in the upper layers of water in the area of the northeast trade winds, but only a very minor contribution in the south Atlantic.

2.3 Manganese nodules

Manganese, nickel, and cobalt are being concentrated in the deep sea floor of the Pacific. The nodules are not restricted to areas of Red Clay deposits, being also found amongst some organic deposits, but they are most often found in inorganic clay. Underwater cameras (Shipek, 1960) have revealed the nodules in the eastern Pacific, where they are forming on a deposit of Red Clay at a depth below 4500 m.

Manganese nodules have aroused considerable interest lately as having possible commercial interest. The nodules range from thin coatings on mineral grains or coccoliths of less than 30 microns to nodules of 850 kg mass. Manganese and iron accrete in concentric layers. Nodule distribution is widespread and especially abundant in the central Pacific. It is found in other areas where low accumulation of calcium carbonate and clay occur, such as the Blake Plateau, which is current scoured, and the Drake Passage. The manganese and iron are probably derived from detrital land sources, and by reaction of sea water with submarine volcanic activity. Rates of growth vary from several centimetres in the present century to 5 mm/million years, the former applying to the continental margin of California and the latter in the deep ocean. In some areas nodules may be covered by sediment; elsewhere they lie on the sediment. They require a nucleus to form around (Turekian, 1968).

The nodules were discovered by the *Challenger* expedition 100 years ago. They include copper, cobalt, and nickel and occur as slabs and crusts as well as large and small nodules.

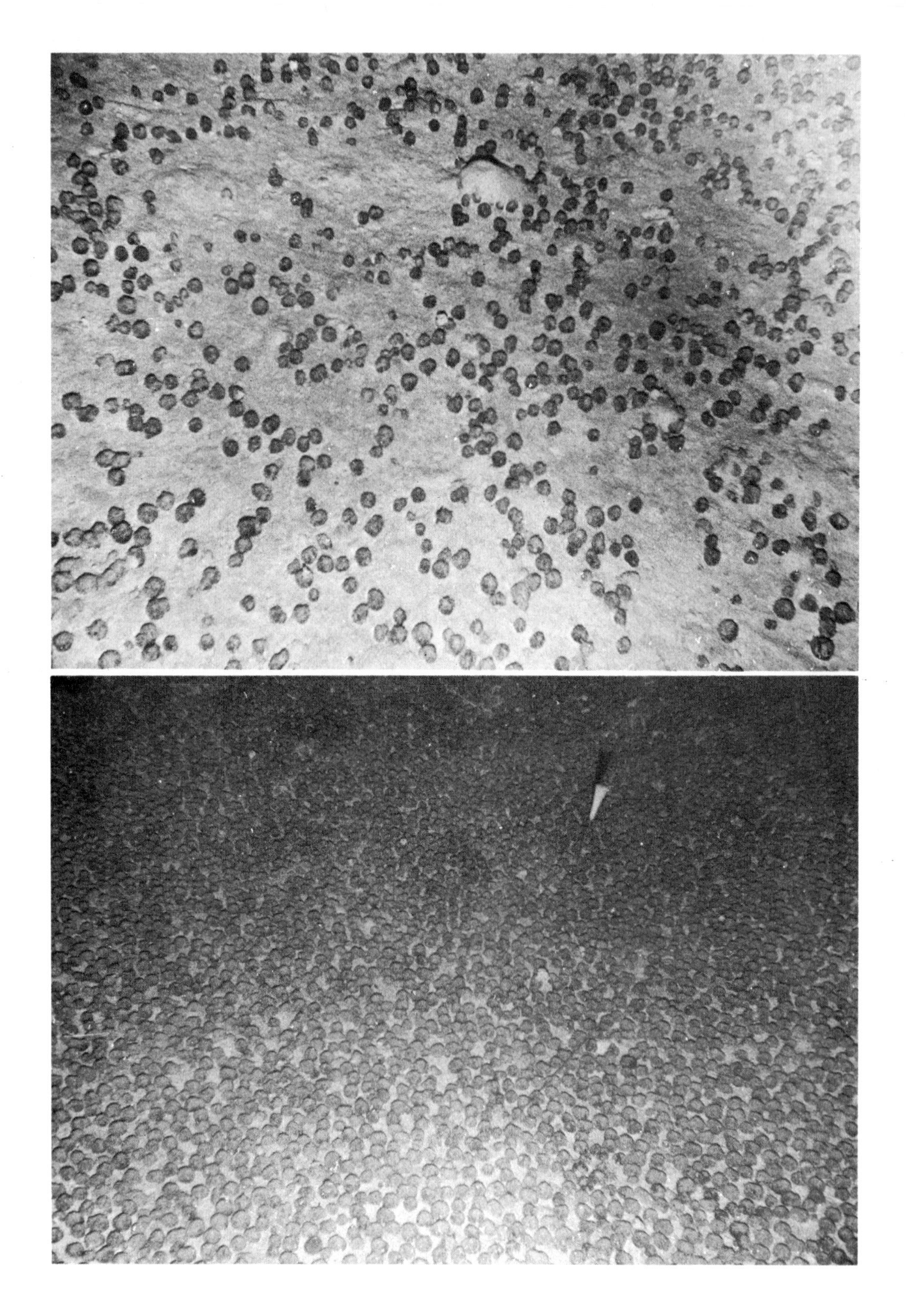

In the larger slabs the coatings are thicker on top than below. On sea mounts crusts can attain 11 cm thickness. Nodules in general accrete more slowly than sediment so the problem arises as to the reason for their non-burial in some areas. In others they are buried, as they are not shown on photographs but are obtained in dredge hauls. In the Pacific they occur on the surface in an east–west belt in 18–20°N and in the southwest Pacific. The latter is an area of very slow sedimentation, while the north Pacific belt lies between the zone of inorganic sedimentation to the north and organic sediments to the south. Nodules were probably more widespread in the late Tertiary when sedimentation rates were generally lower than at present. It is possible that burrowing benthic animals push the nodules to the surface. Table 5.4 shows the relative abundance of different

Table 5.4 Manganese nodule mineral content

Source	Cu	Ni	Mn	Co (grams)
World commercial reserves	10^{14}	$1{\cdot}4 \times 10^{13}$	10^{15}	10^{12}
Pacific nodules	5×10^{15}	$9{\cdot}0 \times 10^{15}$	2×10^{17}	3×10^{15}
Sea water	10^{16}	10^{14}	10^{16}	10^{8}
Top km continental crust	3×10^{19}	3×10^{19}	4×10^{20}	10^{19}
World production 1949–50	$2{\cdot}5 \times 10^{12}$	$1{\cdot}2 \times 10^{11}$	$1{\cdot}4 \times 10^{12}$	7×10^{9}
Remaining world economic reserves	40 years	120 years	700 years	140 years

Nodules in various parts of the Pacific basin in grams weight

Area	Nodules	Mn	Co	Cu	Ni
At surface	10^{17}	10^{17}	10^{15}	10^{15}	10^{15}
Southwest	2·6	0·7	1·0	0·5	1·6
North equatorial belt	2·5	0·5	1·0	1·1	3·5
Buried in top m of sediment					
Southeast area of low Mn	2·8	0·6	0·3	2·5	2·8
Southeast area of high Mn	0·5	0·2	0·02	0·2	0·5
North	1·8	0·2	0·2	0·8	0·5
Southwest	0·8	0·2	0·3	0·2	0·5
Total	10·0	2·4	2·8	5·3	9·4

minerals in the sea and on land. Sea floor deposits contain 50–3000 times the commercial reserves, but the cost of collection is still high. The uneven distribution of land resources could encourage some countries such as Britain and Japan, to exploit undersea sources.

3 Deep sea processes and turbidity currents

Turbidity currents and processes of auto-suspension were mentioned in the discussion on submarine canyons in chapter 3. In this section further information on the evidence for these currents in the sediments of the deep ocean will be mentioned. Deep water sands can be transported for long distances by turbidity currents and form part of turbidites,

Plate 16 Manganese nodules on the foothills of the Carlsberg Ridge in the Indian Ocean (*opposite*). The nodules are 5 cm in diameter and rest mostly on the sediment surface. (*A. S. Laughton, Institute of Oceanographic Science, Wormley.*) *Above,* area of picture 3 × 4 m, depth 4230 m; *below,* area 3 × 4 m, depth 4220 m. (*See Laughton, 1967.*)

although not all deep water sands are necessarily of this origin. Sand moves along the axes of submarine valleys and is found in cores, interbedded with muds. Sand is particularly common where canyons extend close inshore. Most deep water sands are fine or very fine, but a few contain gravel and shells. Many of the sands are graded, and ripple marks also occur. Horizontal laminations, current ripple lamination, convolute laminations, and very fine horizontal laminations all occur within one section 17 cm thick on a small abyssal plain in the southeast Adriatic.

Allen (1970) has summarized the knowledge of turbidity currents and turbidites. The currents are a species of gravity current where a stream of denser fluid displaces a lighter fluid as it moves downslope. Turbidity currents are often initiated by slumping. They have a broad gently rising head; rapidly circulating vortices develop at the rear. The motion of a gravity current is unsteady so that it is not well understood theoretically, while experimental studies suffer from scale difficulties.

In nature, slumps are most likely to start turbidity currents on moderate to steep slopes, where the sediment is poorly consolidated and with excess pore pressures. The upper continental slope provides the best conditions for slumping. Slopes vary between about 1 : 5 and 1 : 50. Cable breaks suggest current velocities of 5–10 m/sec and coarse sand beds can be laid down under these conditions. With sea level at its present height submarine slumps (occurring as often as annually) are common off large rivers. Slumps must have been more frequent on the continental slope during low glacial sea levels.

The assemblage of forms associated with turbidity currents is similar to that found in a semi-arid mountain front area, with its steep slopes, fan deposits, and basin sediments. In the sea, graded beds are more common and coarser in the fans than in the abyssal plain deposits. The similarity of form suggests that underwater rivers are important in the formation of deep sea deposits as they are on land. The turbidity current must lose its momentum when it spreads out on the fan, as the bulk of the sediment is deposited there. The turbidity current must be turbulent and it must carry a well-mixed heterogeneous load. When this is deposited it gradually decreases in thickness away from the point where deposition begins; in each position the sediment becomes finer upwards, giving graded bedding, according to theory. The theory, however, ignores the turbulence and the self-generating capacity of some such currents. A turbidity current may be able to erode even after it has started to deposit, because the deposition takes place behind the head, which may still be eroding. Erosion in mud is normally indicated by flute marks and their amplitude indicates the rate of erosion by the current and its duration. The size of the flute increases with increasing thickness and excess density of the current. It should, therefore, decrease along the path of the current.

A knowledge of modern turbidity current action has helped in interpreting ancient turbidites. These are often a few thousand metres thick, consisting of many coarse and fine-grained beds that alternate vertically on a small or medium scale. The sole of the sand bed is sharp and often has flutes, load casts, or organic markings. Coarse beds, which have a very wide lateral extent with minimal changes in thickness or character, pass up gradually into the fine. Fauna often consist of displaced elements of both deep and shallow species. In the geological record, turbidites occur in the Palaeozoic of Wales and southern Scotland, the flysch of the Alps, and the Ordovician and Devonian of the Appalachians.

A complete turbidite sequence consists of five units. Unit A, the coarsest and lowest, often consists of graded sand, with lumps of mud or small pebbles at the bottom. This unit is massive, without internal layering. Unit B has parallel laminations and is finer-grained. It can attain 50 cm in thickness, and usually consists of fine or very fine sand. Unit C is often thinner than B, and grades up into coarse silt. It is a unit of cross laminations 2–3 cm thick, which increase in angle upwards from 5–15 to 45–55°. It is the unit in which bed load was no longer being carried and the current was waning in power. The laminations change from erosional forms to depositional ones throughout the layer. Ripples also change in form upwards in this layer, from long and low to short, high and symmetrical. The fourth unit D consists of clean laminae of sand, lenses of silt, muddy silt, and clayey silt. The uppermost division, unit E, consists of clayey silts and silty clays, much of which is hemipelagic in type. As the five units are traced along the path of the turbidity current, the upper units increase in proportional thickness to the lower units. The fine deposits may be lacking where the coarse ones are thickest. Ancient turbidites have properties that appear to belong to modern turbidity current sedimentation. Their character is explained by theory, including the auto-suspension hypothesis.

Kuenen (1965) has described experiments in connection with turbidity currents. He used a circular flume in which high velocities could be generated in salt water; various types of sediment mixture were tested. Different rates of current deceleration gave different forms of graded bedding, including ripples, convolute bedding, and horizontal lamination. The latter form develops if the deposition is not too fast, and does not necessarily require current pulsations, and can occur in turbidity currents.

Kuenen also studied the settling of clay suspensions in sea water. With concentrations between 1 and 400 g/l the rate of settling was at first constant and then fell off. Clay that had settled overnight in the circular flume was eroded at a velocity of 30 cm/sec, but when the clay lay on scattered pebbles it was eroded with a velocity of 12 cm/sec and the amount in suspension increased. Below 10 cm/sec all clay settled, while above 48 cm/sec no deposition occurred. When the density exceeds 500 g/l no settling occurred at 4 cm/sec, indicating a 'solid' sliding over the bottom. A current able to transport sand in suspension will not allow clay to settle, hence most sandy turbidites contain little clay. Such clay as there is may be partly of secondary origin.

Once a turbidity current loses sediment it becomes lighter and can transport less sediment and therefore loses more, by positive feedback. A natural turbidity current flowing at 25 cm/sec should come to a halt in less than a week and travel about 100 km, but a large current could take longer and travel up to 500–1000 km if its tail were 12·5 m thick. The period of erosion in a turbidity current is short. During this time flutings can be cut, a process requiring a velocity of over 50 cm/sec. If the surface were rough, inducing local turbulence, a lower velocity would be required. The deposition of sand on the bottom requires a fairly strong current to carry the sand to the site. A problem is raised by the normal grading of sand, which decreases in size upwards; coarser grains are carried more slowly and should, therefore, arrive later than finer ones. Deep sea sand thus cannot be adequately accounted for by normal deep sea currents on the bottom. Settling out from turbidity currents is much more likely in view of the graded bedding.

In order to travel the very great distances required to reach the central parts of abyssal plains, turbidity currents must be able to flow with the minimum loss of energy. Bagnold

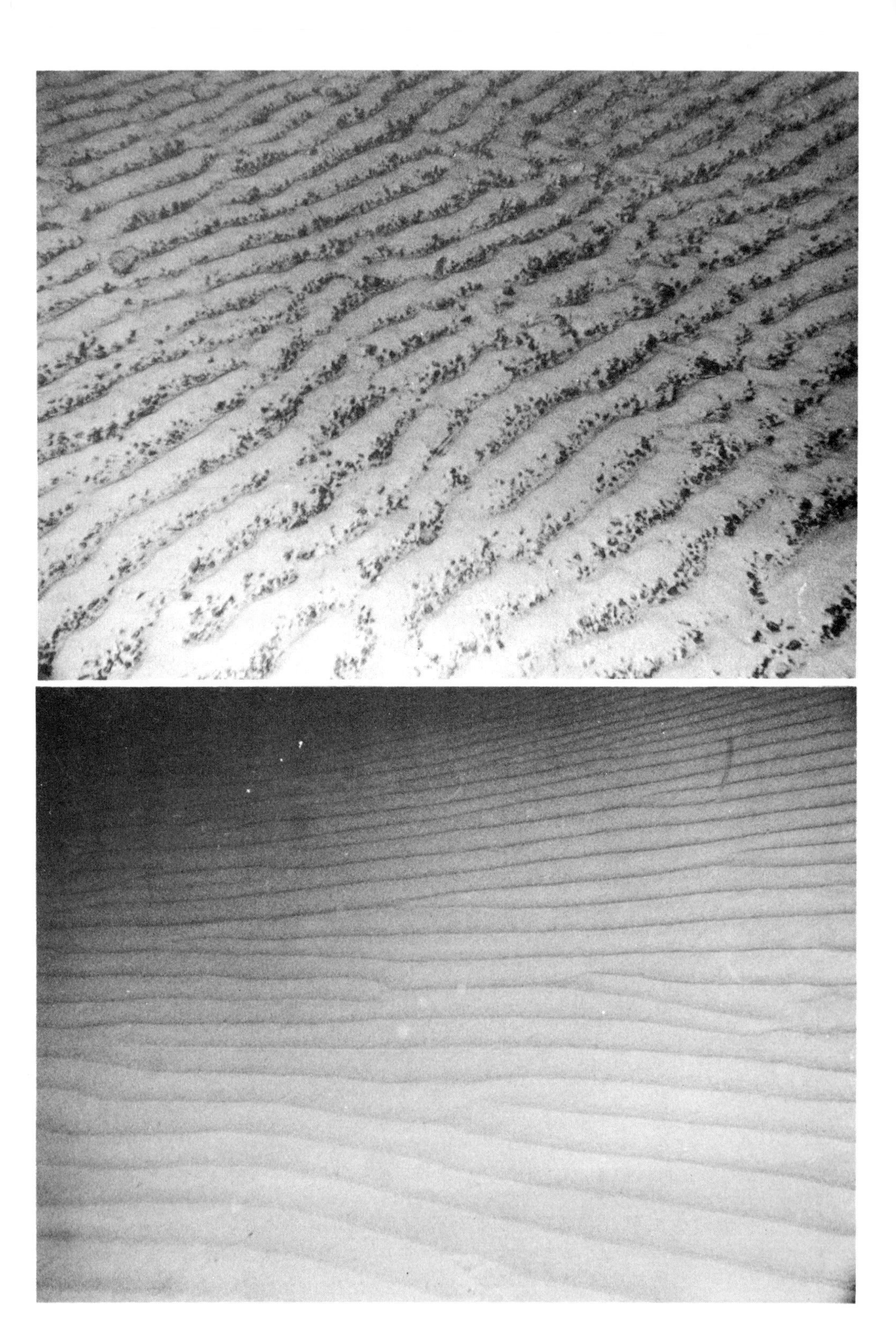

(1963) has shown how a sediment body can be auto-suspended. If tan β (gravity gradient) exceeds $w/\bar{U}_s$ (w is fall velocity of sediment grain, $\bar{U}_s$ is the mean sediment velocity), the suspended sediment has excess density and pushes the fluid in front of it rather than vice versa. The transport rate, i_s, when w, the fall velocity, is less than $\bar{U}_s \tan \beta$, should be limited only by the availability of sediment or the suppression of fluid turbulence at very high concentration in turbulent flow. The sediment should become uniformly distributed in the fluid, and the largest grains should be in the uppermost layers of fluid. A turbulent current should be self-maintaining if

$$\frac{1}{\bar{u}^2}\frac{\rho_s - \rho_w}{\rho_w} gNh \log^2 \frac{13{\cdot}2h}{D_1} \sin \beta - \frac{w}{\bar{u}} < 0{\cdot}03$$

where h is the thickness of the current, N is the concentration, ū is the current velocity, D_1 is the effective size of boundary roughness and could be estimated as 1 cm for a rippled bed, ρ_s and ρ_w are the densities of sediment and water respectively. The value of 0·03 should be increased to 0·045 or 0·06 to allow for the upper flow boundary. If N exceeds 1 per cent by volume, the value should be increased still further to allow for the effect on flow resistance. The initial finite velocity ū sin β must exceed w. The value of ū could increase as the result of instability, allowing more sediment to be picked up. If h were 500 m and N 0·01 and sin β 0·01 for fine sand, with the right-hand value at 0·045, ū could attain 25 m/sec. A considerable initial pressure would be necessary to produce these high velocities. The current velocity might need to be 20–30 times w in the initial disturbance.

Hand and Emery (1964) have studied the evidence for turbidity currents and their deposits in the northern San Diego trough off California. In this area turbidity currents originate in the submarine canyons and cross the subsea aprons at their foot, continuing down the fan-valleys, filling one basin with sediment until they overflow into the next. Levees, aggraded 25 m above the fans, have formed on the sides of the channels, which have smooth concave profiles. The turbidite deposits consist of gravel, sand, and coarse silt in beds with cross bedding, grading, and shallow water fossils. These beds are interlayered with hemipelagic green clayey silt, accumulating at a rate of 11–21 cm/1000 years. The action of turbidity currents was more frequent during the Pleistocene periods of low sea levels. There is considerable variation over short distances in sediment size in any one core, but each bed is fairly uniform, being the product of a single episode. The coarse material is dropped close to the canyon mouths. Only fine sand and some medium sand is carried far beyond. The deposits of turbidity currents form prominent features, including aprons, levees, and channels, with terracing on some aprons. Canyons provide the major route of sediment transport.

Anikouchine and Ling (1967) have examined the turbidite accumulating in some Indo-Pacific trenches. The Java and Mindanao Trenches are receiving mainly turbidites, while the Mariana Trench is receiving mainly pelagic sediments. The Java Trench cores show grading. The sorting is generally poor, owing to much lutite. The sediments contain clay minerals, angular quartz, mica, feldspar, and pyrite, which Kuenen (1954) considers occurs in turbidites. Wood occurs in all the trenches. It and the fish bones and sponge spicules are of neritic origin. Strong laminations occur in the Java and Mindanao

Plate 17 *Above* (*opposite*), ripple-marked sand (wavelength 15 cm) near centre of Carlsberg Ridge in the Indian Ocean (area of picture 3 × 4 m, depth 2410 m). The coarse material in troughs is probably manganese-coated fragments of altered basalt. *Below,* long-crested ripple marks in calcareous sand (wavelength 15 cm) in the same area (picture 3 × 4 m, depth 2450 m). (*A. S. Laughton, Institute of Oceanographic Science, Wormley.*) (*See Laughton, 1967.*)

sediments, ranging from a few mm to 20 cm. Cross bedding, which could be due to ripple marks, was found in the Java Trench cores. All three trenches had mud lumps from 1 mm to 3 cm in size; they were not formed *in situ*. The turbidites ranged in thickness from layers of a few mm to 18 cm. Most of the contacts were sharp. Erosion of the lower deposit is indicated by a lag feature, consisting of mud lumps and silt streaks. In the Java Trench, hemipelagic micaceous lutites are interbedded with graded and ungraded micaceous silts introduced by turbidity currents, which derive their sediment from Sumatra.

Slumping has been observed to occur in several areas, and direct evidence for this is indicated by cable breaks. Slumps can also be inferred from indirect evidence, such as the disposal of sediment through submarine valleys. Slumps have been observed at the mouth of the Magdalena River in Colombia, one causing a deepening of 60 m, involving an area of 10^7 m^2 giving a volume of 3×10^8 m^3 or more. A tsunami caused deepening of 100 m in an area of more than 5 million m^2 on the continental slope west of Suva, Fiji. The volume displaced is estimated at $1{\cdot}5 \times 10^8$ m^3. The slump resulted in a turbidity current that broke a cable 90 km from the source of the current at a depth of 2000 m. The cable was carried 4 km across the level basin floor. Similar slumps occurred in Orkdals Fjord, Norway, and significant changes in level have been observed at the head of Californian submarine canyons.

Slumps generate turbidity currents, which build the fans, cut channels through them, and deposit well-sorted sediments on the abyssal plains. Dill (1969) has observed a slow-moving sand–organic mat producing plucking, gouging, and smoothing, thus eroding the canyon head where it occurred. Sand rivers can carry pebbles and cobbles. Slumping is most common off rivers and at the head of submarine canyons which are areas of rapid deposition, leading to instability. Stability depends on shear strength, which increases with depth of burial, and is dependent on sediment character. Slumps will occur with a sediment thickness of about 40 m, with slopes as low as 1·5 per cent. Slumps that can give rise to turbidity currents have volumes ranging between 10^4 and 7×10^{10} m^3. Speeds of up to 10 m/sec have been recorded from cable breaks. The largest slump known had a volume of 7×10^{10} m^3 and an average depth of 100 m. The slump was 350 km long by the time it had entered the submarine canyon leading to the Japan Trench (Menard, 1965). Slumps change into turbidity currents in the canyons down which they are funnelled. The size of the valley indicates the dimensions of the turbidity current, being commonly 100–200 m deep and several km wide. Turbidity currents spreading across flat abyssal plains are only a few tens of metres thick. Their density probably ranges between 0·004 g/cc and 0·1 g/cc.

Modern turbidites in the San Pedro basin contain sand layers from mere partings to layers more than 1 m in thickness, the mean being 1 cm. The sand has a mean size of 72 microns in one basin and 58 in another. Sands are rather coarser on the fans with values between 52 and 89 microns. Menard concluded that turbidity currents dominate sedimentation in the Pacific at present, in all areas to which they have access, because the rate of sedimentation by this process is so much greater than by pelagic sedimentation. The ratio is as much as 1 : 100 in some areas. Most of the Pacific, however, is not accessible to turbidity currents, so that abyssal hills are the typical landform, and pelagic sediments the most common covering.

The sedimentation in the Cascadia Channel and adjacent abyssal plain off the coast of Washington and Oregon is different from elsewhere in the greater abundance of organic matter in the channel (Griggs *et al.*, 1969). The abyssal plain has an area of 170,000 km^2 and it slopes southwards with depths from 2100 to 2930 m. The sediment is derived from the Columbia and Fraser Rivers, the former providing 12·2 $\times$ 10^6 m^3/year. The Cascadia Channel extends for 950 km across the abyssal plain and is fed via submarine canyons, which have built large fans. The channel extends on through the Blanco Fracture zone into the Tufts abyssal plain. The gradient of the channel across the Cascadia abyssal plain is 1 : 1000. The Cascadia abyssal plain has been receiving pelagic and hemipelagic clays at the rate of 8 to 10 cm/1000 years in the post-glacial period in areas where river sediments can also collect. The rate is 2–3 cm/1000 years where no river deposits reach. Rates in the Pleistocene were much greater, and in places were as much as 170 cm/1000 years. Benthic fauna are prolific in parts of the area; especially in the Cascadia Channel, where burrows are particularly common. The effect of currents is twofold: they provide nutrients for benthic life, and they also bury and thus kill the same life at intervals. Turbidity currents are now confined to the channel.

Winterer *et al.* (1968) show that sediment forming the Delgado fan is not displaced by the Pioneer Fracture zone. The sediments are 1·5 km thick in 3000 m of water at the foot of the continental slope and they mask all the relief on top of the second layer. Fan deposits are 0·5 km thick and they subdue, but do not bury the submarine relief. Abyssal hills and smooth areas continue from the distal edge of the fans and are covered by similar sediment. It is, therefore, inferred that the sediment is pelagic in type and not due to turbidity currents. The Delgado and Monterey deep sea fans overlie and partially merge into an older continental rise. The sediments in the area are probably derived from the north, including the Columbia River, having been transported by the south-flowing Californian ocean current. Sedimentation probably started in the late Tertiary and continued through the Quaternary. The fans are mainly Quaternary in date. The volume of the Delgado fan is about 10^5 km^3, and with a rate of supply of 50 cm/1000 years, it could accumulate in 1 million years.

The Puerto Rico Trench in the Atlantic has also been affected by turbidity currents on the evidence derived from cores. Conolly and Ewing (1967) have found Holocene, Pleistocene, Tertiary, and upper Cretaceous sediments overlying upper Cretaceous andesitic volcanics. Graded beds several mm to 7 m thick can be followed over 200 km, the grain size decreasing towards the high-level abyssal plain and into and along the abyssal plain of the lower trench floor. The deposits were laid down by turbidity currents starting near the Puerto Rico–Virgin Island shelf, and flowing through submarine canyons northwestwards to the high-level abyssal plain. Fine sand 20–50 cm thick was deposited 100 km away from and 10 m above the point where the sediment entered the main trench floor. The elevated abyssal plain slopes from 7500 m to the northwest into the lower main plain at 7920–8000 m. Abyssal plain sand is a heterogeneous mixture of calcareous and terrigenous detritus and deep sea clay, consisting of *Globigerina* and pteropod ooze. The largest single graded bed was 3 m thick and covered 10^4 km^2, giving a volume of 3 $\times$ 10^7 m^3. The sediment was removed from the southern wall of the trench. Erosion of the trench walls is thus still active. The thickness of sediment on the trench floor is 1–1·7 km and consists almost exclusively of turbidity current deposits, including some brown clay, which marks the final stage of current sedimentation.

Van der Lingen (1969) has criticized the too ready correlation of turbidity currents with flysch depositional processes. He considers that flysch deposits do not have the same characteristics as deep sea deposits, and that experimental data do not support this correlation. There are many exceptions and inconsistencies. Alternative theories of deep sea sedimentation, such as ocean bottom current theories for recent sediments in deep water, and tectonically controlled sedimentation under relatively shallow water provide promising alternative hypotheses for flysch deposits.

Turbidity currents are not the only process capable of moving sediment in the deep sea. Heezen and Hollister (1964) show that bottom photographs and cores reveal the influence of strong bottom currents. In areas of steep relief and high elevation, ripples, scour marks, and rock outcrops occur. These features have been observed below the Antarctic bottom current in the western south Atlantic, below the Gulf Stream in the Straits of Florida, and on the Blake Plateau, beneath the outflowing Mediterranean water, and in Drake Passage. Heezen and Hollister recognize three types of deep sea sands and silts: turbidites, accretionary, and residual deposits, the latter two being associated with bottom currents. Velocities necessary to move the sediments vary between 4 and 60 cm/sec, which agree with geostrophic current computations and with observations by direct deep sea current measurements. A variety of ripple marks have been photographed in the deep sea, including asymmetrical, symmetrical, longitudinal, lingoid, and lunate, with wave lengths between 10 cm and several metres, and amplitudes up to 20 cm or more. Current lineations are largely depositional, occurring in the lee of objects on the bottom, similar to crag and tail or sand shadows. Evidence of current activity is found on some continental slopes. Current ripples occur on the south slope of the Grand Banks at a depth of 1010 m. Tidal currents add to the speeds of oceanic currents, giving velocities of 20–25 cm/sec in the Antarctic bottom current, for example. Currents are particularly strong in passages between islands and continents. Evidence of currents is found in 10 per cent of over 1000 deep sea cores containing silt and sand. Features seen in *Globigerina* ooze of about 0·4 mm median diameter require a current speed of about 15 cm/sec. The current marks in mineral sands require higher velocities. Even in the greatest depths speeds higher than those required for the deposition of silt are common. Clean laminae of silt can be explained by deep sea current transport. Photographs have revealed the presence of ripples at depths of 1647 and 1320 m in Calcareous oozes, suggesting that bottom movement is not negligible at these depths. It may be caused by the action of long waves or tsunami or tidal currents.

McCoy and Herzen (1971) have recorded bottom currents between the Lesser Antilles and the mid-Atlantic Ridge by means of bottom photographs. At 19 stations the major bottom current direction was southwards, particularly in the Demarara abyssal plain, while on the flanks of the island arc and ridge no persistent current was observed. Velocities in general were about 1–3 cm/sec or less. Current-controlled sedimentation can produce giant ripples; it is characterized by lack of topographic control, in that sediment is not draped over the relief, nor is it ponded in basins. Changes in thickness are erratic.

A nepheloid layer was recognized by M. Ewing and Thorndike (1965) in the basal layers of the water. It consisted of lutite in suspension on the continental slope and rise. The lutite was in sufficient quantity to produce a downslope flow. This material may represent a major component of deep sea sediment. The nepheloid layer is cloudy and can be

Plate 18 Angular boulders of fractured lava (*above opposite*). Scour pits around the boulders show the effect of currents at a depth of 2000 m on the crest of the Carlsberg Ridge. Massive pillow lavas at 1770 m on the Carlsberg Ridge (*below opposite*). Surface markings due to extrusion and cooling are shown. Area of pictures 3 × 4 m. (*A. S. Laughton, Institute of Oceanographic Science, Wormley.*) (*See Laughton, 1967.*)

studied by optical means. The cloudiness could be due to stirring up of the bottom layer or by material more or less permanently in suspension. The evidence favours bottom stirring as the source because the darkening is greatest near the bottom, and the layer follows the bottom topography. The weight of the suspended sediment could affect the movement of the deep water masses.

The nepheloid layer provides information concerning both processes of sedimentation and bottom currents. Emery and Connary (1970) show that it extends from the bottom to the temperature minimum throughout the Pacific basin in the area north of 10°s. In general its intensity is less than that in the Argentine basin, with the exception of the Alaskan and Tuft abyssal plains, where it is comparable in intensity. The nepheloid layer consists of particulate matter and is recognized by its light-scattering character. The matter is transported to the ocean via the large rivers, and only a very small proportion is wind-borne. It is dispersed in the ocean by lateral mixing and is carried in the large anti-clockwise gyres. The marginal trenches have a strong nepheloid layer of trapped particles, which eventually settle to the trench floor. Little matter comes into the north Pacific Ocean from the south via the deep currents. It appears that the nepheloid layer is a steady-state phenomenon. Over much of the north Pacific Ocean its upper surface is at about 1000 m above the bottom. At this depth the nephelometer, which consists of an underwater camera that records light scattering, records a minimum value.

4 Near shore sediments

4.1 Type and distribution

The variety of nearshore sediments is great, and only a few comments can be made. Shelf sediments vary according to the climate and they also show the effects of recent changes in sea level. Hayes (1967) has attempted to relate the sediment type of the inner continental shelf to the coastal climate. His analysis of the data was based on nautical charts. The findings indicate the percentage of each type and are given in table 5.5.

Table 5.5 Continental shelf sediment types

Mud	Abundant off areas with high temperature and high rainfall, humid tropics 37 per cent
Sand	Abundant everywhere, increasing to a maximum in intermediate zones of moderate temperature and rainfall, and in all arid areas except very cold ones 47 per cent
Coral	Most common off areas with high temperatures 6 per cent
Gravel	Most common off areas with low temperatures—polar and subpolar 3 per cent
Rock	Generally more abundant in cold areas; correlates highly with inner shelf slope 3 per cent
Shell	Distribution not diagnostic of climate 4 per cent

The controls that influence the distribution include climate, which determines the type of weathering and the presence of large rivers to bring the products of weathering to the sea. Coarse deposits are associated with glaciation. Most of the inner shelf sediments at the present time are strongly influenced by Pleistocene events. Apart from climate the shelf sediments are influenced by tectonic activity. Hayes has shown that there is a logarithmic normal distribution of shelf widths and slopes, but only rock correlates strongly with shelf width and slope.

The shelves were divided into three types according to the character of the adjacent land: young mountain range coasts, plateau-shield coasts, and depositional coasts. Each type was associated with a specific distribution of shelf width, with modal values at 4·6–9·2 km, 18·5–37 km, and 74–148 km respectively. Percentages of sand and gravel show little relationship with shelf type; mud increases as the shelf width increases. The relict sediment from the Pleistocene greatly affects the present distribution, so that processes must be considered with this in mind. Relict sediments are probably more important, however, on the outer part of the shelf, as the inner part is more quickly covered by modern sediment.

Swift (1970) has discussed the way in which shelf sediment is becoming adjusted to the abnormal conditions following Pleistocene events in which much of the shelf was exposed during periods of glacially lowered sea level. Curray has suggested that the present belt of fine sediments offshore will gradually prograde seawards until the whole shelf becomes graded and the coarse relict sediments of the outer shelf are covered. In the nearshore zone waves are the dominant force, and these are likely to develop a nearshore modern sand prism in the zone where the wave energy is high and where fine sediments cannot accumulate as a result. Sediment movement beyond this nearshore zone is thought to be small at present. Only cataclysmic events, with about a 100-year recurrence interval, can influence this area markedly. Shelf sedimentation in this zone is limited to extreme storm conditions apart from the effect of wave-drift currents. These currents would result in a net seaward drift of sediment, with a sediment source as the shore and a sink at the shelf break. Tides and other agents of shelf sediment movement should also be considered.

Off the east coast of the United States shelf sediment first becomes graded *in situ*, and the nearshore sediments may be derived in part by winnowing of the relict central shelf sediments. Fine sediments are trapped in the inshore lagoonal areas. In the second stage the inner lagoons become filled and fine sediments will reach the shelf area. Eventually sand will also be by-passed and will reach the outer shelf. Thus mud will first be deposited over the whole shelf and then finally a layer of sand will spread over the whole. At present this shelf consists of modern sediments inside the barrier island and on the outer shelf, with relict sediments on the central shelf. The inner modern sediments also include the barrier sand and the nearshore sand prism, while muds are accumulating on the outer shelf. The Atlantic shelf of the United States is starved of sediment compared with those of the Gulf coast, which is fed by the Mississippi delta, and the Pacific coast, which has a steep gradient and heavily loaded streams. On the Atlantic shelves, where sediment is trapped in the coastal lagoons, sediment sorting is largely autochthonous, while in the latter areas it is allochthonous.

An example of modern shelf sedimentary character in a low latitude area is given by McMaster and Lachance (1969) in their description of the shelf sediments off northwest Africa. Sand is the most common shelf sediment, including biogenic sand, quartz, and glauconite. The former type occurs between Ifni and Cape Blanc (21°N) and south of Cape Verde (15°N), while the sand on the wide Guinea shelf has a low carbonate value.

Glauconite occurs on the outer shelf and upper slope. Silt occurs on the mid-shelf, south of Cape Ghir (30°N), between Gambia and Geba Rivers, and along the narrow Sierra Leone–Liberian shelf. Where the shelf is narrow and sediment supply is abundant, modern sediment is found on the mid-shelf, and the outer shelf is now accumulating fine deposits. Where the shelf is wide, modern fine sediments are accumulating on the inner shelf. Carbonate sediments are collecting in stable areas with a deficient inorganic sediment supply. The Canary and Guinea coastal currents have not prevented shelf sedimentation of silt and clay on this coast. Sediments are low in carbonate (less than 50 per cent) where detrital sediment was deposited during low Pleistocene sea levels and where modern shelf sediments are accumulating.

The North Sea provides an example of shelf gravels derived in part from glacial action (Veenstra, 1969). This shallow sea is characterized by banks and channels. The gravels differ in different parts of the area. Those of the Hinder region near the Thames estuary represent reworked river deposits, while those near the Suffolk coast are fossil beach gravel. Further north there are erratics in the gravels. The Dogger Bank and Inner Well Bank Rough are similar, and the occurrence of till on the sea floor points to their glacial origin. The gravel was brought from the west by ice during the last glacial period. The Cleaver Bank to the east has Scandinavian material but is of the same age. The gravels of the Texel Rough have the same source as the Cleaver Bank, but it belongs to the Saale ice period. The coarser material is found on top of the banks in the central North Sea, with the finer gravels more plentiful in deeper water, due to wave sorting during low sea levels.

Shepard (1948, 1963a) has drawn attention to the deposits characteristic of the continental shelves off glaciated regions, where coarse sediment may be found right to the outer edge of the shelf. Thus many areas in higher latitudes show the influence of glacial deposition in the character of the nearshore terrigenous sediments. In few areas does the material grade in size outwards from shore. Finer material accumulates in hollows on the shelf, while at the edge the material may be coarser or rock may outcrop. Mud is a common constituent off large rivers. Deltaic sediments may accumulate in very great thicknesses, as deposition is accompanied by subsidence of the foundation.

According to Kuenen (1960), the shallow water of the Neritic sediment environment covers only about 10 per cent of the water-covered area. The sediments are, however, very important stratigraphically as most of the rocks now exposed on the earth's surface were once shallow water marine sediments. Most of the terrigenous material is deposited in this zone, where its final resting place is determined by the action of marine processes, such as wave action in shallow water, tidal, and other current activity in deeper water. Waves themselves cannot, however, affect the movement of sand or gravel below a rather limited depth, probably of the order of 10–12 m, according to many observations, although movement in greater depths has also been reported, especially where the bottom is hard. Tidal currents are probably effective to much greater depths, as the sand waves on the edge of the continental shelf off southwest England show. These have been found at a depth of 165 m. There is still too little known of the effect of long waves, seismic waves, and internal waves to assess their part in the movement of material in deeper water, while it has already been suggested that ocean currents are sometimes capable of moving material at great depths, such as the Gulf Stream where it flows over the Blake Plateau.

Another aspect of the neritic zone is the deposition of organic sediment. Much of the limestone in its various forms, which now outcrops on the earth's surface, was originally formed in fairly shallow water, where terrigenous material was not being deposited at the time. Some of the calcium carbonate that now forms calcareous rocks, such as the chalk, may have been deposited by chemical precipitation. Limestone is common in the stratigraphical record, but there are very few areas in the modern shallow water area where it is forming at the present time. One such area is in the Bay of Mont St Michel, where deposits containing 75 per cent calcium carbonate in depths of 10–30 m are accumulating. In other areas the beach material consists of remains of foraminifera with a calcium carbonate content of over 90 per cent, such as some of the beaches of Connemara, west Ireland. Fine calcareous oozes are being deposited in the shallow waters around some of the West Indian islands, but such environments are very restricted. The difference between the present and past distribution of calcareous deposits in shallow water could possibly be explained by the general lowering of sea temperatures at the present time.

Another possibility is that pelagic foraminifera secreting limey skeletons did not exist before the Cretaceous, so that the locking up of large amounts of calcium carbonate as *Globigerina* ooze on the deep sea floor has only taken place since the Cretaceous period, which would lower the concentration of calcium carbonate in sea water. It has been calculated that in another 100–150 million years the pelagic foraminifera will have used up all the available lime, which will then be locked up on the floors of the deep oceans. This would require a considerable modification of oceanic life. The precipitation of lime, in the form of aragonite mud, at present in the shallow water around the Bahamas is taking place because too much carbon dioxide is being assimilated from the water by plants. This leads to oversaturation of lime and the deposition of calcium carbonate. Where the water is saturated with lime, precipitation will be induced either by increase of temperature or decrease of pressure.

Emery (1960) has considered some of the relationships between marine sediments and their environment in the complex region adjacent to the coast of southern California. This area includes many different environments, most of them falling into the relatively shallow zone. They include marshes, gravel beaches, sills of basins, slopes, submarine canyons, basins and troughs, and the deep sea floor. The first three types of sediment need not be discussed, as they are marginal to the sea, occurring at such levels that they are exposed above water level for most of the time. On the mainland shelves the deposits are variable, which can partly be accounted for by the fact that some of the sediment is residual, and therefore, not in equilibrium with modern processes of sedimentation. Most of the shelf is, however, covered by modern sediments, amounting to 95 per cent off Santa Barbara. These deposits are mainly sand and silt, decreasing in grain size offshore. They contain little calcium carbonate, but some clay. The relict sediments were deposited when sea level was much lower than now. Sands are much coarser than the present shelf sediment. Organic deposits consist mainly of shell fragments. In this area shelf sediments form only a thin layer on an eroded shelf surface. There is thus a marked difference between this type of shelf and one of continuous deposition, as off Texas. The island shelves are broadly similar, although local differences are apparent, depending on the character of the island. Figure 5.1 illustrates the character of the sediment in the different zones.

Many of the bank top sediments contain a large proportion of calcium carbonate, which is coarser than the detrital sediment in the vicinity. Another common feature of the bank

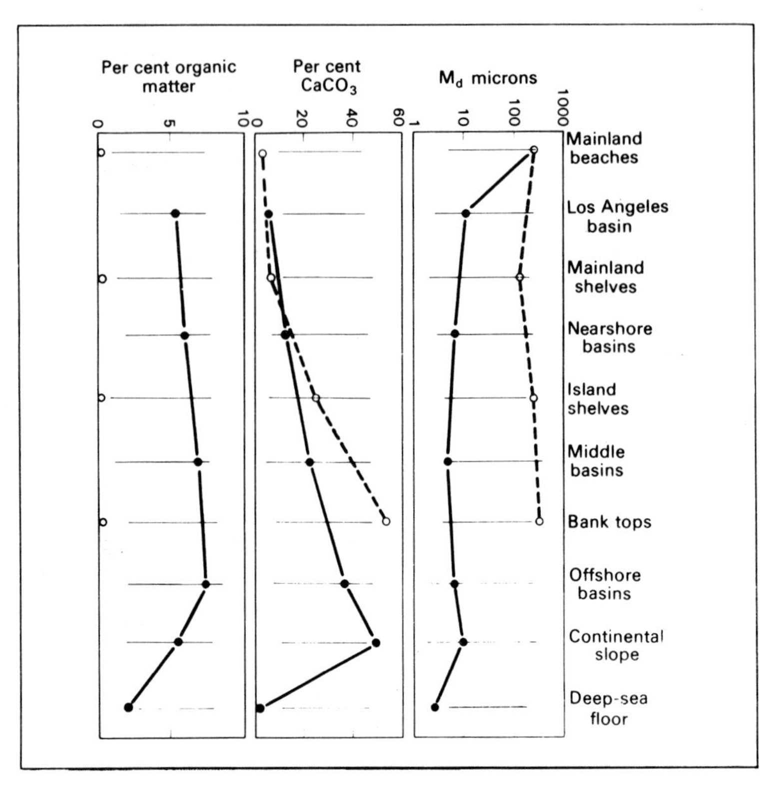

Figure 5.1 Diagrams to illustrate the organic matter, calcium carbonate, and normal detrital matter in different environments on the continental borderland off southern California. (*Modified after Emery, 1960.*)

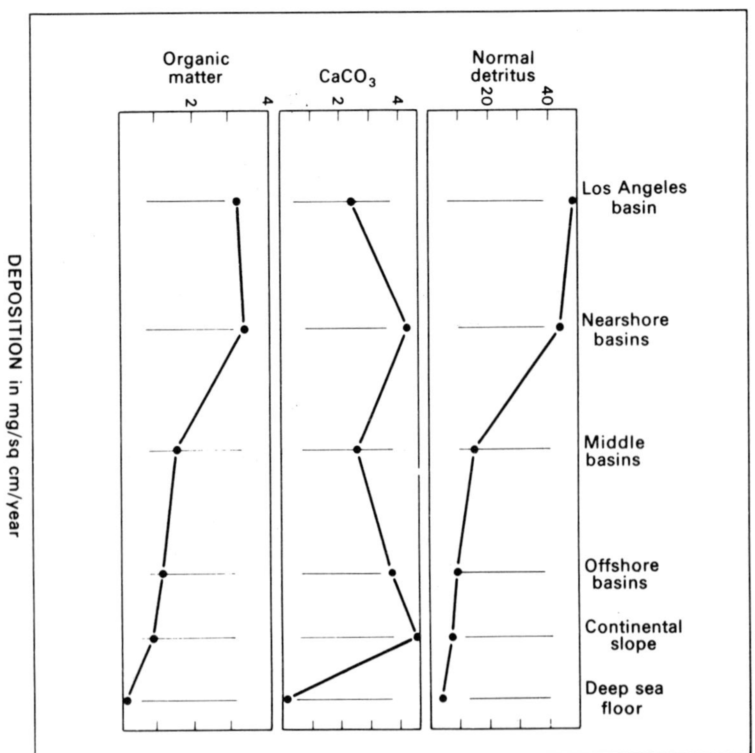

Figure 5.2 Diagrams to illustrate the volume of sedimentation of organic matter, calcium carbonate, and normal detrital matter on the continental borderland off southern California. (*Modified after Emery, 1960.*)

tops is glauconite, often found in areas where detrital sediment is lacking, as it would be on a bank top. It rarely comprises more than 20 per cent of the deposit, but it is coarse grained. The sills of the basins show a very wide variety of sediment. On the slopes, although sediment has been shown to rest on slopes as steep as 70°, the layer of sediment is relatively thin, and absent in places. In others, as one nearshore slope showed, a layer of sediment 6 m thick of median diameter of 22 microns lay on a slope between 9 and 18° steep. These sediments contain much water and can maintain high angles as a result of a certain rigidity which if disturbed leads to loss of strength and subsequent slumping. There seems to be more rock exposed on the continental slope than on the basin slopes. Where slope sediment was found it could be classified as *Globigerina* ooze, as it contained more than 30 per cent calcareous foraminifera. The sediments of the basins and troughs were found to be largely detrital and sedimentary calcium carbonate.

There is a general increase in the lime content in the offshore direction, which can be correlated with the grain size pattern. This decreases from the inner to the middle basins, then increases on the continental slope but decreases again on the deep sea floor. The lime content, consisting of coarser particles than the detrital material, causes an increase in grain size offshore in the outermost basins and on the continental slope. The deep sea deposits in the vicinity are Red Clay.

Various estimates for the rate of sedimentation have been put forward. Shepard (1973) and Revelle (1944) worked out a rate of 13 cm/1000 years. More recent estimates have used radio carbon dating, which can give more accurate results. There is great variability throughout this intricate area, the values varying by a factor of 34. The greatest rate is 123 mg/cm^2/year in the Santa Monica basin, and the smallest is 3·6 mg/cm^2/year on the deep sea floor. The greatest rate of deposition is found in the nearshore basins, which applies to both detrital and calcium carbonate deposition (figure 5.2). The rate of detrital sediment accumulation is at a maximum in the basins now on the shore, such as the Los Angeles basin, and is also fast in the nearshore basins, falling off very markedly in the middle and offshore basins to very low values on the continental slope and deep sea floor.

There is also a wide variation in the rate of deposition of calcium carbonate, which varies by a factor of 45 from 0·2 to 8·9 mg/cm^2/year. The deposition is very high in the nearshore basins; it reaches a peak on the continental slope, but is very low on the deep sea floor. The relative proportions of the different materials deposited is worked out as follows: detrital, 84·2 per cent; calcium carbonate, 13·2 per cent; and organic matter at depth, 2·6 per cent = 100 per cent. These figures show the very important part played by detrital or terrigenous sediments in an area not far from the continental coast, while in the open sea in this vicinity the proportions are very different.

4.2 The formation of petroleum in oceanic sediments

Ocean sediments are worthy of study from the point of view of information concerning the processes of petroleum formation, a product that is becoming of increasing importance and being exploited in increasing quantities in the shelf environment. Petroleum is usually thought to be organic in origin (Emery, 1960). One of the problems of the origin of petroleum is that once formed it may migrate away from its source. The problem of its origin is a chemical one, although its environment of formation is relevant to a study of ocean sediments. Nearly all petroleum is associated with marine fossils and therefore it probably originates in a marine environment, or at least forms from marine deposits. In order to account for the development of petroleum, it is necessary to have an environment in which organic matter can be sealed off from the action of scavengers, where in time it may be converted into petroleum, possibly by the effects of anaerobic bacteria. It is also necessary to have a reservoir rock into which it can migrate and from which it can be extracted.

The conditions associated with turbidity currents can provide this environment (Ericson *et al.*, 1961). The turbidity current can sweep up a large amount of finely divided organic matter in its path, which may be deposited in a relatively restricted area and so become concentrated. It is sealed from the activity of mud-boring creatures at the base of the deposit, as they can only penetrate 10–20 cm, while the deposit of one turbidity current may exceed this thickness. The next turbidity current might provide a layer of sand above the deposit in which the organic matter is isolated and thus form a reservoir for the developing petroleum to pass up into, and from which it can be extracted. The depth at which petroleum forms is greater than that to which corers normally extend, so that the process is difficult to study in operation. The oil deposits of the Los Angeles basin, which were formed in Miocene and Pliocene marine deposits, illustrate the conditions in which oil can form, and which appear to be very similar to those in the basins offshore along the Californian Borderland. The sediments, nearly 4000 m thick, consist of shale interbedded with sand brought by turbidity currents. The shales have 2·6 per cent organic matter and a median diameter of about 4 microns. The organic content was originally higher; it is estimated that it may have been about 6·1 per cent.

Tentative estimates have been made by Emery concerning the total production of organic matter that has any chance of becoming petroleum. He suggests that of the 100 per cent organic matter that is produced by phytoplankton, only 6·4 per cent reaches the bottom of the sea as sediment. At a depth of 3 m this is reduced to 4·2 per cent, while in the shales it is only 2·8 per cent. But the proportion that is likely to be exploited as petroleum is only 0·005 per cent of the original 100 per cent. This can be expressed in a different way, by stating that one barrel of petroleum requires the growth of 19,000 barrels of organic matter. If the hydrocarbons only are considered (and oil consists largely of these), the preservation is a little more efficient, as 1·3 per cent out of 100 per cent is recoverable as petroleum. This source of power, therefore, is a very inefficient method of using the original energy of the sun, which allowed the photosynthesis by which the organic matter was originally created, and which has in time converted it into petroleum.

5 Distribution of sedimentary types

5.1 General

The bulk of the ocean floor is covered by pelagic sediment, although the distribution of shallow water sand and silt by turbidity currents rather blurs the limits of the deep sea sediments. Nevertheless it is possible to delimit the major zones in which each type of deep sea deposit predominates; some of these are very restricted in their distribution. The general pattern is shown in figure 5.3.

Some of the most important facts concerning the distribution of sediments are the

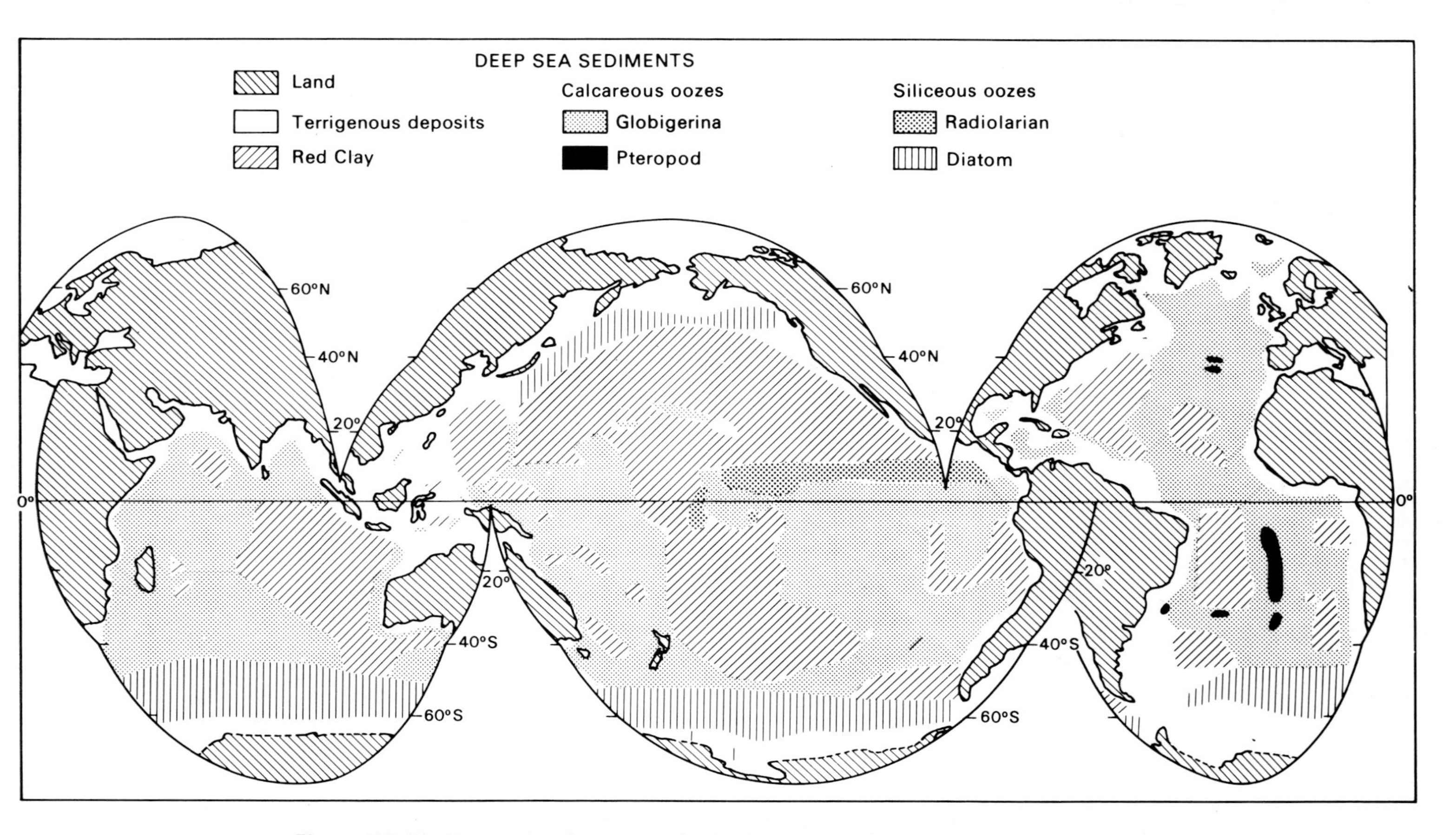

Figure 5.3 World map of sediment types in the deep oceans. (*After Sverdrup* et al., *1946.*)

following: 1) Pelagic sediments are found only in the large ocean basins. 2) The most common types of sediment are the brown or Red Clay and calcareous (*Globigerina*) ooze. The former covers about 38 per cent of the total area, while the calcareous oozes together cover 48 per cent, of which only a very small area is not *Globigerina* ooze. 3) Diatom ooze is restricted to a narrow, almost continuous band around Antarctica, and to a belt across the north Pacific. 4) Radiolarian ooze is found only in the equatorial Pacific. 5) Pteropod ooze is only found in a north–south belt on the mid-Atlantic Ridge around latitude 20°s. 6) The terrigenous belt is variable in width. It tends to be wider in higher latitudes. In lower latitudes calcareous deposits of benthic organisms are more important. In the north polar basin and around the coasts of the north Pacific and Atlantic, terrigenous sediments dominate. 7) Distribution is determined to some extent by depth, with the deepest parts covered by inorganic clay. 8) Although there is some merging of sediment types in the marginal areas, they are on the whole very distinct over wide areas, facilitating and enhancing the value of a study of the types and their distribution.

Distribution of the different types of deposits in the oceans is summarized in table 5.6, according to Sverdrup *et al.* (1946). Table 5.6 and the facts about the distribution given

Table 5.6 Sediment distribution in the major oceans

Sediment	Indian Ocean	Atlantic Ocean	Pacific Ocean
		percentage	
Calcareous oozes	54·3	67·5	36·2
Siliceous oozes	20·4	6·7	14·7
Red Clay	25·3	25·8	49·1

above, bring out several points of importance. Calcareous deposits are most important in the Indian and Atlantic Oceans, which may be accounted for partly by the greater depth of the Pacific, where brown or Red Clay is the most extensive deposit. On the other hand, it seems likely that the more vigorous circulation of the Atlantic Ocean leads to a more rapid replenishment of nutrients, on which foraminifera depend.

The importance of nutrient supply is seen in the pattern of distribution of some of the more restricted oozes. Diatom ooze, which is found only in areas where active upwelling takes place, illustrates this point. It is found in the zone where the Antarctic circumpolar water is formed from upwelling water masses, thus explaining its distribution in a belt around the Antarctic (except where terrigenous material dominates to the southeast of South America).

Another reason why the Pacific may be poorer in calcium carbonate than the Atlantic is the correlation that Trask (1955) found between low salinity and low carbonate values. Observations have shown that the north Pacific, north of 10°N, is very deficient in calcium carbonate, considerably more so than the south Pacific. The Atlantic contains much more than either (figure 5·4). One possible reason for the greater amount of calcium carbonate in the Atlantic might be that the rivers supply more of this material to the Atlantic than the Pacific. A further possible reason could be a net transport of calcium carbonate from the

Pacific to the Atlantic, which is consistent with the higher calcium carbonate content in the subsurface waters of the Pacific.

The reason for the lower calcium carbonate content of the north Pacific is more difficult to explain. It must mean that there is a net transport of calcium carbonate south across the equator, as most of the rivers enter the north Pacific. A possible factor is the greater solubility of calcium carbonate in the north Pacific, on account of the low oxygen content and therefore greater amount of carbon dioxide. Thus calcium carbonate in the water and on the bottom may be more readily soluble. The problem is still far from being finally solved.

The distribution of the most restricted oozes, the calcareous pteropod and the siliceous radiolarian, is mainly determined by the fact that these are species that love warmth and only flourish in restricted areas. These two oozes cover only 1 per cent and 2 per cent of the ocean floor respectively. The present distribution of the oozes has not remained the same permanently, as changing conditions have modified the pattern somewhat. Changes in the pattern of distribution and also in the species of organisms represented and the rate of sedimentation have been used to elucidate the changing climatic conditions and to provide a time scale for these.

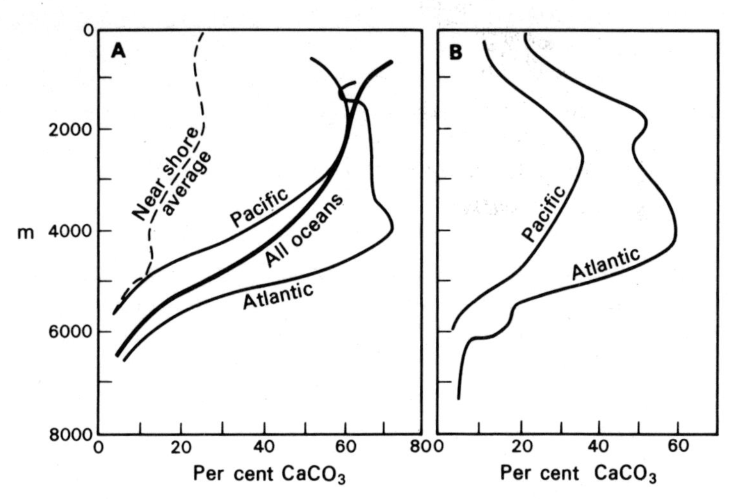

Figure 5.4 The propor- of calcium carbonate in sediments at varying depths. **A**: Average calcium carbonate content of the pelagic deposits. **B**: Average calcium carbonate for all types of sediment. (*After Sverdrup* et al., *1946*.)

5.2 Specific

5.2a Atlantic sedimentary types

In the Atlantic there are two distinct types of sediment. One is truly pelagic in character, but the other carries the terrigenous environment far into the deep ocean and, therefore, provides a link between the two major environments. Ericson *et al.* (1951) comment on these two types. The first, the true pelagic type, is called sediment of continuous accumulation, and consists of *Globigerina* ooze, more or less mixed with inorganic matter of very fine grade, which in places becomes true Red Clay. These deposits are built up by the continuous addition, particle by particle, of matter. The other type is deposited by turbidity currents, growing rapidly for a short time and then remaining static until the arrival of the next turbidity current. This process is more effective in the Atlantic, due to the larger number of rivers (in relation to its area) draining into it, and its smaller size. Only 20 per cent, as against 30 per cent in the Pacific, is more than 1000 km from land. The lack of deep sea trenches allows greater penetration of deep sea turbidity current sediments.

The sediment deposited by turbidity currents differs markedly from that of the other type. It is very much coarser than normal deep water sediment, and it includes well-sorted layers of silt, sand, or even gravel at times. An example is a core taken 170 km southeast of the edge of the continental shelf at a depth of 3470 m at 39°23′N, 70°57′W. Above a gravel layer nearly 300 cm thick, is a layer of lutite with foraminifera. This type of deposit often includes material obviously derived from shallow water, such as mollusc shells of shallow water species. Another characteristic of intermittent deposition is the interbedding of terrigenous material and true pelagic deposits, which accumulate between the turbidity currents.

Ericson *et al.* (1961) have described the different deep water environments in which pelagic and terrigenous deposits may be expected. The most likely areas in which the first type of sediment will be found are the nearly level tops of isolated rises, since these situations are least likely to be disturbed by turbidity current activity. Sediments which are typical of turbidity currents are found in a number of environments. Graded bedding has been found in submarine canyons, supporting the theory that these currents play a part in the formation of canyons. They also occur on the gently sloping plains, which often form the deep water deltaic-like sedimentary area at the mouth of the canyons. Another environment in which they have been found is the deep trenches; the steep sides of these do not allow a great thickness of sediment to accumulate without instability causing slumping, resulting in the stirring of the sediment and subsequent settling in a graded state.

As a result of the localized deposition of terrigenous matter in this way, there is a great variety in the depth at which it is found, and the rate at which it accumulates. Several metres may be deposited during one period of deposition of short duration. In some basins, such as the Sigsbee Deep in the Gulf of Mexico, it can be shown that low-velocity turbidity flow on a broad front has spread a recognizable layer of graded bedding over an area of about 7800 km^2. These beds are relatively fine sediments, supporting the view of low-flow velocity in the turbid water. Thus silts grade up into a thick layer of lutite.

A contrasting type of deposit was laid down at the mouth of the Hudson Canyon, at a depth of 3200 m. The samples, described by Ericson *et al.* (1951) were taken from the submarine delta at the canyon mouth. Sand was a common constituent of these cores, varying from thin films to layers up to 6 m thick; on the average 30 per cent of the cores were sand. This was well graded and bedded and was intercalated with normal deep sea sediments. Its source was the continental shelf, while turbidity currents must have carried it to its present position.

Because deposits can be carried in this way out to the deep ocean basins, great care must be taken when ocean sediment cores are used for estimating the rate of deposition and associated changes of climate. No extraneous matter should be included, or if it is present, it must be allowed for, which can be done because of the characteristic type of bedding and other features. The distribution of cores showing sand and silt in the central Atlantic is shown in figure 5.5.

Since the cores drilled by the JOIDES project have become available, the pattern of sedimentation through a much longer period of time can be assessed. Bostrom, Joensuu, Valdes, and Riera (1972) have used sediment samples from the JOIDES cores to study the geochemical history of the south Atlantic sediments since the late Cretaceous. As a result of their study they conclude that the separation of present South America from Africa began in the Neocomian, which is early Cretaceous, about 120–135 million years ago. The early depressions in the continental rift zone that started the separation were fresh-water lakes. As the rifts widened, marine sediments started to accumulate, and these early marine strata contained evaporites, of early to middle Cretaceous age. These evaporites are widespread and suggest a rather earlier opening of the south Atlantic than that suggested by LePichon (1968). Basalt vulcanism occurred simultaneously with the opening process. Erosion of Brazil was rapid for much of the Tertiary, but this material was diverted south of the Rio Grande Rise or to the north Atlantic. The Brazil basin, therefore, received relatively little sediment. Thus rates of sedimentation in the JOIDES cores 14–22 are lower

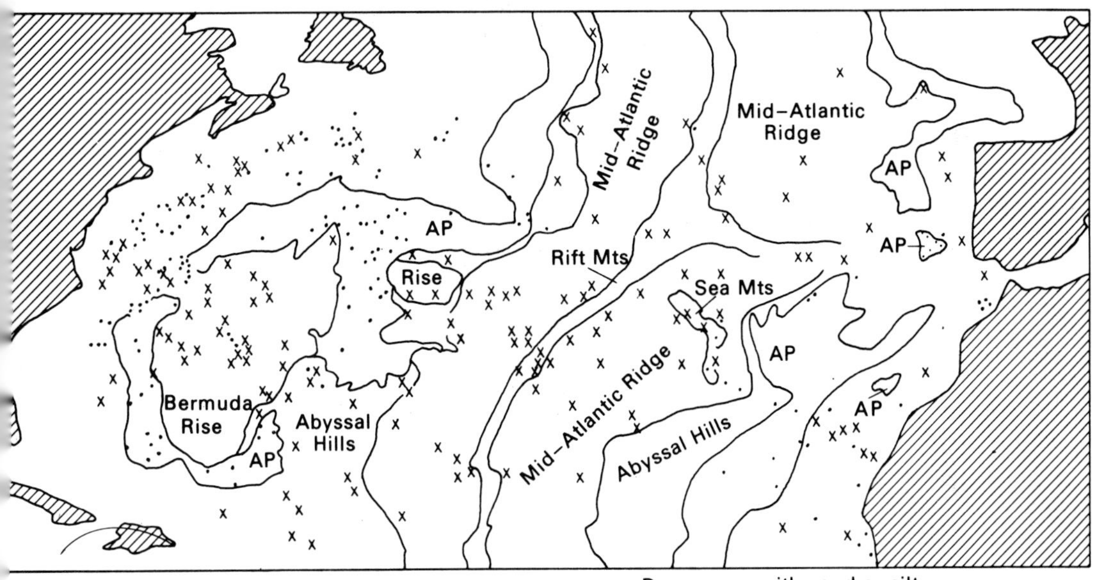

Figure 5.5 Map to show some physiographic and sediment types in part of the Atlantic Ocean. (*Modified after Ericson* et al., *1955, and Heezen* et al., *1959.*)

than those observed along the north coast of South America. Terrigenous sedimentation dominated the basin until the ocean became wide enough for the anomalous sediments of the central volcanic rift zone to become important. These sediments become well developed by the Eocene. These anomalous sediments include iron-enriched deposits. Maximum volcanic activity probably took place around 15–50 million years ago, while the main uplift of the Andean chain in late Miocene onwards provided more terrigenous sediment through enhanced erosion. This terrigenous material reached the central parts of the south Atlantic. The results of this study tend to confirm the views of Maxwell *et al.* (1970) that ocean floor spreading has been fairly constant during the Tertiary. There appears to be little later diagenic alteration of the sediments once they became buried by later material. Manganese does not appear to have migrated on a large scale. Volcanic sediments only occur in significant quantities in the basal layers, the upper being dominated by terrigenous material.

5.2b Indian Ocean

M. Ewing *et al.* (1969) discuss the distribution of sediment in the Indian Ocean and find it thickest in three zones: 1) the Indus and Ganges cones, which together contain 40 per cent of the sediment, 2) basins adjacent to the continents on the east and west, which largely have terrestrial sediment, and 3) an area south of the polar front, which is rich in nutrients and where extensive non-stratified, organic-rich sediments occur, consisting mainly of siliceous oozes. The ridge is a zone of thin sediment up to 100 km from the axis, except in the southwest branch, where a slower spreading rate along this part of the ridge has allowed more sediment to accumulate.

The Indus and Ganges cones are perhaps the most extensive in the world, as they are the product of the most intense denudation. Turbidites associated with the cones reach 10°S at 90°E and 7°S at 81°E. Assuming a mid-Miocene age for the uplift of the Himalayas,

Table 5.7 Relative volume of three sedimentary zones in the Indian Ocean

		Volumes 10^{14} m^3	% of total
Cones:	Indus	2·12	9·9
	Ganges	7·28	34·2
	Total	9·40	44·2
Bordering basins:	Africa and Madagascar	4·81	22·6
	Australia	0·44	2·1
	Total	5·25	24·7
Southern pelagic sediment		2·85	13·4
Other		3·75	17·6
Total		21·25	100·0

giving an age of 15×10^6 years for the cones, the mean sedimentation rate has been 17 cm/100 years. The relative volume accumulating in the three zones mentioned is given in table 5.7. The sediments of the Somali basin are up to 1 km thick and consist of turbidites similar to those of the Ganges and Indus cones. It is suggested that much of the sediment in this basin came from the Indus cone, passing through a gap that may be associated with the Owen Fracture zone, as the depth of the basin is 5100 m compared with 4400 m on the Indus cone. The absence of a known sill lower than 4500 m, however, casts doubt on this idea. The Zambesi Canyon may be the route of sediment supply for the Madagascar basin. In the south of the ocean the Enderby abyssal plain does not contain sediment that can be distinguished from that of other abyssal plains. The conclusion must be that the sediment accumulated before Antarctica was glaciated, or that glacial areas do not provide distinguishable sediment. A small abyssal plain at 45°S, 58°E is isolated by present relief from continental sources. It may derive its thin sediment from the Crozet Plateau; its surface has a dense layer of manganese nodules. The age of the last turbidity current can be estimated from the age of the nodules. Assuming a nodule growth rate of 3 mm/million years and a mean radius of 5 mm, the estimated age is 1·7 million years for the age of the last turbidity current in this area.

The high productivity of the southern part of the ocean results in the diatom ooze in this area. The sediment is homogenous and attained its maximum thickness on the Crozet and Kerguelen Plateaux. There are deep basins on either side of the Ninetyeast Ridge in which sediment is thin. This is because the area is too far north from the rich organic zone; it is below the carbonate compensation depth and too far from the continents to receive terrigenous sediment. It is a Red Clay environment; the material consists of lutite, accumulating at a rate of 3·36 mm/1000 years.

5.2c Arctic Ocean

Only fairly recently has the distribution of sediment in the Arctic Ocean been studied by coring from drifting ice stations. Hunkins and Kutschale (1967) discuss the sediments of the area. The Alpha Rise is one of the ridges in the Arctic basin and it separates the Canada abyssal plain from the Wrangel and Siberia abyssal plains. It is 1000 km wide and has an undulating surface with 1·5 km thick sediments. Layers found in cores correlate from core to core, indicating uniform sedimentation. The sediments are a mixture of normal pelagic sediments and glacial marine sediments. The uppermost layer is a dark brown lutite, with many foraminifera of the species *Globigerina pachyderma* only. This layer is about 10 cm thick; below it is a light brown sandy lutite with few foraminifera. Its base is dated at 70,000 years ago, suggesting that it belongs to the last glacial period, accumulating at about 1·5–2mm/1000 years.

In the Siberian abyssal plain the sediment is deposited by turbidity currents. It is a fine, well-sorted grey lutite supplied by the rivers that enter the East Siberian Sea. The Wrangel abyssal plain contains sediment very similar to that of the Alpha Rise, suggesting that no turbidity currents have reached this plain and the sediment is pelagic in type. In the Canada abyssal plain, 2 km thickness of sediment exist. The uppermost layer is an olive-grey lutite, except for the uppermost 3 cm, which is a brown foraminiferal layer. There is no ice-rafted material in the grey lutite, which is probably a turbidity current deposit and is at least 3 m thick. The findings suggest that Arctic sedimentation has not changed over the last 70,000 years.

5.2d Pacific Ocean

The main sedimentary provinces of the north Pacific Ocean have been discussed by Horn, Horn, and Delach (1970). They suggest that there are three scales of sedimentary provinces in this area. The first order provinces are the main areas of large extent in the ocean: the central north Pacific zone, the Japan–Kurile province, and the Aleutian–Alaskan province (see figure 5.6). These large provinces fit the boundaries of the major water-masses, because there is no large-scale transport of terrigenous sediment towards the basins in the Pacific. This lack of terrigenous sediment in the main basins is due to the presence of bordering trenches along much of the Pacific margin.

The second order features are dominated by the features of positive relief including the Hawaiian Ridge, the Marcus–Necker Ridge, the Shatsky Rise, the Emperor Seamount Chain, and the Ridge and Trough provinces (see figure 5.6). The third order provinces consist of the narrow zone of terrigenous sediment around the rim of the north Pacific and

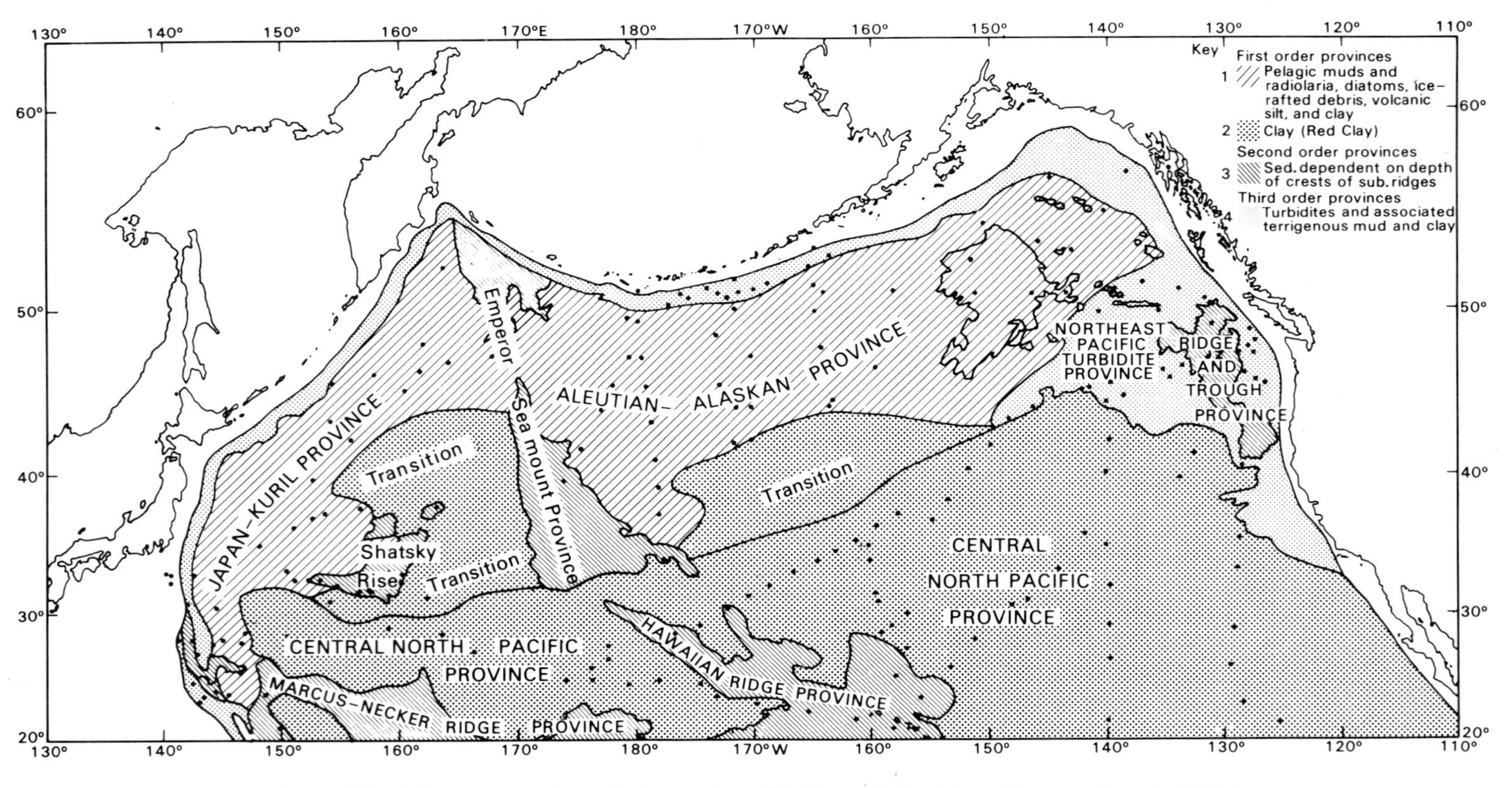

Figure 5.6 Sedimentary provinces in the north Pacific Ocean. (*After Horn, Horn and Delach, 1970.*)

the turbidite province of the northeast Pacific basin, to which terrigenous material has access. The central north Pacific province in the main area of Red Clay accumulation is the north Pacific ocean.

6 Rate of deposition

The rate of deposition plays an important part in the analysis of cores, and is also of interest from the point of view of the age and structure of the oceans. The total thickness of sediments on the sea floor is revealed by seismic surveys and is on the whole small, often amounting to less than 1000 m. About 200 m is the average thickness in the Red Clay-covered areas of the Pacific Ocean and 400 m in the equatorial calcareous belt. Sedimentation rates have been estimated by a number of different methods. They varied greatly in different environments as was to be expected. In one sample taken at the lower edge of the continental slope near the end of the Labrador current, a value of 24 cm/1000 years was recorded. On the edge of the mid-Atlantic Ridge the rate was found to be 11 cm/1000 years, while in the western basin in the Atlantic (in impure *Globigerina* ooze), the rate was 4 cm/1000 years. The same ooze in the Caribbean Sea gave a rate of 0·6 cm/1000 years and an even slower rate was found in Red Clay, 500 km off the coast of California, where the rate was 0·5 cm/1000 years.

Ericson *et al.* (1961) report that in the Atlantic during the last 11,000 years, 108 cores give a range of sedimentation from 0·5 cm to 63·6 cm/1000 years. The latter core was located on the continental rise southeast of Cape Hatteras. The average rate of sedimentation in glacial times was 1·5 times as great as the post-glacial in 46 cores. These 46 cores showed a post-glacial rate of accumulation of 3·5 cm/1000 years, while the glacial rate was 5·1 cm/1000 years. A very high rate of 274·4 cm/1000 years was recorded in the bottom of a canyon northwest of Cape Verde, French West Africa. The high rates of accumulation can probably be accounted for by the special circumstances in which they were found and are not likely to be due entirely to pelagic deposition.

Seismic refraction methods have shown that Pacific sediments reach a thickness between 170 m and 1000 m, with the average a little less than 500 m. An average rate of deposition of 0·5 cm/1000 years would enable this thickness to accumulate in 100 million years. This date goes back into the Cretaceous period. No sediments older than this have been identified in the Pacific (Revelle *et al.*, 1955). Figures for the Atlantic suggest a sediment layer of between 500 and 1000 m. The average rate of deposition is faster in the Atlantic than it is in the Pacific. The 1000 m could accumulate in the Atlantic during the time it would take 200–400 m to accumulate in the Pacific.

There are various possible interpretations of the relatively thin layer of sediment, which at the present rate of accumulation would not go far back towards the beginning of geological time. Either the rate of sedimentation in the past has been slower than now, which could possibly be explained by the more vigorous oceanic circulation to which the cold sources of deep water in high latitudes now give rise. These may well have been absent when there was no ice. Another possibility is that the oceans only developed in their present form during the Cretaceous, which would agree with many of the findings relating to sea floor spreading. Alternatively the earlier sediments may have become consolidated or covered by volcanic lava, with the result that they do not react as unconsolidated sediment to seismic waves. It is possible that all these explanations apply to some part of the ocean, but that none is applicable over the whole area.

In considering the rate of sedimentation and the type of sediment deposited, a number of environmental factors must be taken into account. The most important of these are 1) the general relief pattern of the area, 2) the relation of the area to the sources of inorganic, organic, and terrestrial material, and 3) the physical and chemical conditions, both at the bottom and in the water above. Many minor environmental characteristics also cause local differences.

The rate at which sediment accumulates is extremely variable both in time and space. In time the greatest variability is associated with turbidity currents, which operate irregularly, with long periods between rapid sedimentation, while the pelagic rates are much more uniform, but generally slower. In the Pacific the rate of sedimentation is probably responsible for the distinction between chthonic sediments, a brown clay derived from land, and halmeic sediments, which are mainly formed from solution in sea water. The first forms mainly in the northern and the second in the southern hemisphere. Both are brown clays.

Opdyke and Foster (1970) have calculated sedimentation rates for nine cores from the north Pacific. They use the 0·69 million year boundary for the Brunhes/Matuyama junction.The rate of Red Clay accumulation in the central Pacific is estimated at 3 mm/1000 years. The Aleutian abyssal plain ceased to receive turbidites in the Miocene period. The greatest thickness of sediment located was about 400–500 m near the Shatsky Rise. The deposits in this area consist of pelagic material. In some areas outcrops of Cretaceous age are found. Layer 2 in the Pacific has always proved to be basaltic where it has been studied, so that all the sediments that are there must lie above it. If this is true, then the minimum sedimentation rate must be 1 mm/1000 years. At this rate of deposition the oldest sediments would be 400–500 million years old. They are, however, probably younger than this. The thickness of the sediments and the rate of accumulation is shown in figure 5·7. Evidence of the sediments is consistent with the mobility of the sea floor. Figure 2·6 shows the estimated age of the basement beneath the sediments of the north Pacific. The sedimentary cover must be younger than the age of the basement.

Studies of the diatoms (Donahue, 1970) and the radiolaria (Nigrini, 1970) from the north Pacific cores both suggest a major change in conditions of sedimentation around 700,000 years ago. Between 2 million years ago and 700,000 years ago conditions were stable, but after 700,000 years conditions were generally colder and were more variable. Nine temperature fluctuations based on diatom type were recorded in this interval.

The 700,000 years ago boundary has been recognized elsewhere by different phenomena. It is the time of change from reversed to normal magnetization, and this process is probably related to the same cause that allowed the deposition of tektites in the southeast Indian Ocean. An estimate suggests that nearly 1000 million tons of tektites reached the floor of the Indian Ocean. Most of them are less than 1 mm in size, but a few larger ones have been found (Heezen and Hollister, 1971).

Menard (1965) summarized the data for the Pacific. Dating techniques now allow absolute rates to be calculated from cores. Carbon 14 analysis can give dates to 50,000 years ago on organic sediment, while the ionium–thorium range is 500,000–800,000 years.

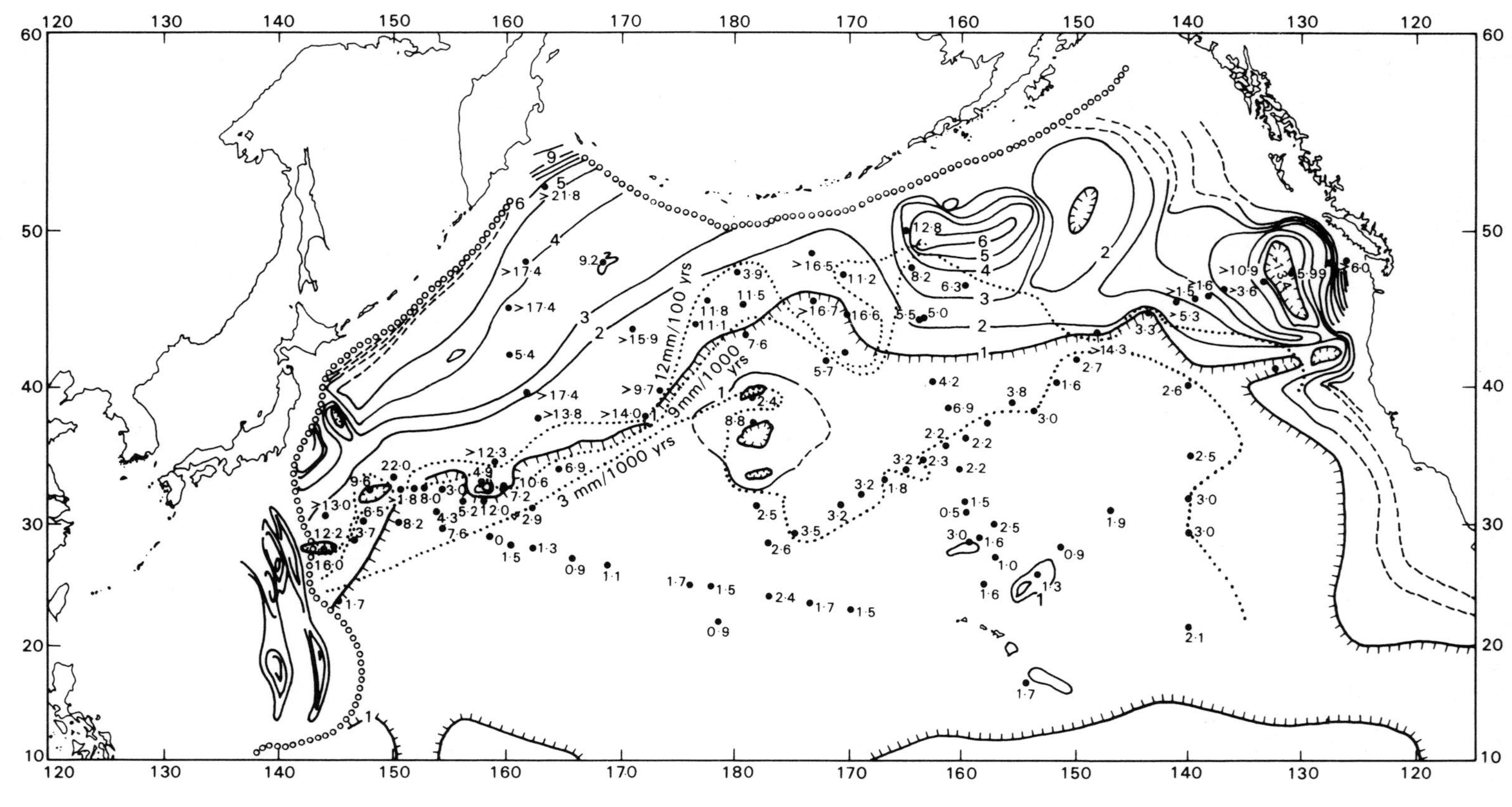

Figure 5.7 Sedimentation rates and sediment thickness in the north Pacific Ocean. Sedimentation rates are shown by dotted lines for the Brunhes epoch. (*After Opdyke and Foster, 1970.*)

Pelagic sedimentation in the abyssal plain off Baja California takes place at 0·9–2·0 cm/1000 years; in the eastern equatorial zones rates are 0·1 cm/1000 years. The more normal Pacific brown clay sedimentation takes place at 0·03–0·16 cm/1000 years. The rates in the north and south differ by a factor of 3, possibly due to more wind-blown dust in the north from the land. The *Globigerina* ooze of the equatorial Pacific accumulates at a rate of up to 1 cm/1000 years or even more, but in the southeast Pacific the rate is only 0·04–0·05 cm/1000 years. The difference is due to organic production and subsequent solution, production being high along the equator. All the carbonate may be destroyed in areas of low productivity and large benthic population. The most rapid rates of sedimentation in the Pacific occur around the Antarctic due to ice rafting, producing glacial marine sediments. Areas of abyssal plain are also subject to rapid sedimentation; the most extensive area of this type of turbidity current sedimentation occurs south of Alaska in the northeast Pacific and in some marginal trenches. The area of abyssal hills is one of variable sedimentation, high values occurring in the hollows and low on the hills, as indicated by outcrops of Tertiary material on the hills in places and topographic smoothing. Very low rates occur in the

Table 5.8 Calcium carbonate accumulation in the glacial and post-glacial

North Atlantic		Post-glacial	1·19	Glacial	0·81 g/cm^2/1000 yrs
Equatorial Atlantic			0·98		2·42
Caribbean			1·37		1·50
Arctic		Foram-rich	0·05	Foram-poor	<0·01
Argentine basin	oceanic rise	Post-glacial	1·50–5·80	Glacial	3·50–7·10
Argentine basin	abyssal plain		0·80–2·50		2·90

south Pacific of around 0·03–0·05 cm/1000 years. This sediment consists mainly of authigenic minerals.

Ku and Broecker (1966) have shown that sedimentation rates in the Arctic have not changed significantly over the last 150,000 years. The mean rate is about 0·2 cm/1000 years. During this period there have been periods rich in foraminifera and ones poor in them, so it is inferred that biological productivity has varied, although there is no simple correlation with changes in the Atlantic. It is also inferred that the Arctic Ocean has not been open during the last 150,000 years. The foram-poor zone extends from 220,000 to 70,000 years ago. The very slow rate of sediment accumulation in the Arctic Ocean may be compared with values given for the Atlantic in table 5·8. In the Antarctic diatom ooze accumulates at about 0·5 cm/1000 years, while in parts of the Bellinghausen Sea sedimentation rates are 3 cm/1000 years in deposits that include detrital material carried to the area by bottom currents. No evidence of turbidity currents was found.

These rates may be contrasted with those recorded in the deep basins of the northeast Indian Ocean, where sediments accumulate at a rate of 15 cm/1000 years in the Andaman basin. Most of the material is supplied by the Irrawaddy, only 14 per cent being carbonate material. These values indicate the great range of sedimentation rates in the oceans, according to the type of sediment and the processes involved. In general the rates are higher in the Atlantic and Indian Oceans. Turekian (1968) has estimated that the ratio of clay deposition rate in the Atlantic to that in the Pacific is greater than 2·6. A mean rate for the Pacific and Indian Oceans is 0·47 g/cm^2/1000 years, the value probably being higher in the Indian Ocean.

Opdyke and Glass (1969) have studied palaeomagnetism in 40 cores from the Indian Ocean and have derived sediment deposition rates from their results. The palaeomagnetic stratigraphy allows correlation from core to core. The calculated rates of sedimentation range from 0·3 to 0·7 cm/1000 years in the deep ocean basins, but values above 2·2 cm/1000 years were obtained. The sites studied covered most of the physiographic provinces. In 32 per cent of the cores, hiatuses were recorded. The central part of the ocean has very thin sediments, especially on the ridge crests, although fairly high sedimentation was recorded of 0·3–0·7 cm/1000 years. This could be accounted for by ocean spreading. The high rate of 2 cm/1000 years was obtained in the Wharton basin northwest of Australia, where sediments are thick. The mean rate recorded was 1·1 cm/1000 years.

Sea floor spreading is one factor affecting the thickness of sediment. J. Ewing and Ewing (1967) show that there is an abrupt change in sediment thickness between the crests and flanks of the mid-ocean ridges, a fact that could be due to intermittent spreading. The present cycle of spreading is estimated to have started 10 million years ago, after a period of quiescence. The sedimentary thickness gradually increases to distances between 100 and 400 km from the axis, beyond which there is a rapid increase in thickness. The rate of sedimentation is assumed to be about 0·3 cm/1000 years on the flanks of the ridge.

The world-wide cruise of the deep drilling vessel *Glomar Challenger* has provided much detailed evidence on the rate and character of sea floor spreading by means of the very long sediment cores obtained. The first three legs of the voyage were situated in the Atlantic Ocean in 1968. The first three sites were in the Gulf of Mexico and 7 holes were drilled. The first site was at the base of the Sigsbee Scarp, the second one on the Challenger knoll, on top of a salt dome, and the third on the southwest side of the Sigsbee abyssal plain. The second site shows the Plio-Pleistocene boundary at a depth of 61 m and an age of 2 million years, giving 3 cm/1000 years sedimentation rate. Other sedimentation rates are summarized in table 5·9. The rate of sedimentation was twice as high at site 3 than site 2, owing to its basin location. Dating was obtained by study of the coccoliths, which are very small. As they are only 7 microns, a simple smear slide can contain 100,000. Sites 4 and 5 were between the Hatteras abyssal plain and the Bahama platform. Site 4 was characterized by chert layers and drilling was hard. Site 5 was in a depth of 5354 m, and entered late Jurassic strata. These are the oldest rocks recorded from the deep ocean. These cores covered nearly all the Cretaceous and Tertiary periods between them at site 4 and 5. Sediment accumulation was 250 m in 125 million years, giving a mean rate of 2 mm/1000 years. The rate is lower than the abundance of turbidites suggests, and is probably the result of small gaps in the sequence, as much sediment was swept away. Carbonate sediment dominates, much of the material being re-sedimented calcareous ooze from the shelves of North America, the Blake Plateau, and the Bahama platform. Calcareous material can survive below the compensation level if it is brought in by turbidity currents. The material is liable to lithification, diagenesis, and cementation. Hard chalk abounds and calcarenites are silicified. Horizon β is a mid-Cenomanian unconformity. Volcanic rocks may occur below horizon B. Some small-scale structural faults also occur.

Sites 6 and 7 were on the Bermuda Rise and site 8 on the rise between the Sohm and Hatteras abyssal plains. Horizons A, β, and B all occur. The topmost member of A corresponds to a mid-Eocene turbidite. The uppermost sediment is brown, deep sea clay, with only spores and pollen. These beds overlie Eocene turbidites, corresponding to the

Table 5.9 Sedimentation accumulation rates, Legs I and II *Glomar Challenger* cruise

Rates in cm/1000 years	Periods in millions of years
Site 2	
3·0	Pleistocene
2·6	Late Pliocene
Site 3	
2·8	Late Pleistocene
7·3	Early Pleistocene
5·0	Late Pliocene
5·0	Early Pliocene
3·8	Late Miocene
Site 6	
0·34	Late Eocene (Red Clay)
2·35	Middle Eocene (diatom ooze and radiolarian clay) 44 m years
Site 8 35°21′N, 67°31′W	
1·0–2·1	down to late Miocene 0–168 m
0·21–0·33	mid-Miocene 168–249 m fine clays and silica plankton
Site 9 32°37′N, 59°10′W	
4·3	Pleistocene–recent
3·6	Basal Pliocene to Pleistocene
1·2	Eocene to basal Pliocene
0·2	Upper Cretaceous to basal Eocene
Pliocene/Pleistocene	78 m in 1·8 m years
Miocene/Pliocene	213 m in 6·0 m years
Eocene	683·2 m in 44 m years
Senonian	764·9 m in 85 m years
Site 10 32°37′N, 52°20′W	
0·8	Mio/Pliocene to Recent
0·12	Eocene/Oligocene to Miocene/Pliocene
0·25	Mid-Eocene to Eocene/Oligocene
1·43	Mid to lower Eocene
0·78	Lower Eocene to upper Maastrichtian
1·20	Upper Maastrichtian to lower Campigian
Miocene/Pliocene	40 m in 6 m years
Eocene/Oligocene	80 m in 38 m years
Upper Mid-Eocene	99 m in 45 m years
Lower Eocene	185 m in 51 m years
Upper Maastrichtian	295 m in 65 m years
Lower Campanian	475 m in 80 m years
Site 11 29°29′N, 45°07′W	
1·4	Lower Pleistocene, 20·1 m in 1·4 years
1·6	to 267·1 m at 16·5 m years
Site 12 19°40′N, 26°01′W	
0·93	Mostly Pleistocene
0·36	Lower Eocene
Pleistocene	41·8 m in 4·5 m years
Lower Eocene	218·0 m in 51 m years

topmost member of horizon A. The change to deep sea clay may be due to uplift of the Bermuda Rise or subsidence of the Hatteras and Sohm abyssal plains, the first being the most likely. A cycle of turbidites followed by clay deposition seems to have occurred. The turbidites have a lower siliceous unit, with sponge spicules and glauconite, and an upper unit of coccolith ooze and chalk, with diatom ooze above as a pelagic deposit of allochthonous origin. These deposits form beds of chert and increase in thickness downwards. Turbidites include redistributed pelagic sediment. Early Cainozoic radiolarian cherts are widespread in the Atlantic. They present an obstacle to deep penetration.

The second leg occupied 5 sites in the north Atlantic. The reflector horizon A is upper Cretaceous turbidite and horizon β does not extend east of Bermuda; it is a lower Cretaceous carbonaceous-rich sediment. Horizon B is the lowest reflector. Igneous rocks were encountered at sites 9, 10, and 11. No turbidites were sampled at site 9, where the deepest calcareous sediment was at 213 m and barren Red Clay at 305 m. Elevation lifted site 9 above the carbon compensation level during the late Miocene, while the level may also have fluctuated. Site 10 was on the lower western flank of the mid-Atlantic Ridge, an area of hills at 4664–4847 m. At a depth of 4697 m calcareous sediment, unaltered and unconsolidated ooze, was drilled down to the basement. Site 11 was on the upper part of the western flank of the mid-Atlantic Ridge in a sedimentary pocket separating steep peaks. Acoustically transparent sediment was drilled to a depth of 800 m. The material nearest to the basalt was soupy and could not be retained in the core as it was so fluid. Site 12 was in the Cape Verde basin. The oldest sediments on leg II were late Cretaceous. Typical oceanic tholeiite basalts and diabases were recovered from three sites, emplaced probably as sills rather than lava flows. The sills were deeply weathered. The spreading rate was calculated to be 1·2 cm/year since the earliest Campanian (Late Cretaceous), based on palaeontological age of the sediments at sites 10 and 11, overlying igneous rock. Red-brown clays were rare in the cores. The sills formed as flows from the ridge crest ploughed through the low-density sediments, and some sediments may be buried below them. The spreading rate from middle lower Cretaceous to early Sampinian was 1·4 cm/year or perhaps a little less. Figures 5·8 shows the sites for the first two legs.

7 Analysis of deep sea sediment cores

Although there are many complications in the analysis of deep sea cores, they have provided valuable evidence on the changing temperatures of the oceans, and these can be correlated with the evidence available from other sources to give a more complete picture of the sequence of climatic change during the Pleistocene period. Favourable deep sea locations can be found where there is an uninterrupted sequence of deposits, such as is rarely available on land.

The necessary preliminary to the development of this type of analysis was to obtain a long undisturbed core of sediment, preferably from an area of slow deposition, so that the lowest layer will be old. In fact some of the cores obtained from the Pacific penetrate right into the Tertiary sediments. It is necessary to avoid, where possible, areas where the strata obtained in the cores have been disturbed by submarine slumping and turbidity currents. The original coring tubes used to obtain long cores were designed by Kullenberg (1947). The corer used a piston, which enables long, undistorted cores up to 20 m long to be recovered.

To illustrate problems of interpretation, two results will be compared. Data obtained from equatorial Atlantic cores showed increasing deposition and calcium carbonate content during periods of higher surface temperature. On the other hand, data from the equatorial Pacific gave equally consistent but opposite results. The rate of calcium carbonate deposition increased during the colder periods in the Pacific, a result that was interpreted as due to the increased deep water circulation during cold periods, leading to a better supply of nutrients. A more numerous foraminiferal population could be supported and so rapid deposition of calcium carbonate took place. In places this led to an alternation of Red Clay with *Globigerina* ooze. These two opposite results are probably both correct, and give warning of the danger of making generalizations, without taking all the relevant environmental variables into account. The changing productivity of the oceans is only one of the means by which the changing climate can be assessed, but it is useful if it can be correlated with differing rates of sedimentation or different types of sediment.

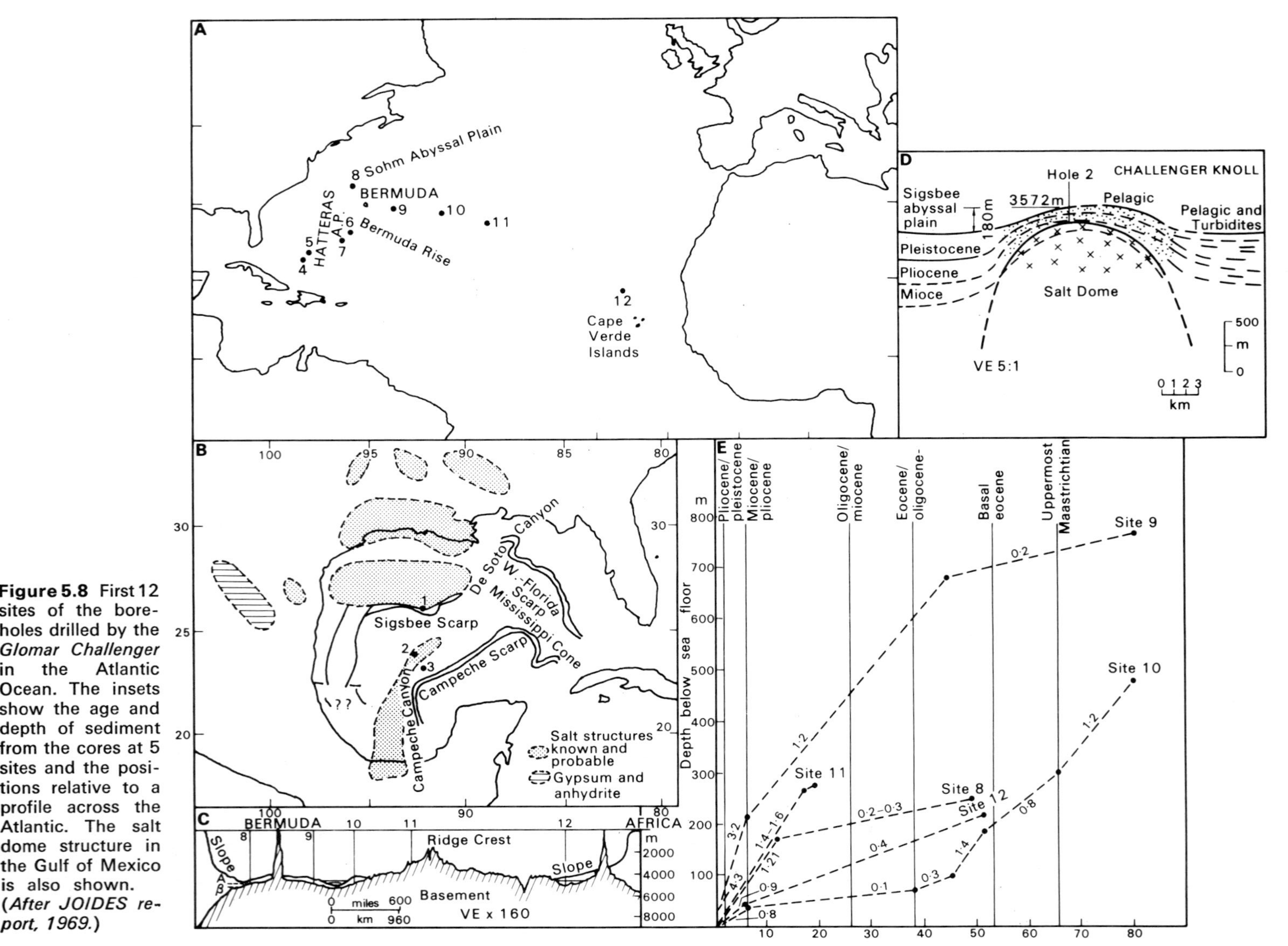

Figure 5.8 First 12 sites of the boreholes drilled by the *Glomar Challenger* in the Atlantic Ocean. The insets show the age and depth of sediment from the cores at 5 sites and the positions relative to a profile across the Atlantic. The salt dome structure in the Gulf of Mexico is also shown. (*After JOIDES report, 1969.*)

Another method of using deep sea cores to determine change of climate is to study the changing species of foraminifera, which are usually related to their optimum temperature. Emiliani (1955) has used the relationship between O^{18} and O^{16} to provide a measure of the water temperature. The actual dating of the different layers of sediment provides the best means of relating the changing conditions of sedimentation to specific climatic variations, which on land may give rise to glacial and interglacial periods. Several dating techniques have been used. The Carbon 14 method is used extensively and the ionium method has been used by Urry (1949).

Hough (1953), working on cores taken in the south Pacific, has given a long record of Pleistocene climatic change. According to the ionium dates of Urry, the core (which was 194 cm long), goes back 990,000 years, although the older part of the curve is extrapolated. The core, from 08°56′s, 92°05′w, consists of alternating Red Clay and *Globigerina* ooze. The latter is thought to have formed during the warm periods. The data from this core and another further south, suggest that the last glacial period started about 64,000 years ago, which agrees fairly well with other estimates. The core shows six colder substages during the last glaciation. The base of the core was thought to penetrate towards the beginning of the first of four major glacial epochs, while the last interglacial was calculated to lie between 268,000 and 64,000 years ago, preceded by a 70,000-year glacial period, which would be the Saale of north Europe and the Illinoian of America.

The work of Broecker *et al.* (1958) relates the changing climate to the rate of deep sea sedimentation. Their work was done with a mid-equatorial Atlantic core, using Carbon 14 dating to measure the rate of sedimentation between the dated layers. The results showed that there was a major change in the rate of sedimentation about 11,000 years ago, the recent rate being considerably less than the earlier rate. The clay deposition rate fell by a factor of 3·7, while the carbonate factor fell by 2·1. Both are related to a change in temperature. There was a uniformly high rate of sedimentation between 25,000 and 11,000 years ago, while an intermediate rate of clay deposition took place before 25,000 years ago. These observations suggest that clay and probably carbonate deposition was highest during the maximum of the glaciation.

The greater rate of clay deposition under cold conditions could be accounted for by various factors. The lowering of sea level would mean that less material would be trapped on the continental shelf. Glacial action enhanced erosion on land, therefore more material, especially of the finer grades, would be available. The increase in calcium carbonate could be due either to different oceanic circulation, or increased supply via the rivers. The organisms are also affected by temperature changes, particularly the coccolithophores, which provide much of the very finely divided calcium carbonate.

The date of 11,000 years ago, at which the deep sea core data suggest a marked change of climate, agrees well with estimates made by other methods. Radiocarbon dates nearly all agree that the mild Allerød, Zone II, period started about 11,000 years ago. This date marks the beginning of the post-glacial climatic improvement, although it was followed by a short deterioration in zone III. Extrapolation back from this date by means of a rate of sedimentation of 5·25 cm/1000 years, makes the whole of the last glacial period run to 64,400 years ago. An intermediate rate of deposition gives a result of 80,700 years ago. A mean value of 70,000 years agrees well with the generally accepted length of the last glacial period. The rates of sedimentation suggested by this study are given in table 5·10.

Table 5.10 Glacial and post-glacial sedimentation rates

	Post-glacial	Glacial gm/cm²/1000 years
Clay	0·22	0·82
Total carbonate	1·34	2·80
Foraminifera	0·40	0·57
Coccoliths	0·94	2·23

Broecker *et al.* (1960) show that a large part of the Atlantic has probably warmed up by 6–10°C over a period of less than 2000 years, centred within 300 years of 11,000 B.P. Further evidence of the effect of this change is the stagnation of water in the Cariaco Trench off Venezuela, where the upper 5–10 m of sediments contain hydrogen sulphide, resting on oxidized clay. There is organic material below this anaerobic layer. The cause of the stagnation is the warming of the water. Water cold and dense enough to allow it to sink to replace the cold water at the bottom of the trench was no longer available.

The advantage of the method of oxygen isotope analysis of Emiliani (1955) is that it obviates the danger inherent in some of the other methods. For example the carbonate content of the deposit and the proportion of species of foraminifera are both liable to subsequent modification as a result of bottom solution, which does not affect the isotopic method. Emiliani used Pacific, Atlantic, and Caribbean cores, the latter giving the best results as the cores were longest. The temperature differences suggested by the analyses were about 6°C. The dates of various levels in the cores have been determined by radioactive isotope methods.

The longest core showed six complete temperature cycles with the oldest minimum 880 cm from the top of the core. If the cycles are of approximately the same duration (which may be justified from their similar thickness), then this level of the core represents 260,000 years. These results also suggest that the rate of sedimentation is falling off owing to a reduction in the number of foraminifera. The results from the Pacific agreed with the earlier conclusion of Arrhenius (1952) that high carbonate phases correlated with low temperature periods. It is suggested that the Plio-Pleistocene boundary may lie at a depth of 610 cm, which is immediately followed by a phase of high carbonate activity and preceded by a period of high temperature. There are about 15 complete carbonate cycles above this level. It is suggested that this represents a total of 600,000 years, which would cover the whole of the Pleistocene. This does not, however, appear to allow sufficient time for all the developments that must be fitted into this period (Gage, 1961).

The cores taken in the Atlantic, which penetrate into the Tertiary, do not show signs of cyclic temperature changes. A study of the benthic foraminifera suggest that the bottom temperatures in the Pacific were about the same today as in the glacial periods, while those in the eastern equatorial and northern Atlantic were about 2·1°C cooler. Interglacial bottom temperatures were not more than 0·8°C higher than the present in the equatorial Pacific. The correlation that Emiliani proposes with glacial events on land seems to fit well as far back as the last major ice advance, which he suggests started about 75,000 years ago; but earlier than this the time scale seems to be very much too short, and the results do not agree with the more recent work of Ericson *et al.* (1961).

Ericson *et al.* use a number of different techniques to get all the available information from a series of cores taken in the Atlantic. The variations in foraminiferal species were studied, as well as variations in the rate of sedimentation. Micro-palaeontology is used to deduce changes in climate, while radiocarbon dates give precision to the results. Beyond the time limit of this dating method, correlations are made on the basis of a study of the coiling direction of planktonic foraminifera. Some of the cores penetrate below the Pleistocene deposition. The oldest extends throughout the Tertiary into the upper Cretaceous, according to the dates indicated by the foraminifera.

The cores used in the analysis cover a very wide area and range of environments, including the mid-Atlantic Ridge, its rift zone, the abyssal plain, submarine canyons, as well as various submarine hills and rises, and the continental slope. Slow continuous type of deposition is the most useful for this type of analysis. Deposits which must have been dropped from floating ice have been found well to the south of the present position of the north Atlantic drift at 46°55′N, 18°35′W, suggesting changes in the past circulation of the north Atlantic. In the older material, which was found in about 1 in 10 cores, no material older than Cretaceous has been found, which also applies to samples taken from the Pacific. In some instances the older sediments are exposed near the surface owing to erosion of the overlying layers, particularly on the steeper parts of the continental slope and other areas of high gradient.

The continuous deposition cores contain both oozes and Red Clay. The latter material accumulates so slowly that a core can cover the whole Pleistocene succession. They are not so useful in other respects, however, as there is no means whereby the changes in climate can be identified. The most useful cores for climatic analysis are those containing foraminifera; these are best preserved in relatively shallow water, in which solution has not destroyed some of the evidence, as happens in the deeper areas. By correlation of a large number of widely separated cores in the Atlantic and Caribbean, it has been possible to establish valid faunal zones, which have been given letters, and which can be related to the average thickness of the sedimentary layers and correlated with the glacial periods.

The results of the analysis show that the period since 250,000 years ago can be divided into six faunal zones. Radiocarbon dates go back to about 20,000 years ago, beyond which the dates are extrapolated on the basis of the rate of sedimentation. The periods covered by the zones and the equivalent glacial epochs are shown in figure 5·9. Zones U, W, and Y contain cold type fauna, while zones V, X, and Z contain the warmer types. In a few cases there is a correlation between the colour of the sediment and the climate, as deduced from the foraminiferal faunal type. The warmer zones are brown, while the cooler ones are various shades of grey. The uppermost layers show a cyclic variation from dark grey to light grey in four cores, nine cycles being indicated in each. So far no explanation for this pattern has been found. The colour change, however, can possibly be related to the periods of desiccation and pluvial phases in Africa. The brown corresponds to the dry periods, since these sediments are most likely to obtain their terrigenous content from the rivers of north Africa, owing to the pattern of ocean currents. The grey colour could be due to carbonaceous matter brought down by the rivers in the pluvial periods, which might correspond with glacial periods further north. The cores showing these colour variations were found in the eastern Atlantic near the equator, on the eastern flank of the mid-Atlantic Ridge. Cores in the western Atlantic and Caribbean did not show this feature.

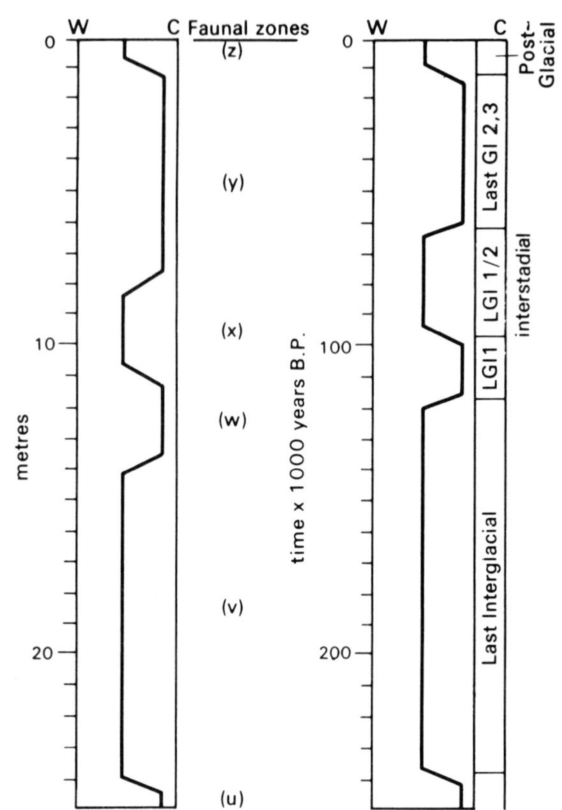

Figure 5.9 Generalized climatic curve and average thickness of sediments in the deep Atlantic and Caribbean. W refers to warm types and C to cold types. (*After Ericson and Wollin, 1968.*)

Zone X shows an assemblage of foraminifera very similar to the present, which suggests that at this time similar oceanographical conditions obtained. This zone is thought to have been deposited during an interstadial in the last glaciation. The temperature fluctuations deduced for the period agree on the whole with those given by Emiliani (1955), although his earlier dating is not reliable. The cores suggest that at no time during the past 50,000 or 60,000 years has the climate been as warm as it is now.

This evidence from the Atlantic can be checked with that obtained by Arrhenius (1952) from the eastern Pacific. He found, partly on the basis of variation in sedimentation rate, that the last cool phase in the Pacific started about 70,000 years ago. Piggot and Urry (1942) give 60,000 years ago for this event, obtained by the ionium method from a core taken from the north Atlantic. A similar date is given by Fisk and McFarlan (1955) from a study of the radiocarbon dates of shells and wood in the Mississippi delta region. The river cut a deep trench at about this time, owing to the falling sea level, resulting from advancing ice. The beginning of this fall is dated at 60,000 years ago, while the maximum fall reached by sea level occurred at least 30,000 years ago. The deposits dated at this time were found in the trench cut by the river at a level of 24 m below sea level near New Orleans. Another wood sample of 28,000 years age was recorded at a depth of 83 m below present sea level near Donaldsville, Louisiana.

More recent and more detailed work on deep sea core analysis has given greater precision to earlier estimates of climatic change and former oceanic conditions. More accurate dating has also been established. The study of changes in climate is mainly based on micro-palaeontological analysis of organic remains in the cores. The coccoliths found in deep sea cores show an abrupt and marked change near the Pliocene–Pleistocene boundary. The

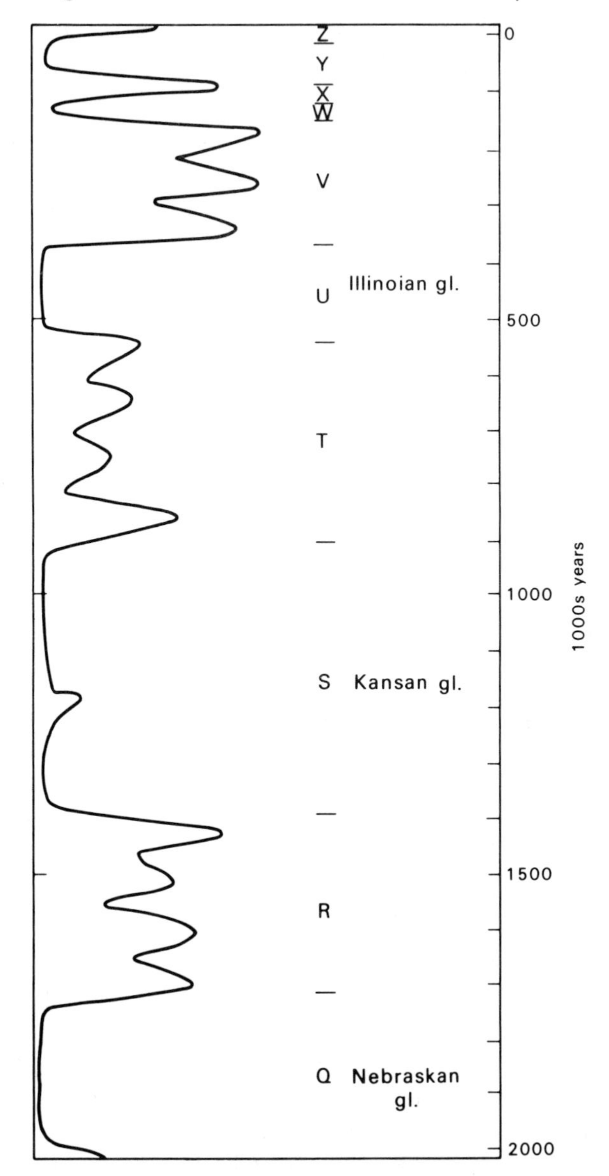

Figure 5.10 Generalized climatic curve derived from deep sea core data. (*After Ericson and Wollin, 1970.*)

main change took place in the lower Pleistocene, probably indicating the Nebraskan–Aftonian transition, the assemblage marking a change from cold fauna below to warm above. Three species disappear at the boundary and one appears out of a total of ten dominant species. Five of the remaining six species change in abundance. At the Pliocene–

Pleistocene boundary, which occurred about 3 million years or more ago, there was a shift from dextral coiling in the Pliocene to sinistral coiling in the Pleistocene populations of *Globigerina pachyderma* in the Los Angeles basin off southern California. At the end of the Pleistocene (about 11,000 B.P.) there was a change from sinistral to dextral population, and *G. subcretacea* and *G. radiolarians* increased rapidly. The boundary between the Pliocene and Miocene in the deep sea can also be recognized palaeontologically. It is marked by a great decrease in the number of discoasters, which continue through the Pliocene in reduced numbers and disappear in the lower Pleistocene. Other species that go out at the end of the Miocene include *Globoquadrina*, *Sphaeroidinellopsis*, and *Globorotalia tumida miocenica*. The Miocene ended about 9 million years ago and the Pliocene about 3 million.

7.1 Atlantic cores

Ericson and Wollin (1968) have analysed cores up to 1300 cm long from the central Atlantic. Their evidence includes the nature of sedimentation, the abundance of *Globorotalia menardii*, the change in coiling direction of *G. truncatulinoides*, palaeontology, and palaeomagnetic reversals. The cores consisted of foraminiferal lutite and were uniform throughout. Figure 5·10 shows the ratio of *G. menardii* to total foraminifera. They estimate the Pleistocene extends back to 2 million years on the basis of *G. menardii* in 10 Atlantic and Caribbean cores, the time scale being based on magnetic reversals. The main faunal difference at the boundary between the Pleistocene and Pliocene is the abundance of discoasters and *G. truncatulinoides*.

The earlier history of sedimentation in the Miocene is discussed by Saito *et al.* (1966) and is based on a study of Miocene fossils in basalt glass in dredge hauls from the crestal area of the mid-Atlantic Ridge near 30°N. Pre-Pleistocene sediment was recognized in 43 cores in the ridge and abyssal hill area, including one Cretaceous and four Eocene. Dredging and cores suggest that sediment is interbedded with basalt flows. Large-scale drifting or spreading is not consistent with the observations. The extruded basalt includes Miocene fauna and this may limit the amount of sediment found. The cores and hauls were recovered from where the Atlantic fracture zones displace the rift on the ridge crest near St Paul's Sea mount. The general increase in age of sediment from the ridge crest may be the result of burial by basalt near the crest. Lower Miocene rocks from the rift-valley were dated at 29 ± 4 million years old, indicating that the last expansion must have taken place at least 20 million years ago. An alternative explanation is that patches of crust escaped movement in the crestal area. The Miocene may have been buried by flows except in windows.

Caribbean cores have been analysed by Ku and Broecker (1966) using Thorium-230 dates to establish the dates of the climatic phases of the glacial period. The results they obtained are as follows:

U/V boundary (Sangamon–Illinoian)	320,000 B.P.
V/W boundary	128,000 ± 13,000 B.P.
W/X boundary	107,000 ± 11,000 B.P.
X/Y boundary	75,000 ± 8,000 B.P.

The results agree with those obtained by extrapolation backwards from the sedimentation rates, indicating a uniform sedimentation rate during the period. These dates also fit with

Plate 19 Photographs of Mount Error (*opposite*), a sea mount lying immediately to the west of the Owen Fracture Zone where it meets the median valley of the Carlsberg Ridge. *Above,* rock outcropping, darkened where there is manganese encrustation and pale where erosion is taking place. There is much coral debris, some live coral and a holothurian. Depth 461 m. *Below,* a vertical dyke some 2 cm wide and 15 cm high protruding through a sediment slope. Depth 572 m. (*A. S. Laughton, Institute of Oceanographic Science, Wormley.*) Area of pictures 1½ × 2 m. (*See Laughton, 1967.*)

evidence given by Broecker *et al.* (1968) for raised coral reefs in Barbados. The island has risen from the sea at a uniform rate, and sufficiently rapidly to separate coral reefs formed at successive high sea levels of the interglacial periods. The three high stands are dated at 122,000, 103,000, and 82,000 B.P. The data show a parallelism, over a period of 150,000 years, between the climate and changes in summer insolation, predicted from the cycles of tilt and precession of the earth's axis, which give the Milankovich cycles.

M. Ewing (1971) has shown that the rate of spreading in the Atlantic has been more rapid in the last 10 million years, although the *Glomar Challenger* results do not confirm this, suggesting a steadier rate of spreading. The relationship between the distribution of land and ice ages suggests that the two are connected. Either an ice-free or an ice-covered polar ocean should be stable, but the mechanism that causes the intermediate unstable state has still to be elucidated. It may be related to lower glacial sea levels, which would influence the flow of water between the Arctic and Atlantic. Thermal isolation of the polar regions seems to favour ice ages. Deep sea cores revealed a large number of temperature fluctuations, beginning in the Miocene, and having a period of between 70,000 and 100,000

Table 5.11 Glacial and interglacial coarse and fine calcium carbonate

	Coarse $CaCO_3$			Fine $CaCO_3$		
Place	Glacial	Interglacial	G/I	Glacial	Interglacial	G/I
North Atlantic	0·4	0·6	0·7	1·5	1·5	1·0
Equatorial Atlantic	0·6	0·9	0·7	1·3	1·2	1·1
Caribbean	0·5	0·7	0·7	0·9	0·6	1·5
Caribbean	0·4	0·6	0·7	1·2	0·9	1·3
Equatorial Pacific	0·25	0·08	3·0	0·75	0·75	1·0
Equatorial Indian	0·4	0·3	1·2	0·3	0·4	0·8

years. The glacial phases are indicated by carbonate maximums in the cores. Glaciations appear to have been synchronous in the north and south.

Broecker (1971) has discussed the calcite accumulation rates in connection with glacial and interglacial conditions. The production of calcium carbonate from plankton depends on the phosphorus present, which is dependent on oceanic mixing to a certain extent. If the rate of calcium deposition can be related to the degree of mixing, then the variation in calcium deposition could be used to determine past mixing patterns. The evidence suggests that calcium accumulation rates were nearly the same in glacial and interglacial deposits in five ocean areas; however, the foraminiferal shells were lower during glacial than interglacial periods in the Atlantic, and higher during glacial times in the Pacific and Indian Oceans, as shown in table 5.11. There are several complications that require further elucidation, such as the separation of ecological effects and solutional processes.

Wollin, Ericson, and Ewing (1971) come to the conclusion from studying cores covering the post-glacial period that there were about 20 climatic fluctuations in post-glacial time since 11,000. Basing their results on the abundance of planktonic foraminifera, they found that warm surface water in some parts of the Pacific were synchronous with cool water in the Atlantic. There was a major change to colder conditions in the Atlantic 3000 years ago, while milder interludes occurred at 25,000, 40,000, and 65,000 years ago. Their results were based on the abundance of *Globorotalia menardii*, coiling ratios of *Globigerina pachyderma* and *Globorotalia truncatulinoides*. All the criteria agreed well.

Imbrie and Kipp (1971) have used quantitative statistical methods to assess the relationship between biological conditions and temperature reconstructions from deep sea core data. Past marine climates are usually estimated on the assumption that pelagic ecosystems have remained unchanged during the Pleistocene. They write quantitative equations that relate the biological side of the ecosystem to the physical parameters of the oceanic environment. The equations are then used on core samples to estimate past marine climates. The analysis is based on a study of planktonic foraminifera. Core top data are first related to present physical conditions. The technique involved the collection of raw data and its transformation if necessary. Varimax factor analysis was carried out on the data. Least squares technique was used to relate the varimax scores to the oceanographical parameters of temperature and salinity. The fossil data from a core was described in terms of the core top varimax assemblages. The palaeo-ecological equations are then used to estimate palaeo-environments.

Fauna were divided into tropical, subtropical, subpolar, and polar assemblages, each group being defined in terms of abundance relative to winter temperature. The abundance of the four climatic groups could then be related to different levels in the core V12–122 from the Caribbean. Reconstructed temperatures could then be related to other evidence of fluctuation over a period of 450,000 years.

Spectral analysis was used to test whether the periodicities present in the data could be related to the Milankovich cycles, the results being based on a sedimentation rate of 2·35 cm/1000 years as a mean value. The orbital eccentricity is matched by the palaeo-temperature curves, but the tilt and precession periodicities were not found. The eccentricity period is 98,000 years, and the others are 40,000 and 22,000, respectively. In the Caribbean the tropical assemblage dominated the core throughout, and the polar one was lacking. Peaks of subpolar assemblages were representative of glacial ages, four such periods occurring during the 450,000 years. The temperature and salinity fluctuations were limited to 5°C and 1·0 ‰ (parts per thousand). The coldest period was the U zone, before 38,000 years ago, and cold spells occurred in the W and Y zones, with warmer periods in the V, X, and Z zones; the warmest part of the V zone, at 220,000 years ago, just missed the maximum warmth, which is shown to be the present as far as surface winter temperature is concerned. Productivity affects *Globigerina menardii* as well as temperature, thus making the correlation imperfect because the figures used depended only upon the temperature. The values for O_{18} also reflected the ice volume changes as well as temperature fluctuations.

7.2 Polar cores

Opdyke *et al.* (1966) measured magnetism in 650 samples from 7 deep sea cores from the Antarctic, and have produced a magnetic stratigraphy based on the reversals. Four faunal zones ϕ, χ, ψ, and Ω, based on the sequential disappearance of radiolaria, have been identified. The ages of the boundaries were ϕ/χ 2 million years, χ/ψ 700,000 years, and ψ/Ω 400,000–500,000 years. Diatom ooze in the cores started to accumulate about 2 million years ago, and the first ice-rafted detritus occurred about 2·5 million years ago. The rates of sedimentation varied from 0·11 to 0·8 cm/1000 years.

Hayes (1967) has shown that the radiolaria in the Antarctic Ocean reflect the northern limit of the pack-ice and the position of the Antarctic polar front. He divides the diatom ooze sediment in long cores into the four faunal zones. The boundaries are given as ϕ/χ 1·6 million years, χ/ψ 900,000 years, ψ/Ω 400,000 years. At the ϕ/χ boundary, clay passes into diatom ooze, and this has been correlated with the Pliocene–Pleistocene boundary. It is thought to mark the large-scale freezing of the sea ice in the Antarctic. This process stimulated the vertical circulation, resulting in the high productivity that exists now. North of the pack-ice, a core revealed an alternation of diatom-rich layers with layers poor in siliceous microfossils. The barren layers indicate a northward migration of the pack-ice. To the north of the polar front, warm and cold fauna alternate, indicating a migration of the position of the front. Two northward migrations occurred in the last faunal zone.

Cores in the Pacific sector of the Antarctic indicate the diatom assemblage—zone ϕ—which is Pliocene in age and represents a warm environment and low phytoplankton activity. Zone χ, which is Pleistocene in age, indicates a cooling of the water and higher productivity. In zones ψ and Ω representing later Pleistocene and recent, further cooling took place. The zonal sequence suggests a continued gradual cooling throughout the Pleistocene in this part of the Antarctic.

In the Arctic Ocean a distinct boundary between two sediment types has been located at a depth of 10 cm in the cores. The upper zone is rich in foraminifera and the lower one poor, although both have ice-rafted material and consist of clay. The boundary between them has been dated at about 70,000 years, although Carbon 14 analysis gives values of 25,000–30,000 years. Pelagic sedimentation has continued throughout in the Arctic at a very slow rate of 0·015–0·03 cm/1000 years. Ice cover has remained throughout the time covered by the cores, which is at least 150,000 years.

The sediments of the Arctic Ocean have been studied by Hunkins *et al.* (1971) and their characteristics related to the physical conditions in the ocean. The Arctic is ice-covered, but the ice margin varies through hundreds of kilometres seasonally. The average ice thickness is 3 m, but it varies about 10 per cent. The removal of the ice would change the albedo over 13 million km^2 from 0·6 to 0·1, and evaporation could lead to major glaciation. The changes in the ice cover affect the sediment characteristics; ice-rafted debris, for example, is at a maximum when the ice is open and can move about freely. Calcium carbonate is also at a maximum when the water is most open. When the ocean is completely open no ice-rafted debris occurs, and the amount is very small when there is a complete and rigid ice cover. There are 231 cores available for the Arctic. These record a very low sedimentation rate of 2 mm/1000 years for pelagic sediment for the last 70,000 years. The rate has been 1·5 mm/1000 years on the Lomonosov Ridge and 2·2 mm/1000 years on the Mendeleyer Ridge during a period of 700,000 years. The water temperature has not changed much over a period of 25,000 years. Four types of sediment are found: 1) normal pelagic sediment, 2) ice-rafted debris, 3) turbidity current deposits, and 4) volcanic glass shards. The ice-rafted material is derived from the ice islands from the Ellesmere ice shelves. The type of sediment does not appear to have changed since the beginning of the Wisconsin, 80,000 years ago to the present. The evidence includes the abundance of *Globigerina pachyderma*, coiling direction of this species, and the amount of insoluble ice-rafted material. The number of calcium carbonate peaks in the Brunhes period vary from 4 to 7, while 8 have been found in the open ocean in this period. The coiling direction changes from left to right as the temperature rises to above 7·2°C for the April isotherm, a line lying south of Iceland. The surface samples of the Arctic show 97 per cent left-coiling. Ice conditions in the Matuyama were more uniform, but evidence for the Brunhes period is conflicting; ice-rafted debris and coiling direction suggest more open water, but low faunal abundance suggests rigid pack-ice. However, the latter observation could be due to greater solution in the Matuyama back to a date of 1·5 million years ago.

7.3 Pacific cores

Riedel *et al.* (1963) have suggested that microfossils in the cores from the Atlantic and Pacific correlate. Fifty cores from the east Pacific rise have been analysed by Burckle *et al.* (1967). They cover the period from Pliocene to lower Miocene up to the present. The sediment was thin near the crest of the rise and only Pleistocene sediment was recovered from here; the sediments thicken and become older on the flanks. The age of the oldest sediment increases away from the axis. Lower Miocene sediment on the lower flanks consisted of Red Clays over calcareous ooze. Only one Oligocene core was found, 1600 km from the rise crest. Eocene sediments were found in the southeast Pacific, the west part of the Tuamotu Ridge, and the area south and southeast of Hawaii. No deep sea Cretaceous deposits were found east of the Pacific Rise. The Pliocene should first appear 90 km from the crest with a spreading rate of 4·5 cm/year and 2 million year Pliocene-Pleistocene boundary; the nearest sediment of this age was found at 100 km. The Pliocene-Miocene boundary is placed at 7 million years, so the upper Miocene should be 315 km from the crest, and the closest point it was found was 350 km from the crest.

Hays *et al.* (1969) have studied a number of cores from the equatorial Pacific around the Clarion and Clipperton Fracture zones. Calcareous sediments lie south of the Clipperton Fracture zone in depths up to 4800 m, and siliceous material occurs between the fracture zones, while clay occurs to the north, where depths are greater. The equatorial sediments are highly stratified east of 170°W, the deeper layers coming to the surface north of the Clipperton Fracture zone, so that Tertiary sediments can be sampled a few degrees north of the equator.

Cores going back to the Palaeocene have been obtained where the age of the material increases to north and south of the equator, and westwards from the crest of the East Pacific Rise. These ages agree with sea floor spreading. Fifteen cores were studied palaeontologically and palaeomagnetically. Some of the fauna do not survive reversals, these probably being the most specialized types. Major changes of fauna occur at times of frequent reversals.

Carbonate peaks correlate with glacial stages. A cooling in the cores was noted at the χ/ψ boundary at 700,000 years, which may relate to the Sherwin till under the Bishop tuff, dated at 710,000 years B.P. Evidence of Pliocene cooling was found in the cores and could be related to glaciation at 3 million years ago in Iceland, 2·7 million years in the Sierra Nevada, at more than 2·5 million years in New Zealand, and 2·5–3 million years in Antarctic cores. Eight glacial stages have occurred during the last 700,000 years. In order to establish these phases, four groups of microfossils have been used. These are diatoms, silicoflagellates, foraminifera, and radiolaria. The period covers the last 4·5 million years of geomagnetic reversals. Two major zones are identified with extinction of species and changes in coiling direction. The first was in the middle of the Gauss normal at 3 million years, and the second near the Olduvai event at 2 million years. These absolute dates allow world-wide correlation. Eight distinct carbonate cycles in the Brunhes series with a periodicity of about 75,000 years in the upper part and 100,000 years in the lower part

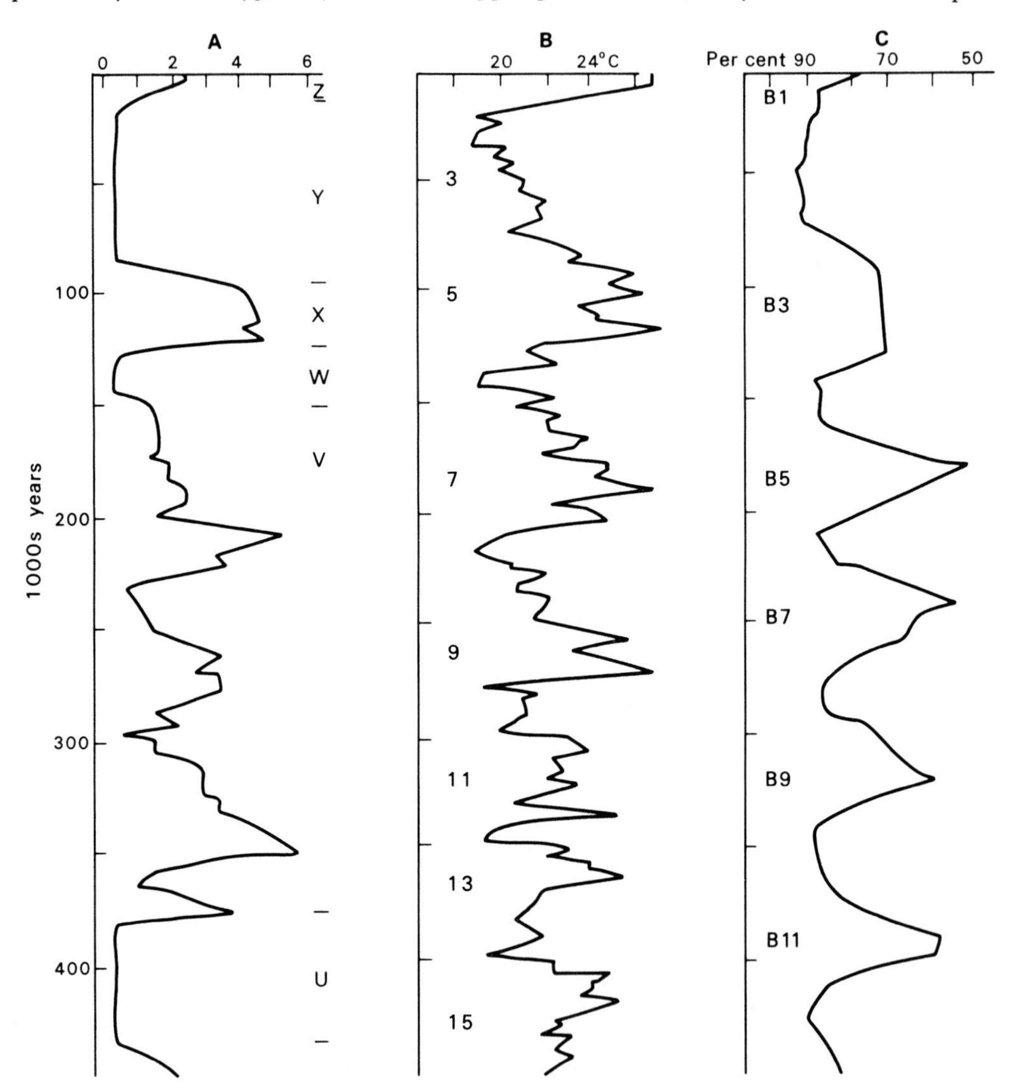

Figure 5.11 Data derived from deep sea cores. **A**: Temperature fluctuations from Ericson and Wollin (1968). **B**: Temperature fluctuations from Emiliani (1960) **C**: Percentage of carbonate in the sediments. (*After Hays* et al., *1969.*)

were established. The peaks indicate the glacial stages, the last high being in the Wisconsinan at 80,000–11,500 B.P. The last 400,000 years of the carbonate cycles correlate in the Pacific and Atlantic. The rates of sedimentation in the Brunhes range between 0·35 cm/1000 years for siliceous ooze to 1·75 cm/1000 years for calcareous sediment. The rate was 44 per cent lower in the Matuyama than the Brunhes, the decrease occurring between the Jaramillo and Olduvai events (figure 5·11).

The boundary between the Pliocene–Pleistocene and the pre-Pliocene lies in the Clipperton Fracture zone. The absence of thick sediment since the Miocene north of the calcareous zone may be due to change in productivity or removal of bottom sediment by bottom currents. It is possible that the wide Tertiary zone of thick sediments is the result of an open Panama Isthmus, so that the Eocene belt of high productivity would have stretched through into the equatorial Atlantic and Caribbean.

Menard (1965) shows that in the Pleistocene the high latitude areas of the oceans were characterized by glacial marine sediments. These can be recognized by their very poor sorting, varied lithology, and erratic stones, especially in areas inaccessible to turbidity currents. Changes also occurred in the position of the diatomaceous sediment in middle latitudes and carbonate sediments in low latitudes. In the Tertiary carbonate-rich sediments were much more extensive than in the Pleistocene. In the early Tertiary carbonate sediments may have extended as far as 45°N in the Pacific, but by Miocene times they only extended to 16°N. During this period bottom water temperatures declined from 10°C to 2°C, according to oxygen isotope analysis. This change in carbonate distribution is probably due to decreased solution in the early Tertiary, as solution is lower at higher temperatures. The abundance of the small coccoliths in the early sediments supports this view. The amount of quartz in the sediments also shows a noted change at about the Tertiary–Quaternary boundary, decreasing in the older deposits. The causes for this change are probably complex, but may be partly related to wind patterns. The paradox of the thinness of marine sediments, in view of the possible great age of the Pacific Ocean basin, may be explained by compaction of the sediment and its consolidation. The second layer of seismology could be shale, with limestone on top. If it is assumed that the second layer is in fact sedimentary in origin, in a consolidated form, then all the estimated erosion from the continents could be accounted for. The average thickness of unconsolidated sediment in the Pacific is 300 m, or an equivalent of 550 m prior to consolidation. At a present rate of accumulation of 0·04 cm/1000 years it seems that this rate must have been at least as small all through the Tertiary. Extrapolation gives an age for the sedimentary layer alone of 1400 million years. Thus it seems possible that the Pacific has been receiving sediment for a large part of geological time, even if the second layer is not wholly sedimentary, and many lines of evidence point to the inclusion of volcanic material in this layer.

Deep sea cores have indicated the complexity of the glacial and Tertiary record. Emiliani (1967) has shown that there have been 16 major temperature oscillations, each one possibly corresponding to a glacial and interglacial, since the extinction of *Discoaster brouweri* in the Pacific. Ice was certainly present before the Pliocene–Pleistocene boundary, which is defined as the first appearance of *Anomalina baltica* in the section at Le Castella, Calabria, although the age of this event is not known. It is not possible, therefore, to indicate the number of glaciations in the Pleistocene.

The more recent studies of the evidence provided by deep ocean sediment cores have revealed increasingly the complexity of events recorded in these cores. The cores confirm, and provide much of the evidence for, the conclusion that the glacial period started much earlier than was thought at one time. The repercussions of these events are widespread. They are concerned with sea level changes (considered in chapter 6 during the Tertiary and Quaternary); these changes themselves have a very important influence on many aspects of geomorphology. Instead of clarifying the situation, much of the recent information has added complications and the number of glacial events has been shown to be large. These significant changes of climate influence ocean currents, fossil faunas, and sediment types. Their results are complicated by other changes such as variations in the distribution of land and sea and in the positions of the land-masses. Faunal changes may result also from magnetic reversals, which affect the environment through their influence on cosmic rays, causing extinction of some species. The magnetic reversals, however, do provide a valuable world-wide chronological tool by means of which correlations can be made between events in the oceans and those on land. Thus the study of deep sea sediments has many aspects of interest and value ranging over a large field of interest.

Further reading

ALLEN, J. R. L. 1970; *Physical processes of sedimentation. Earth Sci. Ser. I.* London: Allen and Unwin. (Includes chapters 5 and 6 on sedimentation in marine conditions, and a discussion of turbidity currents.)

BAGNOLD, R. A. 1963: *Mechanics of marine sedimentation. Autosuspension of sediment.* In M. N. Hill (editor), *The sea.* Vol. **III**, *chap.* 21, *part I.* New York: Wiley. (A technical account of the process of turbidity currents.)

ERICSON, D. B. and WOLLIN, G. 1964: *The deep and the past.* New York: Knopf. (A general account of the study of deep sea cores and sediments.)

ERICSON, D. B. and WOLLIN, G. 1968: Pleistocene climates and chronology in deep sea sediments. *Science* **162,** 1227–34. (Evidence of past climatic changes from sediment cores.)

HEEZEN, B. C. and HOLLISTER, C. 1964: Deep-sea current evidence from abyssal sediments. *Mar. Geol.* **1,** 147–74. (Discussion of the effects of deep-sea currents on sediment deposition and resultant features.)

KUENEN, P. H. 1960: *Submarine geology.* New York: Wiley. (An account of the importance of turbidity currents by one who has advocated the process strongly.)

MERO. J. L. 1965: *The mineral resources of the sea.* Amsterdam: Elsevier. (Includes an account of maganese nodules and other mineral wealth in the oceans.)

SWIFT, D. J. P. 1970: Quaternary shelves and the return to grade. *Mar. Geol.* **8,** 5–30. (An account of shallow water sediment and the effects of changes in sea level on it during the Pleistocene.)

TUREKIAN, K. K. (editor). *Late Cenozoic Glacial Ages.* New Haven: Yale University Press. (A series of papers many of which are concerned with the analysis of deep sea cores and the evidence on changing climate that they provide.)

JOIDES. NATIONAL SCIENCE FOUNDATION. *Reports of the deep sea drilling project.* 1969–: Several volumes starting in 1969 (Sripps Inst. Oceanog., Univ. of Calif.) (These publications give details of the boreholes and cores obtained by the *Glomar Challenger* deep drilling vessel. The first two legs cover the Caribbean and north Atlantic Ocean, and the third the south Atlantic.)

6 The origin of ocean water and changes in sea level

1 The origin of ocean water

2 Changes in sea level during the Tertiary and Quaternary periods: 2.1 Tertiary changes, 2.2 Pleistocene changes; 2.3 Late and post-glacial changes in sea level; 2.4 Present changes in sea level; 2.5 Seasonal fluctuations in sea level

The origin of ocean water is closely bound up with the origin of the salt that gives it the distinctive characteristics associated with sea water as opposed to fresh water found on land. It was the saltness of the sea that led to the early estimates of the age of the oceans in terms of the water they contain, rather than in terms of the ocean basins, which have formed the main focus of the first chapters of this volume. This chapter is concerned essentially with the origin of the water in the ocean, but this also requires a consideration of geological processes, as both the water and the salt must have come originally from within the rocks that form the earth's crust and mantle.

Robert Boyle in the 1670s was the first person to argue that salt in the sea was derived from rocks on land, and he provided the first estimate of sea water salinity, by measuring the density of sea water. Edmund Halley in 1715 first suggested that the age of the ocean waters could be determined by recording the salt carried down to the sea by rivers and estimating the time needed to concentrate the salts to give the observed salinity of sea water. Lack of knowledge of river flows delayed implementation of this idea until it was undertaken by J. Joly in 1899. The result achieved was 90 million years. Joly, however, measured the residence time of the salt that he recorded by dividing the total amount of salt in the ocean by the annual addition from rivers. The value he suggested was about the right residence time for sodium.

One of the important characters of salt in the sea is the fact that all the constituents of sea water bear a constant proportion to each other; it is only the absolute amount of salts that vary in the water. Some of the elements of sea water are present in such small amounts that they can only be detected when they are concentrated, a process that can take place organically or by the process by which manganese nodules grow, for example. Developments in marine chemistry have given added weight to geophysical conclusions, and have themselves become interpretable in these terms. One view is that the earth's core, mantle, crust, ocean, and atmosphere were differentiated about 4000 million years ago, and that during this process ocean water first collected (according to various estimates) between a third and 90 per cent of the present total. The processes that operated to differentiate the original material of the earth are still continuing in volcanic activity, concentrated along the oceanic ridges where new crust is forming and in the island arcs and regions where the crust is being consumed. There are also many isolated volcanoes within the Pacific basin and elsewhere. If only 0·5 per cent of the water released by volcanoes currently active is juvenile water, entering the hydrological cycle for the first time, then the oceans could have accumulated to their present volume in 4000 million years. The evidence suggests that the salinity of the ocean has been more or less constant since the oceans first formed, at least the evidence is definite for the last 200,000,000 years.

The waters that reach the sea from the land have distinctly non-marine characteristics, which would have changed the quality of sea water in the absence of other processes that maintain its character. There appears, however, to be a very complex cyclic process whereby the quality of sea water is maintained through time. These cycles involve the atmosphere, the oceans, rivers, crustal rocks, and oceanic sediment, as well as the ocean crust and mantle. Most of the salt carried by the rivers is derived from the sea as condensation nuclei for rain. a process that accounts for 90 per cent of the chloride and 50 per cent of the sodium in river water. If the ocean water is not changing in type, then the material brought into the sea via the rivers must be locked up in marine sediment. Chemical reactions that take place in the formation of clay minerals, and the locking up of calcium and other elements in marine organisms are methods whereby material can be locked up on the sea floor in sediment. Some of the material is, however, redissolved if the dead organisms sink into very deep water. Eventually marine sediments may be reconverted into hard rocks such as granite. Of all the elements entering the ocean, only sodium and chlorine remain abundantly in solution; less than 1/100 of the other elements remain in solution.

The reactions that take place in the chemistry of sea water are highly complex and involve both organic and inorganic processes. A particularly important organic process is that which ensures that free oxygen remains in the atmosphere. If it were not for this process, the oceans would have caught all the atmospheric oxygen in the form of nitrates in solution or in sediment. Marine bacteria can, however, release nitrogen gas and can convert the oxygen into a form in which it can be liberated by phytoplankton.

The whole complex interactions between the different elements in sea water are such that the sea maintains a balance that ensures that the quality of the water does not change over the ages. The simple methods of measuring the age of ocean water by the amount of incoming salt are thus quite unrealistic, and doubts still remain as to the precise way in which the oceans have accumulated their water supply and of the time scale involved. If the complex salt cycles operate in the ways that chemists have suggested, then evidence of the gradual increase of the ocean volume through the additions of juvenile water via volcanic eruptions gains some weight. The chemical evidence adds to geophysical and other evidence in confirming that recycling processes are fundamental in the operation of global processes. Even the mantle is probably slowly turning over, but with a cycle so long that it has yet to complete one turn (F. Macintyre, 1970).

Recycling and periodicity seem to be important elements of many global processes and the same type of process applies (with perhaps less regularity) to variations in the level of the sea, a matter that in the long term is connected closely with the production of sea water, but in the short term has many other more important causes. Some of these causes of changes in sea level and their results in terms of recorded changes of sea level are considered in the second part of the chapter. Changes of sea level can be considered on a variety of time scales. There are long-term changes that occupy millions of years. There are the changes that can cause rapid oscillation over periods measured in millennia. Still shorter periods of change are the seasonal ones, while short-term changes are associated with events such as submarine earthquakes or meteorological disturbances. At a still shorter time scale are various types of waves.

Of the longer-term changes, such as those occupying millions of years, the effects of earth movements are of outstanding importance. The growth of mountain ranges during the Tertiary period was responsible for deepening some deep sea basins and hence creating a general falling of sea level. The Tertiary curve of sea level has been affected since the Miocene at least, by the development of a major ice-mass in the Antarctic, while the Greenland ice probably started to form soon after. The close connection between ice ages, global tectonics, and mountain growth has been explored in the recent views of the origin of glacial periods, which also seem to have a rough cyclic occurrence over a period of about 250 million years. Sea level can thus be related to the global tectonic scheme through these related occurrences, again indicating the unity of process involved and its many ramifications in such a wide field. The effects of glaciation on the changes of sea level were greatly intensified during the Pleistocene. These changes were extremely rapid in terms of normal sea-level fluctuations during non-glacial geological ages. The rapid changes were brought about by the growth and dissipation of the two major land-based ice-sheets of the northern hemisphere—the Scandinavian and the Laurentide. The magnitude of the changes was of the order of 100 m, compared perhaps with much slower changes, amounting to at least three times that magnitude, during the Tertiary.

The effects of glacial changes of sea level fall into two major groups. There are those associated with the eustatic element of sea-level change, due to the change in volume of the ocean water as part of it becomes locked up on land or is released back into the ocean. The other element is the isostatic depression of the land under the ice load and its subsequent recovery as the ice melts. The resulting isostatic changes are most marked in the previously glaciated areas, where the records of changes of sea level are very complex, as both the eustatic and isostatic factors have influenced them. The oceans are, however, themselves influenced by the changing volume of water that they contain, and this can also complicate the elucidation of sea- level change.

The shorter-term variation in sea-level changes are also often local in occurrence. This applies, for example, to seasonal changes in level, which are of the order of 10s cm in extent and are regular in occurrence, being dependent upon seasonal changes in temperature and pressure. More irregular and still more local changes in sea level are caused by tsunami and meteorological surges. These occurrences can raise sea level up to 10s m above the predicted, although it is rare for such perturbations of sea level to last more than a day or two, and usually they are much smaller and shorter in effect. They can, nevertheless, be disastrous on unprotected low coasts, such as the storm surge of 1953 on the low coast of the southern North Sea, or the tsunami generated by the Alaskan earthquake of 1964.

The longer-term changes of sea level have more lasting and more important effects in most instances. A great deal of work has been devoted to a study of the effects of the Tertiary sea-level changes in the study of denudation chronology. The evidence collected in these studies generally confirms that sea level has fallen during the latter part of Tertiary time, although the evidence is often open to alternative interpretations. It becomes both much more extensive and more complex to decipher in the Pleistocene period when glacial fluctuations dominate the pattern. The evidence is found both above present sea level, in the form of raised shorelines, and below sea level, in the form of drowned beaches and barriers. The effects of changes of sea level are also widely felt in other indirect occurrences. The nature of the sediments on the continental shelf can only be understood in terms of the effects of changing sea level, while modern coastal landforms are intimately related to the fact that sea level only reached its present height about 5000 years ago, following the rapid rise of the Flandrian transgression, during which sea level rose from about −130 m between 15,000 and 5000 years ago. Recent detailed studies have shown that the rise has not been simple (N. A. Mörner, 1971a and b). In fact, sea level responds apparently to very minor changes in climate, but these can only be located from evidence obtained in suitably sited areas, where effects from tidal fluctuations and occasional storm surges do not complicate the pattern too much. The more recent eustatic curves for the post-glacial period indicate that small fluctuations have occurred as the ice has retreated in an oscillatory manner, with minor advances superimposed on the general retreat. The post-glacial rise of sea level is still continuing in many areas, probably due to continued ice retreat in the north.

1 The origin of ocean water

The estimated volume of all the ocean water at present according to Kuenen (1950) is 1370×10^6 km^3. The origin and rate of collection of this water into the ocean receptacles is still an unsolved problem. Water which enters for the first time into the hydrological cycle is called 'juvenile water'. This water has filled the oceans during geological time; its source lies in the igneous rocks of the crust. Juvenile water is now being added to the oceans at the rate of not more than 0·1 km^3/year. This estimate is based on the amount of volcanic activity now in progress. At most, 2 km^3 of igneous rock is produced annually at present. The maximum amount of juvenile water which could be released from this quantity of igneous rock is thought to be not more than 5 per cent. Hence the annual production of 0·1 km^3 is estimated.

Assuming 600 million years since the Cambrian period started, the water in the ocean must have increased by 60×10^6 km^3 during this time. At the beginning of the Cambrian period, therefore, the amount of water in the oceans must already have exceeded 1300 million km^3, if the estimates are approximately correct. The most doubtful point is the amount of volcanic and plutonic activity which has taken place during this long period. It seems reasonable to assume that the oceans have had very nearly the same amount of water in them since the beginning of the Palaeozoic period.

Twenhofel (1932), on the other hand, has suggested that the amount of water in the oceans has been increasing steadily throughout geological time, with an accelerating increase in the Mesozoic (figure 6·1). Walther (1926) has put forward an even more extreme view. He considered that the oceans contained very little water before the beginning of Mesozoic (some 200 million years ago), when they suddenly increased rapidly in volume. His argument is based on the absence of deep sea fauna in the fossil record until after the Paleozoic period, from which he concluded that the deep sea environment did not exist. There are, however, many other ways in which this phenomenon could be explained.

The amount of water increase since the Palaeozoic suggested by Kuenen would be sufficient to raise the ocean level, at its present size, by about 120 m. The Pacific holds about half the present ocean water. If the Atlantic did not exist then the Pacific must have been considerably larger. Although the amount of ocean water has probably only increased very slowly in the Palaeozoic, at times the continental margin has been flooded by shallow seas, geosynclines have developed, but on the whole sea level has apparently not changed very much during the period.

The matter that makes up the hydrosphere, the atmosphere, and the biosphere is not common in the crustal rocks of the lithosphere or the material deeper in the mantle. The materials that make up the outer parts of the earth probably have a common origin and accumulated together. It is still not known precisely how this matter has accumulated or when. The nature of the solar system must be considered in suggesting possible modes of origin. There are some elements that are deficient on the earth compared with the solar system of the sun. This applies particularly to the rare gases, neon, argon, krypton, and xenon. The volatile phases of other abundant elements were also missing in the early stages of the earth's development from planetary matter. These include water vapour,

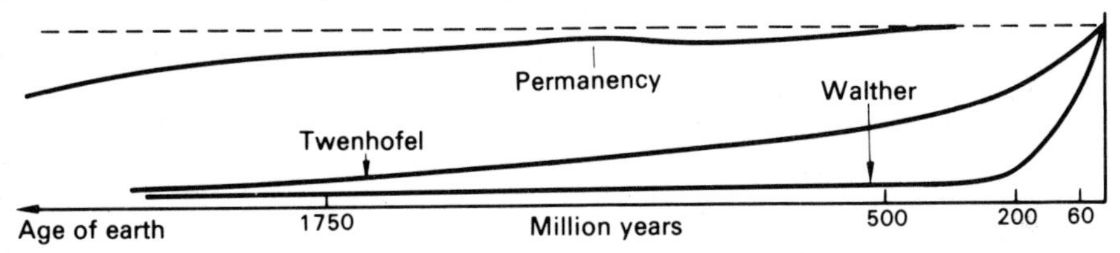

Figure 6.1 Different views on the changing volume of ocean water. (*After Kuenen, 1950.*)

hydrogen, carbon dioxide, and nitrogen. These elements were only present in non-volatile form.

Water would have been included in clay-type minerals and hydrated silicate minerals of different types, while carbon, nitrogen, and chlorine could have occurred in chemical combinations. These compounds have been found in carbonaceous chondrites, a form of meteorite that occasionally reaches the earth. Weathering of crustal rocks will not yield the excess materials that have formed the oceans, atmosphere, and biosphere. These have, therefore, been called excess volatiles, and they must have been derived from the earth's interior by a process of de-gassing after the earth was formed. The main problem is to establish when this process occurred. It could have occurred during some fundamental thermal metamorphosis in the early earth, when the oceans were created in a fairly short period. Alternatively, the volatiles could have accumulated slowly over a long period of geological time through volcanic activity. One reason for the difficulty of deciding between the two alternatives is that there is no record of the first few 1000 million years of the earth's existence, as no rocks survive from this period. The oldest dated rocks are 3200 million years old.

One point in favour of the early accumulation of ocean water is seen in the character of the carbonaceous chondrite meteorites, which could have yielded the necessary volatiles on being heated by gravitational accumulation. It seems likely that the earth started as a cold body and then gradually heated up as the material within it became concentrated. De-gassing on this theory would have taken place in the first 500 million years of the earth's life. This single episode does not conflict with what is known of the constituents of the earth's atmosphere, oceans, and the nature of ancient rocks. The alternative view considers that the volatiles have been produced by volcanic activity throughout the history of the earth. The oceans and atmosphere could have been derived from the mantle throughout the history of the earth by only 1 per cent of the present rate of volatile supply through volcanoes. It is not known how much of the present material ejected from volcanoes does in fact come from the mantle. Information from other planets may enable a decision to be made between the two alternative possibilities for the origin of the ocean.

The level of the ocean can, however, vary even if the total amount of water remains the same. Its level depends in part on the capacity of the ocean basins, and this can vary with time through structural changes. The level of the sea can also vary even if the total amount of water present in the hydrological cycle remains the same and the ocean basins do not change in capacity. Such changes can be brought about by variations in the proportion of water in different parts of the hydrological cycle.

The most massive change, and the best known, is that in the amount of water held in the

Table 6.1 Oceanic volumes and related values

Area of earth, A_e	$5{\cdot}1 \times 10^8$ km^2
Superficial area of oceans, A_{ws}	$3{\cdot}6 \times 10^8$ km^2 = $70{\cdot}8\%$ A_e
Superficial area of continents, A_{cs}	$1{\cdot}5 \times 10^8$ km^2 = $29{\cdot}2\%$ A_e
Area of oceanic crust, A_w	$3{\cdot}0 \times 10^8$ km^2 = 59% A_e
Area of continental crust, A_c	$2{\cdot}1 \times 10^8$ km^2 = 41% A_e
Average thickness of continental crust, T	33 km
Average water depth, D	4·6 km
Volume of continental crust, V_c	$7{\cdot}0 \times 10^9$ km^3
Volume of oceanic crust, V_{oc}	$1{\cdot}5 \times 10^9$ km^3
Volume of oceanic water, V_w	$1{\cdot}37 \times 10^9$ km^3
Age of continents and oceans	$4{\cdot}0 \times 10^9$ years

solid state as ice. This amount has varied markedly in the recent geological past during the Pleistocene ice age. Previous ice ages, such as the Permo-Carboniferous, must also have exerted an effect on the amount of water in the oceans, as evidence suggests very massive ice accumulations during this glacial period. There are also other, less extensive changes that also exert an effect on the amount of water in the oceans, such as the amount of water vapour present in the atmosphere and the amount of ground water. The time period of these different effects varies. Some of the changes are seasonal and the associated variations of sea level are measured in centimetres. Others, such as the glacial and interglacial phases, operate over thousands of years and the variations in level range over at least a hundred metres, while the interval between major ice ages is about 250 million years. The longer period changes will be reviewed first, leading to the smaller, seasonal ones.

The gradual accumulation of water in the ocean basins is supported by Menard (1964). Table 6.1 gives some fundamental values from which the ocean increase in volume throughout its life of 4×10^9 million years is found to be 0·34 km^3/year. The ratio of $V_w/V_c = 20$ per cent, and of $V_w/V_c + v_{oc} = 16$ per cent, are values rather too large to be acounted for by the probable water content of the magmas, which is about 6–8 per cent. Some magmas, however, probably contain more water. Much of the ocean water could have come from differentiation of the upper mantle, which did not yield continental crust as the process operated. The production of sea water probably fluctuated with time, but the available evidence does not throw much light on possible fluctuations, although large random ones appear unlikely. There is only negative evidence to support the hypothesis of uniform water production, but there is little support for Revelle's (1955) view that most of the ocean accumulated in the last 10^8 years at a rate 10 times the mean. His evidence is all open to alternative interpretation. Menard comes to the conclusion that the ocean probably accumulated at a fairly uniform rate throughout geological time.

In table 6·2 some of the causes of variations in sea level are given, some of them being reversible and others not. The table 6.2 emphasizes the very rapid rate of sea-level change in the recent geological past compared with that taking place over much of geological time.

Menard (1971) has suggested that the estimate of ocean water production that Hess (1955) made is too low. Hess estimated a production of 0·4 km³/year, but ocean floor spreading is more rapid than Hess allowed for. On the model of Hess (1955) the rate of

Table 6.2 Causes of sea level fluctuations

Cause	Effect, m		Duration, years	Rate of change, cm/1000 years
Irreversible trends				
Differentiation of continents and mantle	ocean creation		4×10^9	0·1
Accumulation of Pacific submarine volcanics	Raise sea level		20×10^8	0·02
Accumulation of unconsolidated sediments	Raise sea level		160×10^9	0·02
Accumulation of 2nd seismic layer	Raise sea level		180×10^9	0·02
Fluctuations				
Melting of existing ice	Raise sea level	46	—	—
Formation of max. Pleistocene glaciers	Lower sea level	200	$1–2 \times 10^5$	>100
Melting of most recent Pleistocene glaciers	Raise sea level	120	2×10^4	860
Elevation of existing ocean rises	Raise sea level	300	10^8	0·3
Sinking of Darwin Rise	Lower sea level	100	10^8	0·1

production of ocean water would be 3·3 km³/year. This would provide all the ocean water in a period of 420 million years. It is likely that the last 100 million years, and particularly the last 10 million years have produced more active de-gassing and ocean accumulation than the geological average.

2 Changes in sea level during the Tertiary and Quaternary periods

Many different processes can cause changes in sea level, so that the level is unlikely to remain static for long even if the amount of water remains constant. Some of the processes lead to local changes, such as isostatic uplift of areas following deglaciation, or the sinking of geosynclinal areas such as the southern North Sea. Other local movements are associated with tectonic instability, for example, the change in level in Wellington Harbour following the 1851 earthquake. Processes that cause a world-wide change of sea level are called eustatic.

Geodetic changes, such as an increase in the speed of rotation, would alter the geoid by affecting the centrifugal force, causing polar shrinking and equatorial swelling. The hydrosphere will react differently to the lithosphere, resulting in zonal sea-level fluctuations. Polar shift will also cause regions of transgression and regression. A sudden shift of 10° in the position of the pole would cause an equatorial rise of sea level of 2450 m, and a 1° shift would raise the level by 245 m, with equal regression 90° away.

Eustatic changes may be caused by a change in the capacity of the ocean basins. A negative change is caused by the formation of basins, called tectono-eustatism. A positive world-wide change is caused by sedimentation, called sedimento-eustatism. If all the land were peneplaned, sea level would rise 250 m. At the present rate of denudation (estimated at 12 km^3/year), sea level would rise about 3·3 cm/1000 years.

During the last few million years a more obvious and rapid cause of eustatic change has been the glacial-eustatic effect, due to the repeated growth and decay of land-based ice-sheets. Other related effects include that of the change of ocean temperature. A rise of the whole water column of 1°C leads to a rise of sea level of 2 m (Fairbridge, 1961). The changing volume of the ice-sheets has been the most important influence in the Quaternary period, while tectono-eustatism dominated the early Tertiary and carried on throughout the Tertiary and into the Pleistocene, combined with glacio-eustatic changes.

2.1 Tertiary changes

During the Tertiary period much of Britain was land and the changes of sea level there can best be studied by geomorphological analysis. The former base levels can be estimated by studying erosion surfaces, formed either just below sea level by earlier and higher seas, or by subaerial forces which reduced the land to near base level. The results of the study of denudation chronology suggest that sea level fell intermittently during most of the Tertiary period. Surfaces now 600 m or more above sea level are thought to have been formed during this period. The major problem that remains is to ascertain whether the relative fall of sea level was due to the elevation of the land, or the fall of sea level, or by a combination of both movements.

If the main movement was due to a general world-wide tectono-eustatic fall of sea level, then all stable land areas must have been affected similarly. In this case it should be possible to correlate variations of base level by height from place to place. This has been attempted and cross-Channel correlations between Europe and Britain have been suggested, as well as trans-Atlantic correlations. The Tertiary period, however, was one of great tectonic activity, when the main Alpine earth movements took place, influencing many parts of the world. Folding was active in southern Britain, while in the early Tertiary, volcanic activity was widespread in northern Britain. All this tectonic unrest makes it fairly certain that (at least during the early Tertiary) movement of base level was probably a combination of tectono-eustatism and local tectonic effects. It is thus not surprising that the old base-levelled surfaces, which were near sea level during this period are now no longer horizontal.

Many geomorphologists would consider that at least since the Mio-Pliocene, when the 300 m surface was probably formed, base-level changes have been fairly uniform throughout the country, and therefore, likely to be eustatic in origin. Some eastern parts of England cannot be included in this generalization. Deposits of the Pliocene and Pleistocene periods in these areas have been warped down below sea level. West and south of a line running through Braintree in Essex (Wooldridge and Henderson, 1955) there is evidence, in the form of marine deposits of the earliest Pleistocene, that sea level at this time stood 200 m higher than at present. Evidence is found over much of southern Britain for this phase of higher sea level. Over some of the country at least, the land appears to have been stable, and sea level to have fallen since this time.

The reason for the generally falling sea level of the Tertiary period may be related to the very widespread land uplift that took place during this period and the formation of tectonic basins. Parts of New Zealand have been uplifted by up to 3000 m during the late Tertiary and Quaternary period. Material at depth must move beneath the continental areas to make the elevation possible. The material is likely to be derived from beneath the ocean floor, which would deepen. The great elevation of the land in the late Tertiary also helped to initiate the Pleistocene glaciation. This in its turn played a very important part in the fluctuation of sea level.

A number of studies bear on the variation of sea level during the Tertiary. Much of the evidence is conflicting. Certain trends, however, are emerging and certain events are known to have been of over-riding importance, at least in the later stages of the Tertiary. The most important event was the development of ice in the Antarctic. Evidence from the Jones Mountains of western Antarctica indicates that an erosional unconformity has been overlain by 500 m of late Tertiary volcanics. There are grooves, chatter marks, till, and other signs of glaciation on the unconformity beneath the volcanic material. Dates of 22–24 $\pm$ 12 million years have been obtained from basalt flows above the unconformity, and a date of 104 $\pm$ 4 million years was obtained from dykes below it. A Miocene age is considered most likely. Actual evidence of sea level in the Tertiary in Antarctica is limited, but one raised shoreline at 220–250 m above present sea level with Pliocene fossiliferous remains has been found in Cockburn Island, although a Pleistocene age is thought more likely by some. Other remains of the same age occur at 45 m, the difference being the result of tectonic activity.

Further evidence for a fairly early glaciation in Antarctica is provided by deep sea sediments. A glacial maximum is indicated between 2·35 and 3·35 million years ago, and the conclusion is reached that Antarctic glaciation was initiated prior to 5 million years ago. The varying proportion of planktonic fauna (*Globigerina pachyderma*) responding to cold and warmth, which can be identified by their coiling directions, indicate fluctuations of warm and cold conditions, probably associated with glaciations and thus eustatic changes of sea level. Evidence suggests that eustatic falls of sea level due to glaciation may have been as great in the late Miocene and middle Pliocene as those well documented for the Pleistocene, although some evidence suggests rather smaller changes. Mercer (1968) has suggested that the critical factor in the lowering of glacial eustatic sea level was the increase of the eastern Antarctic ice-sheet to form ice shelves, resulting in a lowering of sea level of 60 m. The growth of the western Antarctic ice resulted in a further fall of 4–4·5 m. The growth of the Greenland ice-sheet followed a little later, and caused a further fall of 7 m. This event may mark the boundary of the Pliocene and Pleistocene. The ice-sheets were long-lasting, particularly the eastern Antarctic one, so that their effect on sea level has also been prolonged. Mercer considers that sea level was 72 m above the present level before any of the ice-sheets developed.

Tanner (1968), in summing up the findings on Tertiary sea-level changes, suggests that two rates of change have operated. During the late Cretaceous and early Tertiary, sea level appears to have dropped about 50 m in 70 million years, at a mean rate of 0·7 mm/1000 years. During the Mio-Pliocene period, sea level fell a further 75 m in 25 million years, at a mean rate of 3 mm/1000 years. The first part of the change is thought to be due to the net isostatic rebound due to erosion in the southern Appalachians, while the second part is due to the growth of the Antarctic and Greenland ice-sheets. His curve is shown in figure 6.2. It is difficult to separate the various epeirogenic and tectonic effects, although the operation of different processes in the two periods is probably correct. In the mid-Cretaceous, sea level may have been about 130 m higher than at present, and if glaciation had not occurred sea level would now be about 68 m higher than it is.

The importance of the growth of the ice-sheets in the Antarctic in the later part of Tertiary time is established as one of the most important influences on eustatic sea level during

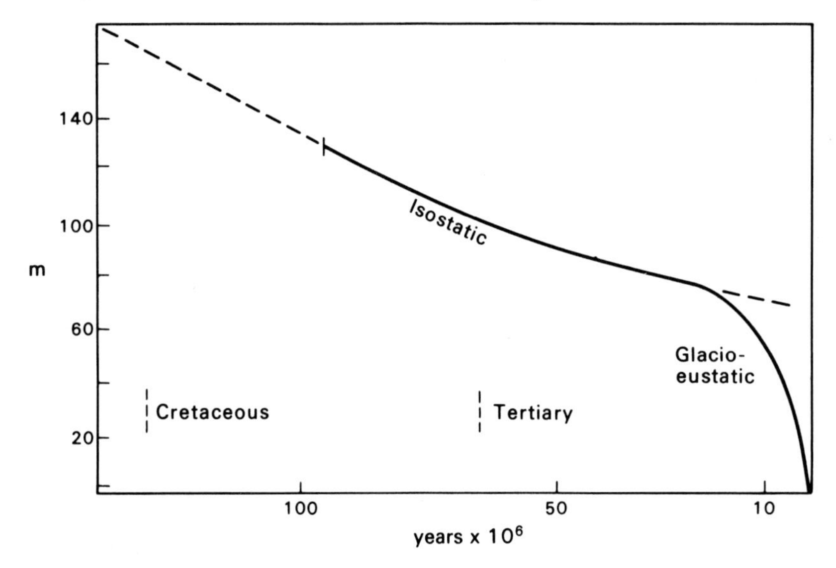

Figure 6.2 Sea-level changes in the Cretaceous and Tertiary. (*After Tanner, 1968.*)

the Miocene and Pliocene periods—an effect that became even more significant in the Pleistocene as other ice-sheets (with much quicker response periods) grew in northern Europe and North America. These ice-sheets resulted in the very rapidly swinging sea levels of the Pleistocene and Holocene periods.

2.2 Pleistocene changes

As the ice-sheets grew in Europe and North America they abstracted more and more water from the oceans. The exact amount of the maximum change of sea level resulting from the growth of ice-sheets is difficult to estimate, because though the area covered at their maximum is well known, the thickness of the ice can only be estimated approximately. Much water was also held on land in temporary lakes, some of which were very large. The sinking of the crust isostatically under the weight of ice displaced deeper-seated material which affected sea level. The ice-sheets were, thus, bi-convex in shape. Another difficulty is that the ice-sheets need not be contemporaneous.

Owing to these uncertainties several different estimates of the maximum fluctuations of sea level have been proposed. Antevs (1928) suggested a value of about 93 m for the last glaciation, and 120–130 m for the maximum. Daly (1934) gives values of 75 m for the last glaciation and 90 m for the maximum, while Penck (1933) suggested 100 m for the maximum. Other estimates are almost double. It seems that about 100 m is a reasonable estimate. This maximum low level was probably attained during the last glaciation, owing to the superimposition of the tectono-eustatic curve on the glacio-eustatic one. This more than compensated for the smaller amount of ice volume during the last glaciation (see figure 6.3).

The relationship between Pleistocene ice volumes and sea-level lowering has been calculated by Donn, Farrand, and Ewing (1962). Their values are given in table 6.3.

Table 6.3 Ice volumes and sea-level lowering

Volumes in 10^6 km^3	Estimate A Wisconsin			Estimate B Wisconsin		
	Classical	Early	Maximum	Classical	Early	Maximum
	30,000	30,000 years		30,000	30,000 years	
Total glacier ice	70·78	74·36	83·57	84·25	88·38	98·65
Existing glacier ice	28·45	28·45	28·45	34·75	34·75	34·75
Pleistocene ice excess	42·33	45·91	55·12	49·50	53·63	63·90
Sea-level lowering in m (Surface area of world seas 361 × 10^6 km^2)						
	105·5	114·5	137·4	123·4	133·7	159·3

Estimates A and B refer to different values of the ice thickness in the Antarctic. Estimate A is based on a thickness of 2·0 km, and estimate B on a thickness of 2·5 km.

Bloom (1971) has considered the eustatic changes of sea level resulting from the melting of the last major ice-sheet. He also has drawn attention to the part played by the formation of new ocean floor by plate movement and sea floor spreading. The water equivalent of the last ice-sheet is estimated at about 43 × 10^6 km^3 by Flint (1971) and the spreading of the ocean floor since the last interglacial period could account for about 6 per cent of the returned water, so that post-glacial sea levels would be almost 8 m lower than the interglacial one of 100,000 years ago. The maximum lowering of sea level in the last glaciation probably occurred about 15,000 years ago when sea level was 130 m below its present level —lower than that of the earlier glaciations, and hence the lowest during the Pleistocene and probably the whole Tertiary period. Since 15,000 years ago sea level has risen at a decelerating rate.

Sea level was higher than at present during the warm interglacial periods. At times of low sea level, during the glacial maximums, much of the shallow continental shelf was dry land or covered by ice-sheets. Rivers extended their courses across the dry land. In many of the major river valleys of the British Isles and elsewhere there is evidence of the lower sea level in the form of buried channels. The Thames cut down its bed to form a deep and narrow channel, subsequently filled with later deposits during a phase of higher sea level. Elsewhere overdeepened channels may be the work of glacial scour, which can take place well below sea level where the ice is thick, as for example in the fjords of Norway and elsewhere, and in the tunnel valleys in the North Sea. The straits of Dover and the southern

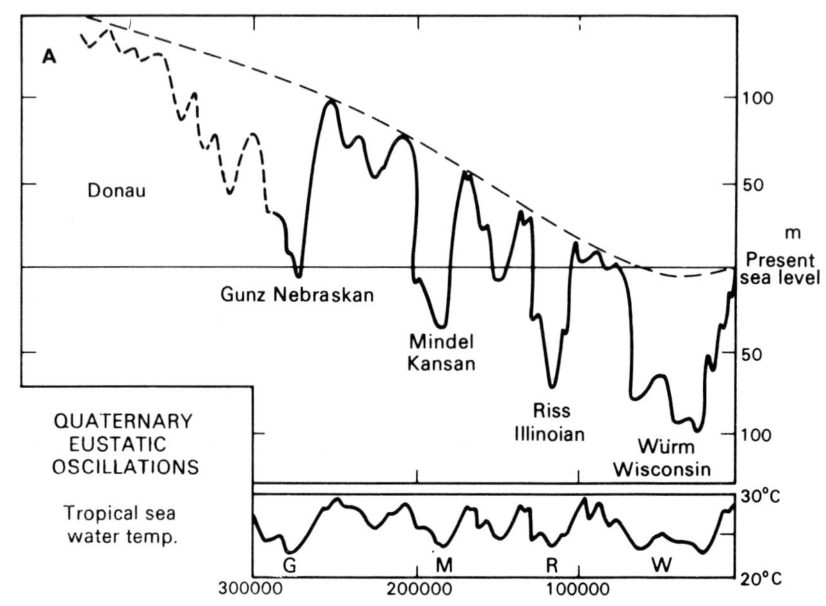

Figure 6.3 A (*above*) and **B** (*opposite*): Changes of sea level during the Pleistocene period. (*After Fairbridge, 1961.*)

North Sea were dry land during low glacial sea levels and Britain was linked to the continent, an important fact with regard to the immigration of men, animals, and plants. Glacial drift was widely deposited on the continental shelf around Britain during the low sea-level phases. Since the rise of sea level, this material now forms the coastline for considerable distances, and because of its non-resistant character, these coasts are liable to rapid modification.

The higher interglacial sea levels also left their mark in many coastal areas, and extended their influence far inland in the form of river terraces, aggraded to the higher sea level. Terraces can often be associated with the interglacial periods by their associated fauna and flora. A number of high interglacial sea levels have been recognized. In the Hoxnian interglacial, sea level was about 30 m higher than it is now.

The last major interglacial, between the Riss and Wurm, is correlated with the Eemian Sea deposits of the European continent. The evidence shows that there has been a steady fall in the high interglacial levels and of the low glacial levels.

Sea levels during the Pleistocene have been affected by both eustatic and isostatic effects resulting from the growth of large ice-sheets. The isostatic effects have been most marked in the high latitudes in which the major ice-sheets developed—northern Europe and northern North America. Both causes of sea-level change must be taken into account in considering changing sea levels in these areas. The isostatic effect is not, however, confined to the previously glaciated areas, as has been pointed out by Bloom (1967). He points out that as the ice-sheets grew, so the amount of water in the ocean was reduced. At present nearly 3 per cent of the land is covered by ice, while the value was nearly 8 per cent at the maximum of the last glacial advance between 20,000 and 16,000 years ago. Three-fifths of the area that has lost its ice is in North America and one-fifth in northern Europe. The load

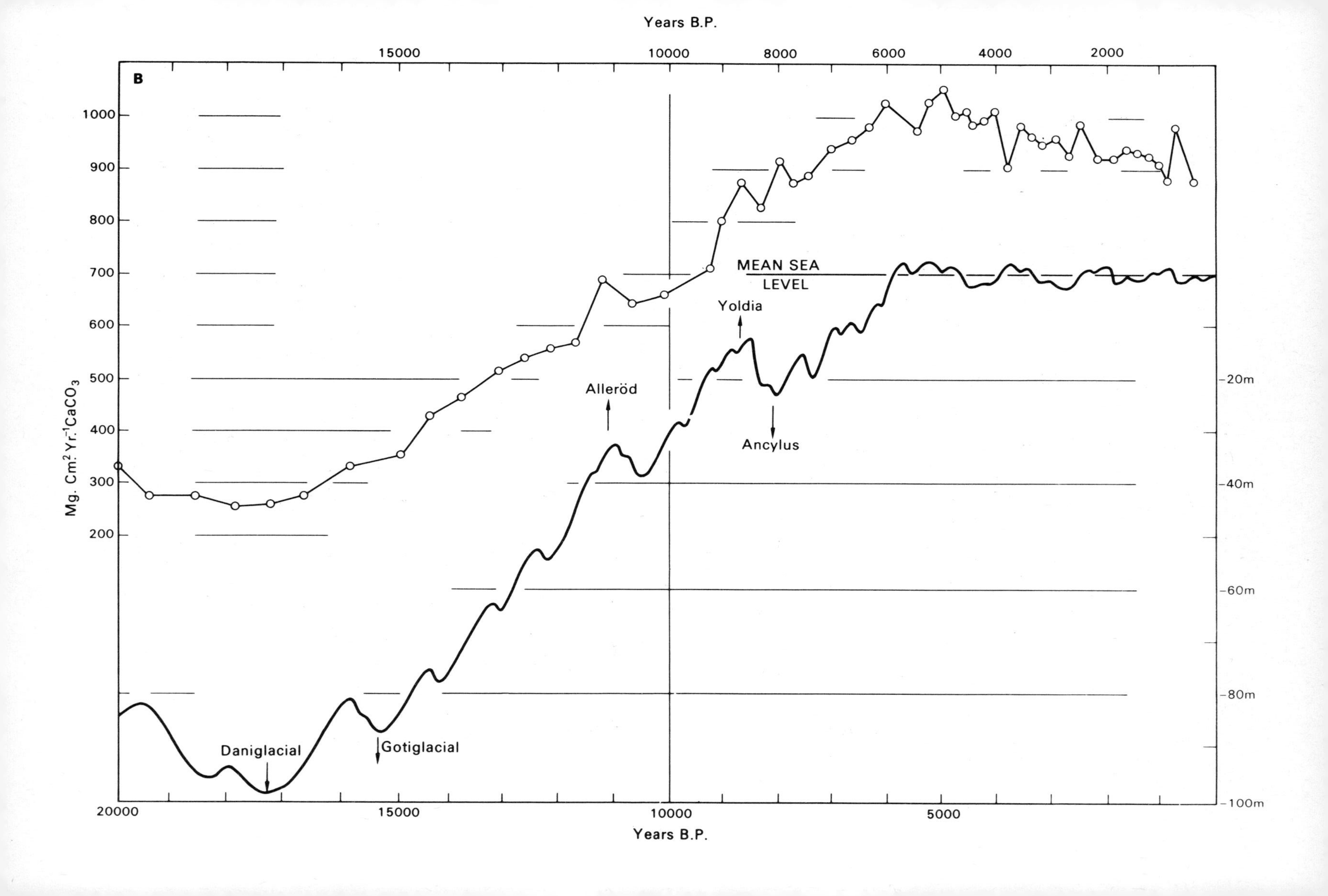

Years B.P.
15000
10000
8000
6000
4000
2000
B
1000
900
800
700
600
500
400
300
200
Mg. Cm.^{2} Yr.$^{-1}$CaCO$_3$
MEAN SEA
LEVEL
Yoldia
Alleröd
Ancylus
Daniglacial
Gotiglacial
–20m
–40m
–60m
–80m
–100m
20000
15000
10000
5000
Years B.P.

of ice is estimated to have averaged 140–170 bars in the ice-covered areas. During interglacial periods this load is transferred to the oceans, which cover 70 per cent of the earth's surface. The excess load over the oceans would be 10–12 bars. The change in position of the load during glacial and interglacial periods must cause isostatic compensation and flow in the deeper layers of the crust, or in the mantle, to maintain the isostatic balance. The rapid rate of compensation suggests that deep-seated movement takes place in the low-viscosity asthenosphere at a depth of a few hundred kilometres.

A pressure of 10 bars is usually the limit at which compensation takes place, so the ocean should also respond to the addition and subtraction of mass during glacial and interglacial phases. Bloom has worked out the theoretical effects of these responses of the ocean in both coastal and oceanic island localities. The assumption is made that the deformation extends 50 km from the edge of the load. At the coastline a sea-level fall of 100 m is followed by a sea floor rise of 30 m, due to isostatic response. This results in a rise of sea level along the coast. The reverse applies when sea level rises 100 m, a change that is followed by a smaller fall. The total change could exceed the eustatic change by one-third. The coast is assumed to be stable while the ocean floor responds to the variation of load. A single oscillation of eustatic sea level could result in the formation of coastal features at four distinct levels, owing to the lag in the isostatic response. The isostatic effect in the areas beyond the ice-sheets is small compared with those in glaciated areas; it is nevertheless, enough to complicate the evidence of sea-level change derived from a study of coastal landforms. Studies made along the eastern coast of the USA support Bloom's theory.

2.3 Late and post-glacial changes in sea level

Much more evidence of sea-level changes in the later periods of the Pleistocene and in recent time is available, partly because radiocarbon dating covers this period. An increasing number of dates are available on organic materials associated with a specific sea level. The literature on this topic is now very large, so only a brief summary will be given of the results of studies of eustatic changes of sea level during the period since about 30,000 years ago. A few comments concerning the complex isostatic effects on sea-level variations in glaciated areas also seem appropriate. These studies relate to the isostatic uplift of glaciated regions.

The last major glacial phase started about 74,000 years ago, giving the Wisconsinan period in North America, the Wurm and Weichsel in Europe, and the Devensian glaciation in Britain. The fluctuations of sea level during the period between the start of this ice advance and the present are closely associated with glacial advances and retreats.

There are several estimates of the changes of sea level during the last 35,000 years. It is generally agreed that between about 35,000 and 30,000 years ago sea level was not very different from its present level, although data are scarce and Mörner (1971a) finds no evidence for a high sea level at 30,000 B.P. This view is given by Shepard (1963b), but McFarlan (1961) has suggested that the sea was at a level of −76 m between 18,000 and 35,000 years ago. The evidence for a rapid fall of sea level between 25,000 and 18,000 is limited. Milliman and Emery (1968) use 80 radiocarbon dates to arrive at a curve for sea level based on data from the Atlantic continental shelf of the USA. They conclude that sea level was at its present height 35,000 years ago, and fell to a minimum level of −130 m 15,000 years ago. The envelope of possible values for the minimum level lies between 132 and 60 metres below present sea level, but both the upper and lower curves give the same time for the maximum fall of sea level. Sea level appears to have fallen slowly from 35,000 to 21,000 B.P., and then more rapidly to 16,000 B.P. A rapid rise started at 14,000 B.P. continuing until 7000 B.P., and thereafter at a slower rate until about 4000 B.P., since which time the change has probably been very small. The data on sea level agree with evidence of a rapid increase of surface temperature at 15,000 B.P. in the Caribbean deep sea cores.

The contribution of Shepard and Curray (1967) is basically similar. They have some evidence for a similar or slightly higher level to the present in the Sangamon interglacial at 100,000 to 150,000 years ago. A fall to −120 m took place during the early advance of the

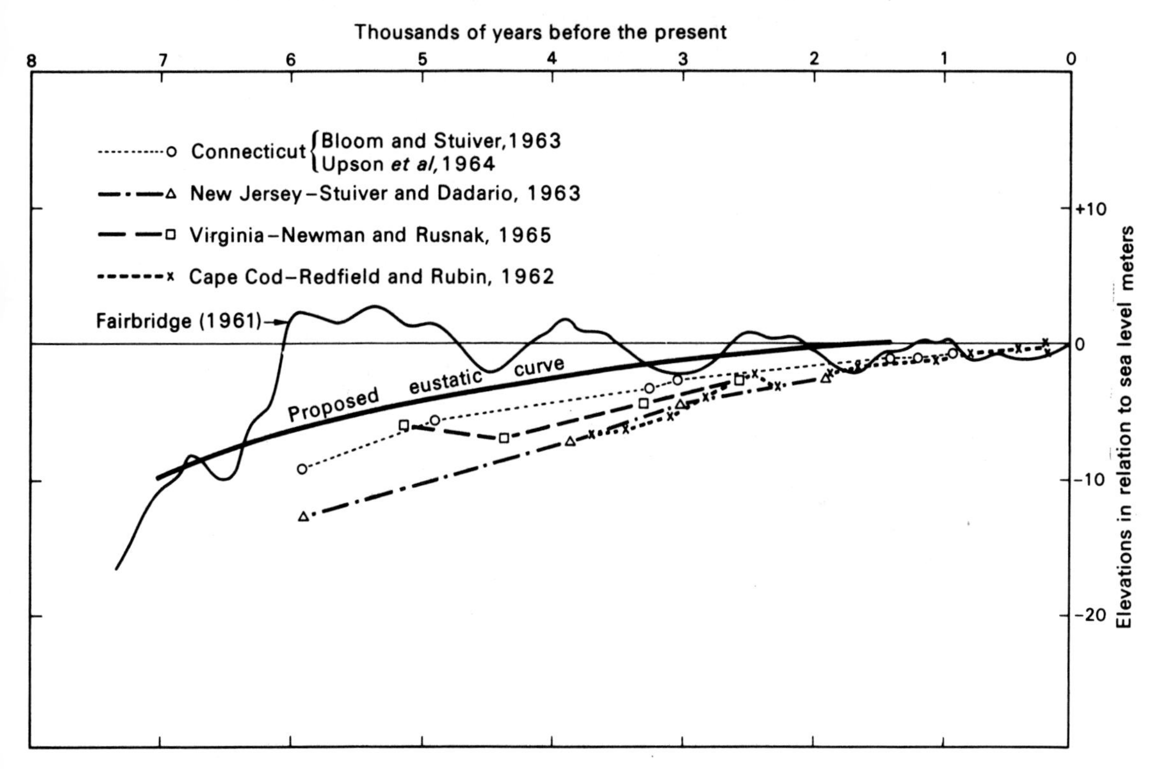

Figure 6.4 Sea-level changes over the last 7000 years. (*After Shepard and Curray, 1967.*)

last glaciation about 50,000–60,000 years ago. They suggest a rise to about −20 m at 30,000 B.P., followed by a renewed fall to −120 m between 10,000 and 20,000 B.P. The data are extremely scattered for the period between 10,000 and 20,000 years ago, with a minimum value of 160 m below present at 19,000 B.P. Their curve is shown in figure 6.4.

The Flandrian transgression which constitutes the rise of sea level from its low position in about 15,000 B.P. is the best documented phase, owing to an abundance of radiocarbon dates. The main problem of this period is whether sea level has exceeded its present height or not. The detailed analysis of Fairbridge (1961) suggests a very complex system of high and low levels, some of which exceeded the present sea level. Fairbridge considers that sea level has oscillated about its present height since 6000 B.P. by about 3–4 m above and below.

Most other observers consider that the higher sea levels of Fairbridge can be explained by local effects and are not eustatic in origin. Fairbridge gives positive values at 5700, 4900, 3700, 2400, 2200, and 1000 years B.P. Some of his data are derived from raised coral reefs in the Pacific. The data on these have been reviewed by Curray, Shepard, and Veeh (1970) in their study in Micronesia. Several examples of elevated reefs were found. Guam provided clear evidence of raised coral reefs, at an elevation of +0·6 m being dated at 2880 ± 110 years B.P., giving a sea level 1·4–1·8 m higher than present. On the other hand 33 of the islands visited failed to show any elevated reefs. These islands are considered more stable than Guam. The low terraces that have been taken to indicate higher sea levels were found to be cemented rubble ramparts, associated with present high tide level. No raised corals in the position of growth were found. The ramparts dated from 2500–3000 years B.P. The conclusion was reached that sea level had not stood higher than present in the Holocene in the Caroline and Marshall Islands that were visited, but sea level must have been near the present when the rubble ramparts were formed. They draw a distinction between the tectonically active and inactive areas, only the active areas showing evidence of sea levels higher than the present.

Other evidence that appears to indicate higher levels, such as raised shoreline barriers in eastern Australia, can also be explained equally well by local tectonic instability, variations in wave refraction, or availability of beach material. A particularly complete record of late glacial and subsequent changes in sea level is available in the deposits of the Dutch coast, which have been studied by Jelgersma (1961, 1966). The chronology is based on radiocarbon dating of peat deposits accumulating in salt-marsh conditions. One difficulty in the interpretation of these records is the subsidence that this area has been undergoing for a very long period, dating back to the Mesozoic at least. Compaction of the sediment must also be taken into account. Jelgersma considers that a slowly rising sea level during the last 5000 years is most likely. This slow rise followed a period of much more rapid rise, which Jelgersma considers covered the period from 20,000 to 5000 years ago. Similar evidence for a slow rise of sea level during the last 5000 years also comes from observations in south Florida, made by Scholl and Stuiver (1967). Their work suggests that in this area sea level was 4·4 m below the present 4400 years ago. During the next thousand years, sea level rose to 1·6 m below present fairly fast, but in the last 3400 years it has risen only 1·6 m. The data were obtained from mangrove swamps. It is thought that this area is stable, so that true eustatic changes are indicated. The curves for the Netherlands and Florida agree closely.

Andrews (1970, p. 23) has collected several estimates of sea-level changes in an attempt to reach the most likely eustatic curve for the last 13,000 years. Table 6·4 gives his data. These values indicate the range of estimates that have been suggested and illustrate the many uncertainties that enter into the establishment of a eustatic sea-level curve, even when many data are available. Some of the variations may be the result of the water load effect to which Bloom (1967) has drawn attention. Andrews uses Shepard's curve, which lies in a central position within the data scatter. Moran and Bryson have shown, in their analysis of the rate of ice wastage in the Laurentide ice-sheet, that the eustatic change of sea level was very closely associated with changes in ice volumes in the continental ice-sheets, which

Plate 20 The submerged forest at Clarach Beach, near Aberystwyth, Wales (*above opposite*). Corrugated *Scrobicularia* clay is exposed below the modern storm beach. (*J. A. Taylor.*) *Below*, the submerged forest at Ingoldmells, Lincolnshire. The tree stumps are rooted in the till of the last major ice advance, overlain by Neolithic peat. (*C.A.M.K.*)

Table 6.4 Estimates of eustatic sea level changes

1	2	3	4	5	6	7	Years B.P.
−0·5		+1	+1		−1		1000
−1		+2	−2		−2		2000
−2		+3	−3		−3	−3	3000
−3		+5	+2		−4	−4	4000
−4		−2	+3	0	−5	−8	5000
−7	0	−0·5	0	−4	−7	−10	6000
−10	−8	−4	−6	−9	−10	−19	7000
−16	−21	−19	−16	−17	−19	−30	8000
−22	−32	−33	−14	−28	−35	−42	9000
−31	−38	−36	−32	−35		−58	10000
−40				−44		−70	11000
−48				−52		−90	12000
−58				−62		−110	13000

1 Shepard (1963)
2 Moran and Bryson (1969)
3 Schofield (1964)
4 Fairbridge (1961)
5 Godwin *et al.* (1958)
6 Jelgersma (1961)
7 Milliman and Emery (1968, inferred from their figure 1)

were wasting very rapidly during the period of the Flandrian transgression. Nearly 75 per cent of the total rise was contributed by the Laurentide ice-sheet, while the European one produced much of the remainder.

Scholl, Craighead, and Stuiver (1969) have slightly revised their earlier eustatic curve and related it to actual years, which differ from Carbon 14 years. The new curve (shown in figure 6·5), agrees with the older one from 4800 years to the present, but before this date the newer curve shows a slower rise. The last 4000 years submergence has been at 3·5 cm/100 years. The rate of biological coastal sedimentation has been at 3·0 cm/100 years. There has probably been a slight increase of sedimentation rate with age, which may reflect

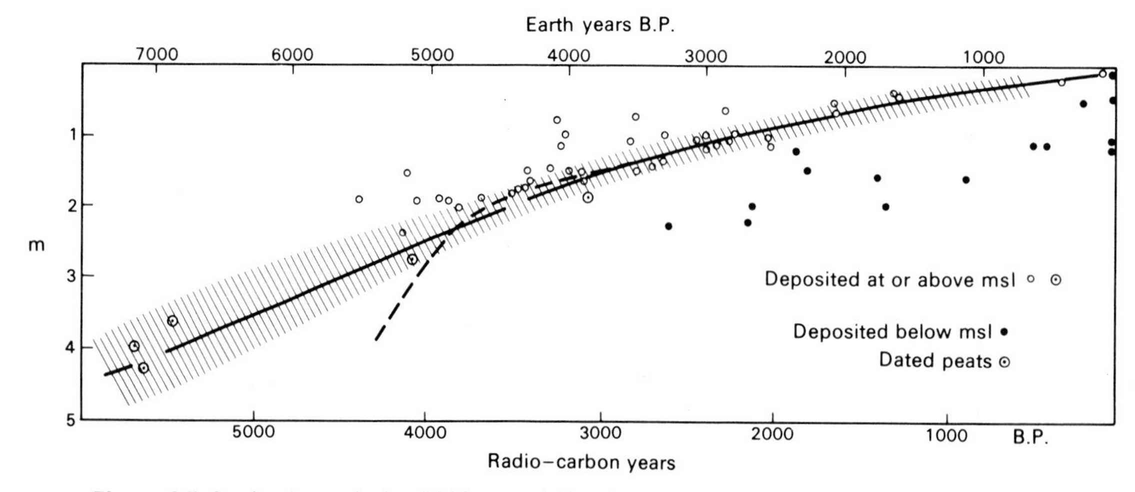

Figure 6.5 Sea level over the last 7000 years. (*After Scholl, Craighead, and Stuiver, 1969.*)

the slowing down of the rate of sea-level rise in the last 4000 years. Scholl *et al.* consider that the rise of sea level has been continuous with no halts or rises above the present level, according to the evidence derived from Florida. The associated sedimentation rates in the fresh water swamps of calcite mud are calculated to have been 2·8 cm/100 years from 4000 to 5000 years B.P., 1·9 cm/100 years from 3000 to 4000 years, 1·7 cm/100 years from 1000 to 3000 years, and 1·2 cm/100 years from the present to 1000 years ago.

Emery, Niino, and Sullivan (1971) have used Carbon 14 dates to assess changes of sea level on the continental shelf of the east China Sea. They found that the levels agreed fairly well with those calculated for the east coast of America and in the Gulf of Mexico.

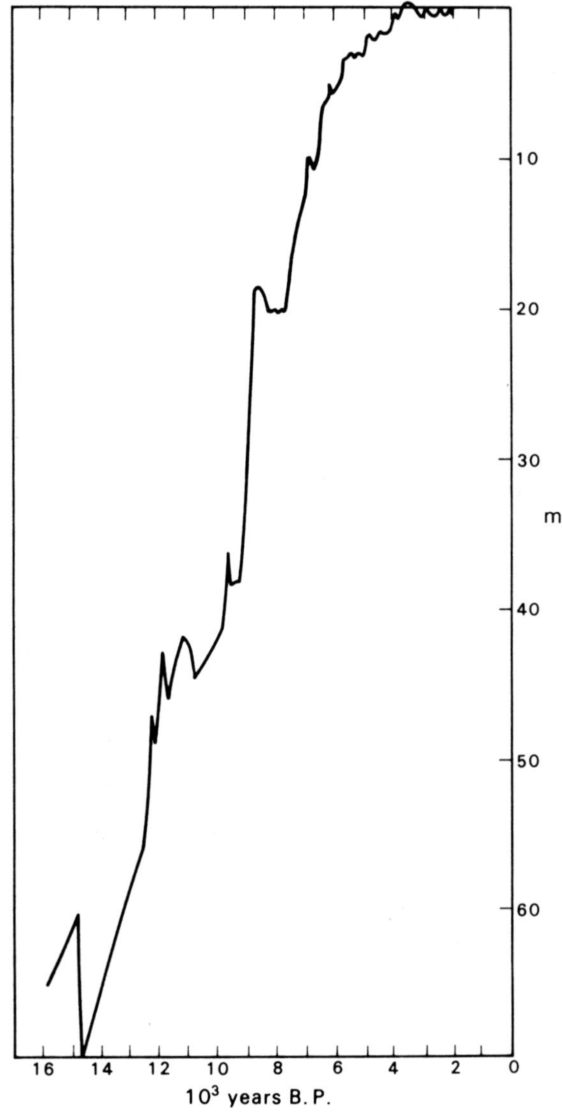

Figure 6.6 Eustatic sea-level curve over the last 16,000 years. (*After Mörner, 1971b.*)

Mörner (1971b) has calculated a new eustatic curve for the last 20,000 years from a very detailed study of the raised shoreline features of southern Sweden. This is an area that has undergone uplift isostatically and, therefore, according to Mörner, provides the best situation in which to establish the eustatic curve, as the isostatic one is well known in the area. Mörner's results (figure 6·6 and table 6·5) show that there is very high crustal sensitivity to changes both of ice load and to water load. He considers that the hydroisostatic effect could account for the differences between his curve and the others that have been mentioned. This factor affects the maximum lowering of the eustatic sea level during the maximum glaciation of the late Weichsel around 15,000 years ago. Mörner's curve shows a

Table 6.5 Post-glacial sea-level and maximum late glacial lowering of sea level

Climatic division	Pollen zone	Date B.C.	Ice load	Sea level	Tilting	Eustatic level, m
Agard interstadial		10700	–	+	rapid	−56·0
Fjaras stadial	I a	10350	+	–	slow	−47·0
Bolling interstadial	I b	10300	–	+	rapid	−49.0
Older Dryas stadial	I c	9950	+	–	none	−42·0
Allerød interstadial	II	9800	–	+	rapid	−46·0
		9000				−42·0
						−45·0
Younger Dryas, Ra, and Salpausselka	III	8050	• +	–	slow	−42·0
Early Preboreal interstadial	IV a	7800	–	+	rapid	−36·0
Preboreal stadial		7700		–		−38·5
Maximum late Weichselian sea-level lowering						
Mörner	Ocean, land		−85, −90 m	stable		
Brittany	Shoreline		−90, −89 m	stable		
Bassa Versilia	Sediment		−90 m	stable		
Flint–ice volume			−132 to −120 m			
Garrison and McMaster	Terrace		−145 m	} Subsided eustatic + hydroisostatic lowering		
Shepard	Shelf break		−132 m	}		
Milliman and Emery	Sea level		−130 m	}		

number of oscillations and sea levels above the present during the last 4000 years, although not as high as Fairbridge suggested.

Mörner considers that sea level has fluctuated during the post-glacial rise, but has never exceeded 0·4 m above the present level on the evidence derived from southern Sweden. Late Pleistocene changes have been world-wide in character, from correlations between Scandinavia, North America and New Zealand, South America and Central Africa. The response to a climatic change is almost immediate, both in a change in the ice volume and the change in sea level related to it. These changes are indicated by Carbon 14 dates and can be clearly identified in the late glacial period up to 8000 years ago. The evidence is more difficult to decipher in the recent period, but there is still much evidence of climatic oscillation in the pollen sequences and other data such as glacier fluctuations, changes in mollusc faunas, and bog development. The periods of glacier advance correlate with eustatic sea-level regressions, and of glacier retreat with eustatic transgressions during the post-glacial period. The temperature changes recorded in the Camp Century ice core in central Greenland, and the deep sea cores discussed by Ericson, Wollin, and Ewing (1970) correlate well with eustatic fluctuations. An abrupt amelioration is indicated between 9000 and about 8600 years B.P., and this correlates with a large transgression between 9300 and 8800. A climatic deterioration between 2000 and 2200 B.P. correlates with an eustatic regression about 2150 B.P. The curves for temperature derived both from the Greenland ice core and the deep sea sediment cores, as well as the well-dated evidence of eustatic sea level, all show marked fluctuations, which appear therefore to be real.

The evidence for sea levels above the present is partly derived from Thorium–Uranium dates supplied by Veeh (1966). He has dated corals in the Pacific and Indian Oceans by the ratios Th^{230}/U^{238} and U^{234}/U^{238}. The first ratios provide dates in the range 90,000 $\pm$20,000 to 160,000 $\pm$40,000, and the second in the range 80,000 $\pm$50,000 to 180,000 $\pm$60,000. The results of the analyses are internally consistent and suggest a eustatic sea level higher than the present at about 120,000 $\pm$20,000 years, possibly during an interglacial period. The age is indicated by the amount of Th^{230}, because marine carbonates contain uranium but are free of the decay product Th^{230}.

In previously glaciated areas the changes in sea level are more complex, as both eustatic and isostatic changes must be taken into account. The isostatic recovery of the land leads to a relative fall of sea level, which may lead either to an absolute fall of sea level or a slower rise, according to its rate of fall relative to the eustatic rise.

A detailed study of the relationship between post-glacial isostatic uplift and sea level changes has been made by Andrews (1970). He has derived curves for post-glacial uplift from data obtained at coastal sites on heights and dates of former sea levels. These data may be presented as uplift–emergence curves for one site, or as patterns of emergence over a large area at one time.

The emergence curve is the sum of the uplift, which is an isostatic rebound, and the effect of the eustatic rise. The post-glacial uplift can be expressed in the form $U'_p = A(I - i^t)/I - i$. U'_p is the post-glacial uplift in time t since glaciation; t is the time measured in increments of 1000 years; i is a constant with a value of 0·677 for Arctic Canada; A is the amount of rebound between t = O and t = 1. The isostatic uplift accelerates slowly at first as the ice is thinning. At the time of deglaciation, uplift is rapid, while subsequently the rate decreases exponentially. Thus the period during which sea level is falling is one of decelerating uplift. The rebound appears to have been faster in Arctic Canada than in Britain or Fennoscandia. In the two latter areas uplift started rather earlier, when eustatic sea level was rising faster, so that in many areas transgressions as well as regressions have occurred. Tentative values for i are 0·770 and 0·742, and those for A values are 25 per cent and 27 per cent for a 10,000-year recovery period, for Britain and Fennoscandia respectively. The A value for Arctic Canada is 33 per cent.

The three-dimensional analysis of isostatic recovery in terms of former sea levels can best be based on the mapping of synchronous shorelines. Isobases, which are lines of equal post-glacial emergence over the same length of time, are shown. The method of trend surface analysis is useful in generalizing the pattern of emergence. Trend maps of rates of recovery can also be calculated. The isobase map for eastern Arctic Canada for 8000 B.P. shows a maximum emergence since this date of 280 m in southeast Hudson Bay, with a second maximum of 240 m over northwest Hudson Bay. The maximums for emergence since 6000 B.P. are 120 and 100 m respectively, illustrating the very rapid uplift during the period immediately following deglaciation. Maps for the rates of uplift show the same two centres rising at over 5 m/century at 6000 B.P., while the present rate of rise is rather over 1·3 m/century in southeast Hudson Bay and not quite 1 m/century in the northwest of the Bay. The value is under 0·3 m/century along the extreme eastern part of Baffin Island and southern Newfoundland and Nova Scotia. The uplift still remaining is 160 m in eastern Hudson Bay and 120 m in northwest Hudson Bay. It is 40 m in eastern Baffin Island and Nova Scotia. Uplift is also still going on in Fennoscandia, at an estimated rate of 9·3 m/century at the head of the Gulf of Bothnia (Donner, 1969).

2.4 Present changes in sea level

The evidence of the tide gauges around the British Isles shows that sea level is rising slowly in southeastern England. The most accurate tide gauges at Newlyn in Cornwall and Felixstowe in Suffolk show rises of 2·3 mm/year and 1·7 mm/year respectively (Valentin, 1954). The major part of this rise is due to the present eustatic rise of sea level, resulting from continued deglaciation, which has been particularly active in the 1930s and 1940s. The line of no change, according to Valentin, runs north–south between Aberdeen, which is rising at 0·5 mm/year, and Dundee, which is sinking by the same amount. It then runs south-southwest to Anglesey and then probably turns north around Scotland. The evidence for the rise of central Scotland is doubtful, and Hafemann (1954) considers that Scotland has not risen isostatically since the first century A.D.

A somewhat similar pattern is found on the coast of northwest Europe. Rapid uplift is still active around the Gulf of Bothnia. The line of no change runs through the extreme southwest of Norway, across north Denmark and southern Sweden. The maximum rate of sinking occurs in south Denmark and north Germany on the North Sea coast.

The present eustatic rise of sea level has been estimated at 1·12 mm/year by Fairbridge (1961) and 1·18 mm/year by Wexler (1961). The latter discusses some of the factors on which this estimate depends. The basis of his argument rests on the estimated changes in volume of the Antarctic ice-sheet, which is by far the largest. Its variation will have a dominant effect on sea level. It has been estimated that if all the ice melted, sea level would rise by 100 m, which would flood many of the major cities and all the ports.

Various estimates of its state of balance have been suggested; the majority consider that the Antarctic ice is growing in volume. A general climatic warming, which is taking place, in the northern hemisphere glaciated areas, could cause an increase in the Antarctic ice because a slight rise of temperature would enable the air to hold more moisture, without raising the temperature above freezing point.

Some values of the current changes in sea level are given in table 6.6 on page 276.

Glacio-eustatic effects will continue to operate until all the ice, now on the land, finally melts. There are two alternatives for the near future—either the ice will continue to melt and sea level to rise, or it is possible that the present is an interglacial period. Sea level would then fall as ice again builds up to form new ice-sheets over northwest Europe and North America.

Plate 21 Raised beaches, Baffin Island, NWT, Canada (*above opposite*). Small limestone pebble beaches have formed as a result of isostatic recovery, causing a falling sea level. Where the gradient is steep enough a sequence of chevron-shaped ridges, pointing seaward, have developed. On the low-gradient area in the distance there are no beaches, but circular thaw ponds, and beyond them the shallow, ice-covered Foxe Basin. *Below,* an example of a raised rock coastal platform is shown, backed by an abandoned cliff, north of Port Donain, Isle of Mull, Scotland. (*J. M. Gray.*) (*See Gray, 1974.*)

Table 6.6 Current changes in sea level

+1 mm/year	Poland, Latvia, and Germany
+0·5	Denmark
+1·5	Holland
+1·5	Portugal
+1·4	France
+1·7	Italy
+0·6	North Africa
+2·2	Formosa
+1·0	Japan
+2·7	Portland, Maine
+0·9	Boston, Massachusetts
+2·4	New York
+3·0	Portland, Virginia
+1·8	Miami, Florida
−13·7	Juneau, Alaska
−6·1	Yakutat, Alaska
+2·1	Seattle, Washington
+1·5	San Francisco, California
−0·3	Los Angeles, California
+4·0	Galveston, Texas
+1·0·	Key West, Florida
+9·1	Eugene Island, Louisiana

2.5 Seasonal fluctuations in sea level

Seasonal changes provide an example of fairly small-scale cyclic changes of sea level, which differ entirely from the longer-term fluctuations that have been considered so far. Seasonal changes in sea level are partly caused by variations in heat storage in the ocean. The warmer the water, the less dense it is and the higher a given column will stand. Other variations are due to changes in atmospheric pressure. This acts as an inverted barometer—the higher the pressure, the lower the water level. A pressure change of 34 mb is equivalent to 30 cm of water elevation. Other variables causing seasonal changes are wind regimes, changes in precipitation, and in water characteristics.

The pattern of seasonal changes has been described by Pattullo (1963). In March the level is lower than the mean in the northern hemisphere and higher in the south, apart from the Arabian Sea and the Gulf of Siam, and in the zone between 40°N and 60°N. The only negative values in the southern hemisphere occur along the south coast of Australia. The largest deviation of −40 cm occurs in the Bay of Bengal, although values of −19 cm occur off Mexico, central America, and northeast Siberia. Values are negative also in the Arctic. The pattern in September is very similar to that in March, but the signs are changed. The maximum deviation is still in the Bay of Bengal, where it reaches +54 cm, while it is +13 cm and +27 cm off west Mexico and northeast Siberia. The main exceptions to the reversal occur off southeast USA and Iceland. There are negative deviations off south Australia at both seasons.

In June and December the patterns are not so regular. In June negative deviations tend to occur in the centres of the oceans. Values are positive in the northern Indian Ocean, the western Pacific on either side of the equator, and around south Australia. The deviations are smaller on the whole, in June the largest being −18 cm in the Gulf of Siam, −13 cm in north Norway, +30 cm in the Bay of Bengal, and +14 cm in south Australia. In December the largest positive values of +20 cm occurs in the Gulf of Siam, +16 cm occurs along the coast of Norway, −26 cm in the Bay of Bengal, and −10 cm in south Australia. Again the pattern is almost reversed at the opposite season. In the Arctic Ocean negative values occur in June along the north coast of Greenland, Europe, and Asia, while in December the negative values are restricted to north Alaska and northeast Siberia.

The largest changes occur in the Bay of Bengal, in which there is a major change in wind and current direction with the seasons due to the monsoon effect. In the subtropics the changes are mainly due to seasonal variations in heat storage, and amount to 10–15 cm. Heat variations are also probably responsible for the changes in the tropics. Atmospheric pressure changes are likely to be more important in the subarctic north of 40°N. In some coastal areas the effect of the wind is important, such as an increase in the onshore wind component.

The annual variation of sea level along the eastern seabord of the United States could be accounted for in part by variation in river runoff according to Meade and Emery (1971). A regression analysis shows that between 7 and 21 per cent of the total variation in the annual sea-level variation could be accounted for by variation in river flow. The secular rise of sea level accounts for between 29 and 68 per cent, leaving between 10 and 50 per cent unaccounted for.

Rossiter (1962) has discussed periodic and secular long-term changes in sea level, covering periods of a day upwards. The causes of these changes can be divided into oceanographical factors, climatological factors, geodetic factors, and long-period astronomical tides. The first set of changes are concerned with gradients set up in water bodies due to dynamical relationships between various types of currents. Climatological effects include many of the seasonal changes that have already been mentioned, including the changes discussed by Pattullo and the changes resulting from glaciation. Local events include the generation of surges, which will be considered under wave action in book II. Geodetic changes include the effects of glacial isostatic responses, already mentioned. There are also the effects of sudden movement of the earth's crust along faults, which, for example, suddenly elevated part of Wellington Harbour as the result of an earthquake. Major changes of sea level would result from movement of the axis of rotation of the earth, but these are very long term and their effect is hypothetical and difficult to discern. There are also long-term tidal effects. These include the lunar fortnightly, luni-solar fortnightly, lunar monthly, solar semi-annual, solar annual, and the nodal tide of 18·62 years. The theoretical value of the nodal tide at Newlyn is 4·8 mm, its maximum value in European waters being 9·1 mm in Oulu in Finland, while Newlyn has the minimum value. There should be a pole tide of 5 mm resulting from the 14-month minor fluctuation cycle of the pole of rotation. These variations of sea level provide a serious obstacle to the use of mean sea level as a reference plane for geodetic levelling. The shorter-term variations due to the tides and waves will be considered in more detail in book II. Occasional changes, due to surges and tsunamis, can be locally and occasionally significant. Dynamic gradients resulting from currents also provide complications in using sea level as a datum.

Recent work has shown that both the structure of the ocean basins, and the water within them, are much less static than was thought at one time. Neither the position of the oceans and continents, nor the level of the sea, are constant. Detailed observations reveal their movements increasingly clearly.

Further reading

ANDREWS, J. T. 1970: A geomorphological study of post-glacial uplift with particular reference to Arctic Canada. *Inst. Brit. Geog. Sp. Pub.* **2.** (A detailed discussion of isostatic recovery in a previously glaciated area, including a long bibliography.)

BLOOM, A. L. 1967: Pleistocene shorelines: a new test of isostasy. *Bull. Geol. Soc. Amer.* **78,** 1477–94. (A discussion of the effect of the variation in water volume in the oceans on the isostatic response of the ocean floor.)

FAIRBRIDGE, R. W. 1961: Eustatic changes of sea level. *Physics and chemistry of the earth.* Vol. **4.** Oxford: Pergamon Press, pp. 99–185. (An account of the causes of eustatic changes of sea level; a curve of these changes is presented.)

JELGERSMA, S. 1966: Sea level changes during the last 10,000 years. Roy. Met. Soc. Sym. on World Climate from 8000 to 0 B.C. London: Roy. Met. Soc. (An account of sea-changes based on data from the Netherlands.)

MACINTYRE, F. 1970: Why the sea is salt. *Sci. Amer.* (A fairly detailed account of the chemical cycles that operate in the ocean water, with reference to the collection of water in the ocean basins.) *Sci. Amer.* readings introduced by J. R. Moore (San Francisco: Freeman). Chap. 13, 110–121: *Oceanography.*

MÖRNER, N. A. 1971a: Eustatic changes during the last 20,000 years and a method of separating the isostatic and eustatic factors in an uplifted area. *Palaeogeog. Palaeoclim. and Palaeoecol.* **9** (3), 153–81. (Provides a new eustatic curve based on a very detailed observations on the south coast of Sweden, which has been uplifted.)

SCHOLL, D. W., CRAIGHEAD, F. C., and STUIVER, M. 1969: Florida submergence curve revised: its relation to coastal sedimentation rates, *Science* **163,** 562–4. (A revised eustatic curve based on observations in the Gulf of Mexico, covering the last 7000 years.)

SHEPARD, F. P. and CURRAY, J. R. 1967: Carbon 14 determinations of sea level changes in stable areas. In M. Seara (editor), *Progress in Oceanography* **4.** Quaternary history of the ocean basins. Oxford: Pergamon, 283–91. (An eustatic curve is presented that has been used often but which has since been modified by later work.)

TANNER, W. F. 1968; Tertiary sea level symposium—introduction and multiple influences on sea level changes in the Tertiary. *Palaeogeog. Palaeoclim, and Palaeoecol.* **5,** 7–14, 165–71. (The special number contains a number of papers on Tertiary sea level and the evidence for its variation, including the introduction and summing up by Tanner.)

THOM, B. G. 1973: The dilemma of high interstadial sea levels during the last glaciation. *Progr. Geogr.* **5,** 167–246. (An examination of various hypotheses about and the evidence for high interstadial sea levels.)

WEXLER, H. 1961: Ice budgets for Antarctic and changes of sea level. *J. Glaciol.* **3,** 867–72. (An estimate of the effect of changes of Antarctic ice on sea level and glaciation.)

References

AGASSIZ, A. 1899: The islands and coral reefs of Fiji. *Bull. Mus. Comp. Zool. Harv.* **33,** 1–167.
1906: General report of the expedition steamer *Albatross* from October, 1904 to March, 1905. *Mem. Mus. Comp. Zool.* **33,** xiii, 77.

ALLEN, J. R. L. 1970: *Physical processes of sedimentation. Earth Sci. Ser.* **1.** London: Allen and Unwin, chaps. 5, 6.

ANDREWS, J. T. 1970: A geomorphological study of postglacial uplift with particular reference to Arctic Canada. *Inst. Brit. Geog. Sp. Pub.* **2.**

ANIKOUCHINE, W. A. and LING, H. Y. 1967: Evidence for turbidite accumulation in trenches in the Indo-Pacific region. *Mar. Geol.* **5,** 141–54.

ANTEVS, E. 1928: The last glaciation. *Amer. Geog. Soc. Res. Ser.* **17.**

ARRHENIUS, G. 1952: Sediment cores from the eastern Pacific. Exped. 1947–48. *Swedish Deep Sea Report* **5,** fasc. 1.

BAGNOLD, R. A. 1963: Mechanics of marine sedimentation—autosuspension of sediment. In M. N. Hill (editor), *The sea.* Vol. **III.** New York: Wiley, chap. 21, pt I.

BANDY, O. L. 1967: Foraminiferal definition of the boundaries of the Pleistocene in southern California, U.S.A. In M. Sears (editor), *Progress in oceanography.* Vol. **4:** Quaternary history of the ocean basins. Oxford: Pergamon.

BANDY, O. L. and WADE, M. E. 1967: Miocene–Pliocene–Pleistocene boundary in deep-water environments. In M. Sears (editor), *Progress in oceanography.* Vol. **4:** Oxford: Pergamon.

BELDERSON, R. A. and STRIDE, A. H. 1966: Tidal current fashioning of a basal bed. *Mar. Geol.* **4,** 237–57.
1969: Tidal currents and sand wave profiles in the northeast Irish Sea. *Nature* **222,** 74–5.

BENIOFF, H. 1954: Orogenesis and deep crustal structure—additional evidence from seismology. *Bull. Geol. Soc. Amer.* **65,** 385–400.

BLACKETT, P. M. S., BULLARD, E. C., and RUNCORN, S. K. (editors) 1965: A symposium on continental drift. *Phil. Trans. Roy. Soc.* **A 258,** 228.

BLOOM, A. L. 1967: Pleistocene shorelines: a new test of isostasy. *Bull. Geol. Soc. Amer.* **78,** 1477–94.
1971: Glacial–eustatic and isostatic controls of sea level since the last glaciation. In K. K. Turekian (editor), *Late Cenozoic Glacial ages.* New Haven: Yale University Press, chap. 13.

BLOOM, A. L. and STUIVER, M. 1963: Submergence of the Connecticut coast. *Science* **139,** 332–4.

BOSTROM, K., JOENSUU, O., VALDES, S., and RIERA, M. 1972: Geochemical history of south Atlantic sediments since Late Cretaceous. *Mar. Geol.* **12** (2), 85–121.

BOTT, M. H. P. 1971: *The interior of the earth.* London: Edward Arnold.

BOURCART, J. 1950: La théorie de la flexure continentale. *Compte Rendu Inter. Géog. Un.* Lisbon, **16,** 167–90.

BOYCE, R. E. and SMITH, E. L. 1968: Geomorphology of Erben Guyot. *Mar. Geol.* **6,** 179–83.

BROECKER, W. S. 1961: Discussion of radio-carbon dating of late Quaternary deposits, S. Lousiana. *Bull Geol. Soc. Amer.* **72,** 159–61.

1971: Calcite accumulation rates and glacial to interglacial changes in oceanic mixing. In K. K. Turekian (editor), *Late Cenozoic glacial ages.* New Haven: Yale University Press, chap. 9.

BROECKER, W. S., EWING, M., and HEEZEN, B. C. 1960: Evidence for an abrupt change in climate close to 11,000 years ago. *Amer. J. Sci.* **258,** 429–48.

BROECKER, W. S., THURBER, D. L., GODDARD, J., KU, T., MATTHEWS, R. K., and MESOLELLA, K. J. 1968: Milankovich hypothesis supported by precise dating of coral reefs and deep-sea sediments. *Science* **159,** 297–300.

BROECKER, W. S., TUREKIAN, K. K., and HEEZEN, B. C. 1958: The relation of deep-sea sedimentation rates to variations in climate. *Amer. J. Sci.* **256,** 503–17.

BUDINGER, T. F. 1967: Cobb seamount. *Deep Sea Res.* **14,** 191–201.

BUFFINGTON, E. C. and MOORE, D. G. 1963: Geophysical evidence on the origin of the gullied submarine slopes, San Clemente, California. *J. Geol.* **71,** 356–70.

BUGH, H. W. 1971: Sea-floor spreading in the southwest Indian Ocean. *J. Geophys. Res.* **76** (26), 6276–82.

BULLARD, E. C. 1954: The flow of heat through the floor of the Atlantic Ocean. *Proc. Roy. Soc.* **A 222,** 408–29.

1968: Reversals of the earth's magnetic field. *Phil. Trans. Roy. Soc.* **A 263,** 481–524.

1969: The origins of the oceans. In *The ocean, Scientific American* special issue, reprinted 1969, San Francisco: Freeman, 16–25.

BURCKLE, L. H., EWING, J., SARTO, T., and LEYDON, R. 1967: Tertiary sediments from the east Pacific Rise. *Science* **157,** 537–40.

BYE, J. A. T. 1971: The slope of abyssal plains. *J. Geophys. Res.* **76** (18), 4188–94.

BYRNE, J. V. and EMERY, K. O. 1960: Sediments of the Gulf of California. *Bull. Geol. Soc. Amer.* **71,** 983–1010.

CARLSON, R. O. and BROWN, M. V. 1955: Seismic refraction profiles in the submerged Atlantic coastal plain near Ambrose Lightship. *Bull. Geol. Soc. Amer.* **66,** 969–76.

CARRUTHERS, J. N. 1963: History, sand waves and nearbed currents of La Chapelle Bank. *Nature* **197,** 942–7.

CARSOLA, A. J. 1954: Submarine canyons on the Arctic slope, *J. Geol.* **62,** 605–10.

CARTWRIGHT, D. E. 1959: Submarine sand waves and tidal lee-waves. *Proc. Roy. Soc.* **A 253,** 218–41.

CARTWRIGHT, D. E. and STRIDE, A. H. 1958: Large sand waves near the edge of the continental shelf. *Nature* **181,** 41.

CASTON, V. N. D. and STRIDE, A. H. 1970: Tidal sand movement between some linear sand banks in the North Sea off northeast Norfolk. *Mar. Geol.* **9,** M38–M42.

CHARLESWORTH, J. K. 1957: *The Quaternary era.* London: Edward Arnold.

CHAVE, K. E. 1967: Recent carbonate sediments—an unconventional view. *J. Geol. Educ.* **15** (5), 200–1.

CHESTER, R. and ELDERFIELD, H. 1970: Dust over the oceans. *New Sci.* **47,** 27 Aug., 432.

CHESTER, R. and JOHNSON, L. R. 1971a: Atmospheric dusts collected off the Atlantic coasts of north Africa and the Iberian Peninsula. *Mar. Geol.* **11,** 251–60.

1971b: Atmospheric dusts collected off the west African coast. *Nature* **229,** 105–7.

CHRISTENSEN, N. I. 1970: Composition and evolution of the oceanic crust. *Mar. Geol.* **8,** 139–54.

CLOET, R. L. 1954a: Sand waves in the southern North Sea and in the Persian Gulf. *J. Inst. Navig.* **7,** 272–9.

1954b: Hydrographic analysis of the Goodwin Sands and the Brake Bank. *Geog. J.* **120,** 203–15.

COLLINSON, D. W. and RUNCORN, S. K. 1960: Palaeomagnetic observations in the United States. *Bull. Geol. Soc. Amer.* **71,** 915–58.

CONOLLY, J. R. 1968: Submarine canyons of the continental margin, east Bass Strait (Australia). *Mar. Geol.* **6,** 449–62.

CONOLLY, J. R. and EWING, M. 1967: Sedimentation in the Puerto Rico trench. *J. Sedim. Petrol.* **37,** 44–59.

COTTON, C. A. 1958: The rim of the Pacific. *Geog. J.* **124,** 223–31.

COX, A. 1969: Geomagnetic reversals. *Science* **163**, 237–45.

COX, A. and DOELL, R. R. 1960: Review of palaeomagnetism. *Bull. Geol. Soc. Amer.* **71,** 645–768.

CREASE, J. 1962: Velocity measurements in the deep water of the western Atlantic Ocean, summary. *J. Geophys. Res.* **67,** 3173–6.

CRONAN, D. S. and TOOMS, J. S. 1969: The geochemistry of manganese nodules and associated pelagic deposits from the Pacific and Indian Oceans. *Deep Sea Res.* **16,** 335–59.

CURRAY, J. R. 1964: Transgressions and regressions. In R. L. Miller (editor), *Papers in marine geology: Shepard commemorative volume.* New York: Macmillan.

CURRAY, J. R., MOORE, D. G., BELDERSON, R. H., and STRIDE, A. H. 1966: Continental margin of western Europe: slope progradation and erosion. *Science* **154,** 770–71.

CURRAY, J. R., SHEPARD, F. P., and VEEH, H. H. 1970: Late Quaternary sea-level studies in Micronesia. Carmarsel Expedition. *Bull. Geol. Soc. Amer.* **81,** 1865–80.

DALY, R. A. 1934: *The changing world of the ice age.*

1936: *The changing world of the ice age.* New Haven: Yale University Press.

1948: Coral reefs—a review. *Amer. J. Sci.* **246,** 193–207.

DANA, J. D. 1885: Origin of coral reefs and islands. *Amer. J. Sci.* **3,** 30, 89–105, 169–91.

DARWIN, C. R. 1842: *The structure and distribution of coral reefs.* London: Smith, Elder and Co.

DAVIES, J. L. 1964: A morphogenetic approach to world shorelines. *Zeit. für Geomorph.* **8,** Sp. Nu. 27*–42*.

DAVIS, W. M. 1928: *The coral reef problem.* New York: Amer. Geog. Soc.

DEACON, M. 1971: *Scientists and the sea,* 1650–1900. *A study of marine science.* London: Academic Press.

DE BOER, G. 1964: Spurn Head: its history and evolution. *Trans. Inst. Brit. Geog.* **33,** 71–89.

DELANY, F. M. (editor) 1970: *Geology of the eastern Atlantic continental margin.* 2. Europe working party 31st symposium, Cambridge. Report No. 70/14 I.C.S.U./S.C.O.R. London: H.M.S.O.

DEMENITSKAYA, R. M. and HUNKINS, K. L. 1970: Shape and structure of the Arctic Ocean. In E. A. Maxwell (editor), *The sea.* Vol. **4,** Pt II. New York: Wiley.

DIETZ, R. S. 1952: The geomorphic evolution of the continental terrace shelf and continental slope. *Bull. Amer. Assn Petrol. Geol.* **36,** 1802–19.

1961: Continents and ocean basins evolution by spreading of the sea floor. *Nature* **190,** 854–7.

DIETZ, R. S., MENARD, H. W., and HAMILTON, R. L. 1954: Echograms of the mid-Pacific expedition. *Deep Sea Res.* **1,** 258–72.

DIETZ, R. S. and SPROLL, W. P. 1970: Fit between Africa and Antarctica. *Science* **167,** 1612–1614.

DILL, R. F. 1969: Earthquake effects on fill of Scripps submarine canyon. *Bull. Geol. Soc. Amer.* **80,** 321–8.

DINGLE, R. V. 1965: Sand waves in the North Sea mapped by continuous reflection profiling. *Mar. Geol.* **3,** 391–400.

DONAHUE, J. G. 1967: Diatoms as indicators of Pleistocene climatic fluctuations in the Pacific sector of the Southern Ocean. In M. Sears (editor), *Progress in oceanography.* Vol. **4.** Oxford: Pergamon.

1970: Pleistocene diatoms as climatic indicators in north Pacific sediments. J. D. Hays (editor), *Geol. Soc. Amer. Mem.* **126,** 121–38.

DONN, W. L., FARRAND, W. R., and EWING, M. 1962: Pleistocene ice volumes and sea-level lowering. *J. Geol.* **70,** 206–14.

DONNER, J. J. 1959: Late and postglacial raised beaches in Scotland. *Suolmal. Tiedeak.* (Tom. Ann. Acad. Sci. Fenn.) *Ser. A III, Geol. Geog.* **53,** 5–25.

1969: A profile across Fennoscandia of late Weichselian and Flandrian shorelines. *Comm. Phys.-Math. Ed. Soc. Sci. Fennica* **36,** 1–23.

DONOVAN, D. T. (editor) 1968: *Geology of the shelf seas.* Edinburgh: Oliver and Boyd.

EMERY, K. O. 1950: Continental slopes and submarine canyons. *Geol. Mag.* **87,** 102–4.

1960: *The sea off southern California.* New York: Wiley.

1965: Geology of the continental margin off eastern USA. In W. F. Whittard and R. Bradshaw (editors), *Submarine geology. Proc. 17th Colston. Res. Soc.,* Bristol. London: Butterworth.

1970: Continental margins of the world. In F. M. Delany (editor), *The geology of the east Atlantic continental margin.* Inst. Geol. Sci. Report 70/13 1: General and economic papers. London: H.M.S.O.

EMERY, K. O. and CONNARY, S. D. 1970: Nepheloid layer in the north Pacific. J. D. Hays (editor), *Geol. Soc. Amer. Mem.* **126,** 41–82.

EMERY, K. O., HIROSHI, NIINO, and SULLIVAN, B. 1971: Post-Pleistocene levels of the east China Sea. In K. K. Turekian (editor), *Late Cenozoic glacial ages.* New Haven: Yale University Press, chap. 14.

EMERY, K. O., UCHUPI, E., PHILLIPS, J. D., BAWM, C. O., BUNCE, E. T., and KNOTT, S. T. 1970:
Continental rise off eastern North America. *Bull. Amer. Assn Petrol. Geol.* **54,** 44–108.

EMILIANI, C. 1955: Pleistocene temperature. *J. Geol.* **63,** 538–78. 1966: Palaeotemperature analysis of the Caribbean cores P 6304–8 and P 6304–9 and a generalized temperature curve for the past 425,000 years. *J. Geol.* **74,** 109–26.

EMILIANI, C. 1967: The Pleistocene record of the Atlantic and Pacific sediments, correlation with the Alaskan stages by absolute dating and the age of the last reversal of the geomagnetic field. In M. Sears (editor), *Progress in oceanography*. Vol. **4.** Oxford: Pergamon.

ERICSON, D. B., EWING, M., and HEEZEN, B. C. 1951: Deep-sea sands and submarine canyons. *Bull. Geol. Soc. Amer.* **62,** 961–8.

ERICSON, D. B., EWING, M., HEEZEN, B. C., and WOLLIN, G. 1955: Sediment deposition in the deep Atlantic. *Geol. Soc. Amer. Sp. Pap.* **62,** 205–20.

ERICSON, D. B., EWING, M., WOLLIN, G., and HEEZEN, B. C. 1961: Atlantic deep-sea sediment cores. *Bull. Geol. Soc. Amer.* **72,** 193–286.

ERICSON, D. B. and WOLLIN, G. 1964: *The deep and the past.* New York: Knopf.

1968: Pleistocene climates and chronology in deep-sea sediments. *Science* **162,** 1227–34.

1970: Pleistocene climates in the Atlantic and Pacific Oceans. A comparison based on deep-sea sediments. *Science* **164,** 1488.

ERICSON, D. B., WOLLIN, G., and EWING, M. 1970: Late Pleistocene climate recorded in Atlantic and Pacific deep-sea sediments. In K. K. Turekian (editor), *Volume dedicated to R. F. Flint*. New Haven: Yale University Press.

EWING, G. C. (editor) 1965: *Oceanography from space.* Woods Hole, Mass.: Woods Hole Oceanographic Institution.

EWING, J. and EWING, M. 1967: Sediment distribution on the mid-ocean ridges with respect to spreading of the sea floor. *Science* **156,** 1590–92.

EWING, M. 1971: The late Cenozoic history of the Atlantic Basin and its bearing on the cause of ice ages. In K. K. Turekian (editor), *Late Cenozoic glacial ages.* New Haven: Yale University Press, chap. 21.

EWING, M. *et al.* 1953: Exploration of the northwest Atlantic mid-ocean canyon. *Bull. Geol. Soc. Amer.* **64,** 865–8.

EWING, M. and CONNARY, S. D. 1970: Nepheloid layer in the north Pacific. *Geol. Soc. Amer. Mem.* **126,** 41–82.

EWING, M., EITTREIM, S., TRUCHAN, M., and EWING, J. I. 1969: Sediment distribution in the Indian Ocean. *Deep Sea Res.* **16,** 231–48.

EWING, M. and HEEZEN, B. C. 1955: Puerto Rico trench, topographic and geophysical data. *Geol. Soc. Amer. Sp. Pap.* **62,** 255–68.

EWING, M. and PRESS, F. 1955: Geophysical contrasts between continents and ocean basins. *Geol. Soc. Amer. Sp. Pap.* **62,** 1–6.

EWING, M. and THORNDIKE, E. M. 1965: Suspended matter in deep ocean water. *Science* **147,** 1291–4.

EWING, M. and WORZEL, J. L. 1954: Gravity anomalies and structure of the West Indies, Part I. *Bull. Geol. Soc. Amer.* **65,** 165–74.

EWING, M., WORZEL, J. L., STEENLAND, N. C., and PRESS, F. 1950: Geophysical investigations in the emerged and submerged Atlantic coastal plain. *Bull. Geol. Soc. Amer.* **61,** 877–92.

FAIRBRIDGE, R. W. 1961: Eustatic changes of sea-level. *Physics and chemistry of the earth.* Vol. **4.** Oxford: Pergamon.

(editor) 1968: *Encyclopedia of oceanography.* New York: Reinhold.

FAIRBRIDGE, R. W. and NEWMAN, W. S. 1968: Postglacial crustal subsidence of the New York area. *Zeit. für Geomorph.* **12,** 296–317.

FAIRBRIDGE, R. W. and STEWART, H. B. 1960: Alexa Bank, a drowned atoll on the Melanesian border plateau. *Deep Sea Res.* **7,** 100–116.

FISHER, J. J. 1968: Barrier island formation: discussion. *Bull. Geol. Soc. Amer.* **7** (4), 1421–6.

FISHER, R. L. 1961: Middle America trench: topography and structure. *Bull. Geol. Soc. Amer.* **72,** 703–20.

FISK, H. N. and MCFARLANE, E. 1955: Late Quaternary deltaic deposits of the Mississippi River. *Geol. Soc. Amer. Sp. Pap.* **62,** 279–302.

FLINT, R. F. 1971: *Glacial and Quaternary geology.* New York: Wiley.

FOX, P. J., HEEZEN, B. C., and HARIAN, A. M. 1968: Abyssal antidune. *Nature* **220,** 470–72.

FOX, P. J., LOWRIE, A., and HEEZEN, B. C. 1969: Oceanographer fracture zone. *Deep Sea Res.* **16,** 59–66.

FUNNELL, B. M. 1967: Foraminifera and Radiolaria as depth indicators in the marine environment. *Mar. Geol.* **5,** 141–54.

GAGE, M. 1961: New Zealand glaciation and the duration of the Pleistocene. *J. Glaciol.* **3,** 940–43.

GARTNER, S. 1970: Sea-floor spreading, carbonate dissolution level and the nature of horizon A. *Science* **169,** 1077–9.

GASKELL, T. F. 1954a: Seismic refraction work by H.M.S. *Challenger* in the deep oceans. In a discussion on the floor of the Atlantic Ocean. Part II: Deeper structure, 356–60. *Proc. Roy. Soc.* **A 222,** 287–407.

1954b: A discussion on the floor of the Atlantic Ocean. Part II: The deeper structure of the Atlantic, 341–407. *Proc. Roy. Soc.* **A 222,** 287–407.

1960: *Under the deep oceans.* London: Eyre and Spottiswoode.

GASKELL, T. F., HILL, M. N., and SWALLOW, J. C. 1958: Seismic measurements made by H.M.S. *Challenger* in the Atlantic, Pacific and Indian Oceans and in the Mediterranean Sea, 1950–53. *Phil. Trans. Roy. Soc.* **A 251,** 23–83.

GERSHANOVICH, D. E. 1968: New data on geomorphology and recent sediments of the Bering Sea and the Gulf of Alaska. *Mar. Geol.* **6,** 281–96.

GIBSON, W. M. 1960: Submarine topography in the Gulf of Alaska. *Bull. Geol. Soc. Amer.* **71,** 1087–1108.

GIERLOFF-EMDEN, H. G. 1961: Nehrungen und Lagunen. *Petermanns Geog. Mitt.* **105** (2), 81–92; **105** (3), 161–76.

GODWIN, H. 1940: Studies of the postglacial history of British vegetation. III and IV. *Phil. Trans. Roy. Soc.* **B 230,** 239.

GODWIN, H., SUGGATE, R. P., and WILLIS, E. H. 1958: Radio-carbon dating of the eustatic rise in ocean level. *Nature* **181,** 1518–19.

GOODELL, H. G., WATKINS, N. D., MATHER, T. T., and KOSTA, S. 1968: The Antarctic glacial history recorded in sediments in the Southern Ocean. *Palaeogeog. Palaeoclim. and Palaeoecol.* **5,** 41–62.

GORSLINE, D. S. 1970: Submarine canyons: an introduction and papers following. *Mar. Geol.* **8,** 181–291.

GRAY, J. M. 1974: The rock platform of the Firth of Lorn, western Scotland. *Trans. Inst. Br. Geogr.* **61**, 81–100.

GRIGGS, G. B., CAREY, A. G., and KULM, L. D. 1969: Deep-sea sedimentation and sediment–fauna interaction in Cascadia Channel and on Cascadia Abyssal Plain. *Deep Sea Res.* **16**, 157–70.

GROOT, J. J. and GROOT, C. R. 1966: Marine palynology: possibilities, limitations and problems. *Mar. Geol.* **4**, 387–98.

GUILCHER, A. 1958: *Coastal and submarine morphology*. Translated by B. W. Sparks and R. H. W. Kneese. London: Methuen; New York: Wiley.

1963: Continental shelf and slope. In M. H. Hill (editor), *The sea*. Vol. **III**. New York: Wiley.

1964: Present time trends in the study of recent marine sediments and in marine physiography. *Mar. Geol.* **1**, 4–15.

1966: Les grandes falaises et mégafalais des côtes sud-ouest et ouest de l'Irlande. *Ann. de Géog.* **75**, 26–38.

HAFEMANN, D. 1955: Zur Frage der jungen Niveauveranderungen an den Kusten der Britischen Inseln. *Abh. Akad. Wiss. u. Lit.* Mainz. Math.-Nat. Kl. Jg. 1954 Nu. 7, 253–312.

HAILS, J. R. 1965: A critical review of sea-level changes in east Australia since the last glacial. *Aust. Geog. Stud.* **3**, 65–78.

HAMMOND, A. L. 1971: Plate tectonics: the geophysics of the earth's surface. *Science* **173**, 40–41.

HAND, B. M. and EMERY, K. O. 1964: Turbidites and topography of the northern end of the San Diego trough, California. *J. Geol.* **72**, 526–42.

HARLAND, W. B. 1967: Testing a theory of continental drift. *Uchennye Zapiski, Regional'naya Geologiya* **10**, 71–98.

1969: Contributions of Spitzbergen to understanding of the tectonic evolution of the north Atlantic region. In Marshall Kay (editor), *North Atlantic—geology and continental drift—a symposium.* Int. Conf. on Stratigraphy and structure bearing on the origin of the north Atlantic Ocean, Gander, Newfoundland, 1967. Tulsa, Oklahoma: Amer. Assn Petrol. Geol.

HARVEY, J. G. 1966: Large sand waves in the Irish Sea. *Mar. Geol.* **4**, 49–56.

HAYES, D. E. 1966: A geophysical investigation of the Peru–Chile trench. *Mar. Geol.* **4**, 309–52.

HAYES, D. E. and PITMAN, W. L. 1970: Magnetic lineations in the north Pacific. *Geol. Soc. Amer. Mem.* **126**, 291–314.

HAYES, M. O. 1967: Relationship between coastal climate and bottom sediment types on the inner continental shelf. *Mar. Geol.* **5**, 111–32.

HAYS, J. D. 1967: Quaternary sediments of the Antarctic Ocean. In M. Sears (editor), *Progress in oceanography*. Vol. **4**. Oxford: Pergamon.

HAYS, J. D., SAITO, T., OPDYKE, N. D., and BURCKLE, L. H. 1969: Pliocene–Pleistocene sediments of the equatorial Pacific: their palaeomagnetic, biostratigraphic and climate record. *Bull. Geol. Soc. Amer.* **80**, 1471–80.

HEEZEN, B. C. *et al.* 1967: The Quaternary history of the ocean basins. INQUA 1965: A symposium. In M. Sears (editor), *Progress in oceanography*. Vol. **4**. Oxford: Pergamon.

HEEZEN, B. C. and DRAKE, C. L. 1964: Grand Banks. slump. *Bull. Amer. Assn. Petrol Geol.* **48,** 221–5.

HEEZEN, B. C., and EWING, M. 1952: Turbidity currents and the 1929 Grand Banks earthquake. *Amer. J. Sci.* **250,** 849–84.

1963: The mid-ocean ridge. In M. N. Hill (editor), *The sea.* Vol. **III.** New York: Wiley.

HEEZEN, B. C. and HOLLISTER, C. 1964: Deep-sea current evidence from abyssal sediments. *Mar. Geol.* **1,** 147–74.

1971: *The face of the deep.* New York and London: Oxford University Press.

HEEZEN, B. C., HOLLISTER, C. D., and RUDDIMAN, W. F. 1966: Shaping the continental rise by deep geostrophic contour currents. *Science* **152,** 502–8.

HEEZEN, B. C. and LAUGHTON, A. S. 1963: Abyssal plains. In M. N. Hill (editor), *The sea.* Vol. **III.** New York: Wiley.

HEIRTZLER, J. R., DICKSON, G. O., HERRON, E. M., PITMAN III, W. C., and LE PICHON, X. 1968: Marine magnetic anomalies, geomagnetic field reversals, and motions of the ocean floor and continents. *J. Geophys. Res.* **73,** 2119–26.

HEIRTZLER, J. R. and LE PICHON, X. 1965: Crustal structure of the mid-ocean ridges. Magnetic anomalies over the mid-Atlantic ridge. *J. Geophys. Res.* **70,** 4013.

HERZER, R. R. 1971: Bowie Seamount. A recently active, flat-topped seamount in the northeast Pacific Ocean. *Can. J. Earth Sci.* **8** (6), 676–87.

HESS, H. H. 1954: Geological hypotheses of the earth's crust under the oceans. A discussion on the floor of the Atlantic Ocean. Part II: The deeper structure of the Atlantic. *Proc. Roy. Soc.* **A 222,** 341–48.

1955: Serpentine, orogeny and epeirogeny. *Geol. Soc. Amer. Sp. Pap.* **62,** 391–408.

1965: Mid-ocean ridges and tectonics of the sea floor. In W. F. Whittard and R. Bradshaw (editors), *Submarine geology and geophysics.* Proc. 17th Colston Res. Soc., Bristol. London: Butterworth.

HILL, M. N. 1957: Recent geophysical exploration of the ocean floor. *Physics and chemistry of the earth* **2,** 129–63. London: Pergamon.

(editor) 1963: *The sea.* Vol. **III.** The earth beneath the sea. Section 1: Geophysical exploration. New York: Wiley.

HILL, M. N. and KING, W. B. R. 1953: Seismic prospecting in the English Channel and its geological interpretation. *Quart. J. Geol. Soc.* **109,** 1–20.

HILL, M. N. and LAUGHTON, A. S. 1954: Seismic observations in the eastern Atlantic, 1952. In a discussion on the floor of the Atlantic Ocean. Part II: Deeper structure. *Proc. Roy. Soc.* **A 222,** 287–407, 348–56.

HILLS, G. F. S. 1947: *The formation of the continents by convection.* London: Edward Arnold.

HOLTEDAHL, O. 1950: Supposed marginal fault lines in the shelf area off some high northern lands. *Bull. Geol. Soc. Amer.* **61,** 493–500.

1952: A comparison of a Scottish and a Norwegian shelf area. *Trans. Edin. Geol. Soc.* **15,** and southeast 214–20.

HOLTEDAHL, H. 1958: Some remarks on geomorphology of continental shelves off Norway Alaska. *J. Geol.* **66,** 461–71.

HORN, D. R., HORN, B. M., and DELACH, M. N. 1970: Sedimentary provinces of the north Pacific. J. D. Hays (editor), *Geol. Soc. Amer. Mem.* **126,** 1–21.

HOSKINS, H. 1967: Seismic reflection observations on the Atlantic continental shelf, slope and rise, southeast of New England. *J. Geol.* **75**, 598–611.

HOUGH, J. L. 1953: Pleistocene climatic record in a Pacific Ocean core sample. *J. Geol.* **61**, 252–62.

HOYT, J. H. 1967: Barrier island formation. *Bull. Geol. Soc. Amer.* **78**, 1125–36.

HOYT, J. H., OOSTDAM, B. T., and SMITH, D. D. 1969: Offshore sediments and valleys of the Orange River. *Mar. Geol.* **7**, 69–84.

HUDSON, J. D. 1967: Speculation on the depth relation of calcium carbonate solution in recent and ancient seas. *Mar. Geol.* **5**, 473–80.

HUNKINS, K., BE, A. W. H., OPDYKE, N. D., and MATHIEU, G. 1971: The late Cenozoic history of the Arctic ocean. In K. K. Turekian (editor), *Late Cenozoic glacial ages.* New Haven: Yale University Press, chap. 8.

HUNKINS, K. and KUTSCHALE, H. 1967: Quaternary sedimentation in the Arctic Ocean. In M. Sears (editor), *Progress in oceanography.* Vol. **4**, Oxford: Pergamon.

IMBRIE, J. and KIPP, N. G. 1971: A new micropalaeontological method for quantitative palaeoclimatology: application to a late Pleistocene Caribbean core. In K. K. Turekian (editor), *Late Cenozoic glacial ages.* New Haven: Yale University Press, chap. 5.

INGLE, J. C. 1966: The movement of beach sand. *Developments in sedimentology* **5**. Amsterdam: Elsevier.

INMAN, D. L. and NORDSTROM, C. E. 1971: On the tectonic and morphological classification of coasts. *J. Geol.* **79** (1), 1–21.

ISAACS, J. D., REID, J. L., SCHICK, G. B., and SCHWARTZLOSE, R. A. 1966: Near-bottom currents measured in 4 km depth of Baja California coast. *J. Geophys. Res.* **71**, 4297–4303.

ISACKS, B., OLIVER, J., and SYKES, L. R. 1968: Seismology and the new global tectonics. *J. Geophys. Res.* **73**, 5855–99.

ISACKS, B. L., SYKES, L. R., and OLIVER, J. 1969: Focal mechanism of deep and shallow earthquakes in the Tonga-Kermadec region and the tectonics of island arcs. *Bull. Geol. Soc. Amer.* **80**, 1443–70.

JEFFREYS, H. 1950: *Earthquakes and mountains.* 2nd edition. London: Collins.

JELGERSMA, S. 1961: Holocene sea-level changes in the Netherlands. *Med. Geol. Sticht.* Ser. **C 6** (7), 1–101.

1966: Sea-level changes during the last 10,000 years. *Roy. Met. Soc. Sym. on World Climate from 8000 to 0 B.C.* London: Roy. Met. Soc.

JOHNSON, D. W. 1919: *Shore processes and shoreline development.* New York: Wiley.

JOHNSON, G. L. and HEEZEN, B. C. 1967: Morphology and evolution of the Norwegian and Greenland Sea. *Deep Sea. Res.* **14**, 755–71.

JOHNSON, M. A. 1964: Turbidity currents. In H. Barnes (editor), *Oceanog. and Mar. Biol. Ann. Rev.* **2**. London: Allen and Unwin.

JOHNSON, M. A. and STRIDE, A. H. 1969: Geological significance of sand transport in the North Sea. *Nature* **224**, 1016–17.

JOINT OCEANOGRAPHIC INSTITUTES DEEP EARTH SAMPLING (JOIDES) 1965: *Science* **150**, 709–16.

JOIDES NATIONAL SCIENCE FOUNDATION 1969: Reports of the deep sea drilling projects.

JONES, E. J. W., EWING, J., and TRUCHAN, M. 1971: Aleutian plain sediments and lithospheric plate motions. *J. Geophys. Res.* **76** (33), 8121–7.

JONES, N. S., KAIN, J. N., and STRIDE, A. H. 1965: The movement of sand waves on Warts Bank, Isle of Man. *Mar. Geol.* **3,** 329–36.

JONES, O. T. 1941: Continental slopes and shelves. *Geog. J.* **97,** 80–99.

1954: The characteristics of some lower Palaeozoic marine sediments. In a discussion on the floor of the Atlantic Ocean. Part I: Sediments. *Proc. Roy. Soc.* **A 222,** 287–407, 327–33.

KARIG, D. E., PATERSON, M. N. A., and SHOR, G. C. 1970: Sediment-capped guyots in the mid-Pacific mountains. *Deep sea. Res.* **17,** 373–8.

KAY, G. MARSHALL (editor) 1969: *North Atlantic—geology and continental drift—a symposium.* Int. Conf. on Stratig. and Structure bearing on the origin of the north Atlantic Ocean, Gander, Newfoundland, 1967. Tulsa, Oklahoma: Am. Assn Petrol. Geol.

KELLEY, J. C. and MCMANUS, D. A. 1969: Optimizing sediment sampling plans. *Mar. Geol.* **7,** 466–71.

KENYON, N. H. and STRIDE, A. H. 1970: The tide-swept continental shelf sediments between the Shetland Isles and France. *Sedimentology* **14,** 159–73.

KIDSON, C. 1963: The growth of sand and shingle spits across estuaries. *Zeit. für Geomorph.* NF **7** (1), 1–22.

KIDSON, C., STEERS, J. A., and FLEMMING, N. C. 1962: A trial of the potential value of aqua-lung diving in coastal physiography on British coasts. *Geog. J.* **128,** 49–53.

KING, C. A. M. 1964: The character of the offshore zone and its relationship to the foreshore near Gibraltar Point, Lincolnshire. *East Mid. Geog.* **3,** 230–43.

1972: *Beaches and coasts.* 2nd edition. London: Edward Arnold.

1973: The dynamics of coastal accretion in south Lincolnshire. In D. R. Coates (editor), *Coastal geomorphology.* Binghamton: State University of New York, 73–98.

KING, W. B. R. 1954: The geological history of the English Channel. *Quart. J. Geol. Soc.* **110,** 77–102.

KNOTT, S. T. and HOSKINS, H. 1968: Marine geological evidence of Pleistocene events in the structure of the continental shelf off northeastern United States. *Mar. Geol.* **6,** 5–43.

KRAUSE, D. C. 1965: East and west Azores fracture zone in the north Atlantic. In W. F. Whittard and R. Bradshaw (editors), *Submarine geology and geophysics.* Proc. 17th Colston Res. Soc., Bristol. London: Butterworth.

KRAUSE, D. C. and MENARD, H. W. 1965: Depth distribution and bathymetric classification of some sea-floor profiles. *Mar. Geol.* **3,** 169–94.

KU, T. L. and BROECKER, W. S. 1966: Atlantic deep-sea stratigraphy: extension of absolute chronology to 320,000 years. *Science* **151,** 448–50.

1967: Rates of sedimentation in the Arctic Ocean. In M. Sears (editor), *Progress in oceanography.* Vol. **4.** Oxford: Pergamon.

KUENEN, P. H. 1948: Turbidity currents of high density. *Inter. Geol. Cong.* 18 Sess. G.B. **8,** 44–52.

1950: *Marine geology.* New York: Wiley.

1953: Significant features of graded bedding. *Bull. Amer. Assn Petrol. Geol.* **37,** 1044–66.

1954: Eniwetok drilling results. *Deep Sea Res.* **1,** 187–9.

1960: *Submarine geology.* New York: Wiley.

1965: Experiments in connection with turbidity currents and clay suspensions. In W. F. Whittard and R. Bradshaw (editors), *Submarine geology and geophysics.* Proc. 17th Colston Res. Soc., Bristol. London: Butterworth.

KUENEN, P. H. and MENARD, H. W. 1952: Turbidity currents, graded and non-graded deposits. *J. Sedim. Petrol.* **22,** 83–96.

KULLENBERG, B. 1947: The piston core sampler. *Svenska Hydrog. Biol. Komm. Skr.* **3.**

LANDES, K. K. 1952: Our shrinking world. *Bull. Geol. Soc. Amer.* **63,** 225.

LARSON, R. L. 1971: Near-bottom geologic studies of the east Pacific rise crest. *Bull. Geol. Soc. Amer.* **82** (4), 823–42.

LAUGHTON, A. S. 1954: Laboratory measurements of seismic velocities in ocean sediments. *Proc. Roy. Soc.* **A 222,** 336.

1959: The sea floor. *Sci. Prog.* **186,** 230–49.

1965: Bathymetry of the northeast Atlantic Ocean and recent geophysical studies. In W. F. Whittard and R. Bradshaw (editors), *Submarine geology and geophysics.* Proc. 17th Colston Res. Soc., Bristol. London: Butterworth.

1965–6: The Gulf of Aden. *Phil. Trans. Roy. Soc.* **A 259,** 150–71.

1967: Underwater photography of the Carlsberg Ridge. In J. B. Hersey (editor), *Deep-sea photography,* Baltimore: Johns Hopkins University Press, 191–206.

1968: New evidence of erosion on the deep ocean floor. *Deep Sea Res.* **15,** 21–30.

LAUGHTON, A. S. and MATHEWS, D. H. 1965: Geology and geophysical structure of the northwest Indian Ocean. In W. F. Whittard and R. Bradshaw (editors), *Submarine geology and geophysics.* Proc. 17th Colston Res. Soc., Bristol. London: Butterworth.

LEES, G. M. 1953: The evolution of a shrinking earth. *Quart. J. Geol. Soc.* **109,** 217–57.

LE PICHON, X. 1968: Sea-floor spreading and continental drift. *J. Geophys. Res.* **73,** 3661–3697.

LE PICHON, X. and FOX, P. J. 1971: Marginal offsets, fracture zones and the early opening of the north Atlantic. *J. Geophys. Res.* **76** (26), 6294–6308.

LE PICHON, X. and HAYES, D. E. 1971: Marginal offsets, fracture zones and the early opening of the south Atlautic. *J. Geophys. Res* **76** (26), 6283–93.

LE PICHON, X. and TALWANI, M. 1964: Gravity survey of a seamount near 35°N 46°W in the north Atlantic. *Mar. Geol.* **2,** 262–77.

LINTON, D. L. 1951: Problems of Scottish scenery. *Scot. Geog. Mag.* **67** (2), 65–85.

LUDWIG, W. J., HAYES, D. E., and EWING, J. I. 1961. The Manila trench and West Luzon trough. I: Bathymetry and sediment distribution. *Deep Sea Res.* **14,** 533–44.

MCCAVE, I. N. 1971: Sand waves in the North Sea off the coast of Holland. *Mar. Geol.* **10,** 199–225.

MCCOY, F. W. and VON HERZEN, R. P. 1971: Deep-sea camera photography and piston coring. *Deep Sea Res.* **18,** 361–73.

MCFARLAN, E. 1961: Radio-carbon dating of late Quaternary deposits, southern Lousiana. *Bull. Geol. Soc. Amer.* **72** (1), 129–58.

MACINTYRE, F. 1970: Why the sea is salt. In *Oceanography, Sci. Amer.* readings, introduced by J. R. Moore. San Francisco: Freeman, chap. 13.

MACINTYRE, I. G. 1970: Sediments off the west coast of Barbados, West Indies: diversity of origins. *Mar. Geol.* **9,** 5–24.

MCKENZIE, D. P., 1967: Some remarks on heat flow and gravity anomalies. *J. Geophys. Res.* **72,** 6261–74.

MCKENZIE, D. P. and PARKER, R. L. 1967: The north Pacific: an example of tectonics on a sphere. *Nature* **216,** 1276–80.

MCMASTER, R. L. and LACHANCE, T. P. 1969: Northwest African continental shelf sediments. *Mar. Geol.* **7,** 57–68.

MALFAIT, B. T. and DINKELMAN, M. G. 1972: The development of the Caribbean plate. *Bull. Geol. Soc. Amer.* **83,** 251–72.

MASON, R. G. 1960: Geophysical investigations of the sea floor. *Liverpool and Manchester Geol. J.* **2,** 389–410.

MAURY, M. F. 1855: *The physical geography of the sea.* Reprinted 1963, J. Leighley (editor). Cambridge, Mass.: Belknap Press of Harvard University Press.

MAXWELL, A. E. (editor) 1971: *The sea.* Vol. **IV,** Parts 1 and 2. New York: Wiley.

MAXWELL, A. E., HEEZEN, R. P., HSII, UON, JINGHWA, K., ANDREWS, J. E., TSUNEMASA, SAITO, PERCIVAL, S. F., MILOW, E. D., and BOYCE, R. E. 1970: Leg 3 of *Glomar Challenger. Science* **168** (3935), 1047–59.

MEADE, R. H. and EMERY, K. O. 1971: Sea level as affected by river runoff, eastern United States. *Science* **173,** 425–7.

MENARD, H. W. 1958: Development of median elevations in ocean basins. *Bull. Geol. Soc. Amer.* **69,** 1179–86.

1959: Geology of the Pacific sea floor. *Experientia* **15,** 205–13.

1964: *The marine geology of the Pacific.* New York: McGraw-Hill.

1965: The world-wide oceanic rise-ridge system. *Phil. Trans. Roy. Soc.* **A 258,** 109–22.

1967a: Transitional types of crust under small ocean basins. *J. Geophys. Res.* **72,** 3061–73.

1967b: Sea-floor spreading. *Trans. Amer. Geophys. Un.* **48,** 217.

1969a: The growth of drifting volcanoes. *J. Geophys. Res.* **74,** 4827–37.

1969b: *Anatomy of an expedition.* New York: Reinhold.

1971: The late Cenozoic history of the Pacific and Indian Oceans. In K. K. Turekian (editor), *Late Cenozoic glacial ages.* New Haven: Yale University Press.

MENARD, H. W. and LADD, H. S. 1963: Oceanic islands, seamounts, guyots and atolls. In M. N. Hill (editor), *The sea.* Vol. **III.** New York: Wiley.

MENARD, H. W. and SMITH, S. M. 1966: Hypsometry of ocean basin provinces. *J. Geophys. Res.* **71** (18), 4305–25.

MERCER, J. H. 1968: The discontinuous glacio-eustatic fall and Tertiary sea level. *Palaeogeog. Palaeoclim. and Palaeoecol.* **5,** 77–86.

MERO, J. L. 1965: *The mineral resources of the sea.* Amsterdam: Elsevier.

MEYERHOFF, A. A. 1970: Continental drift: implication of palaeomagnetic studies, meteorology, physical oceanography and climatology, *J. Geol.* **78,** 1–61.

MEYERHOFF, A. A. and TEICHERT, C. 1971: Continental drift III: late Palaeozoic glacial centers, and Devonian–Eocene coal distribution. *J. Geol.* **79** (3), 285–321.

MIDDLETON, G. V. 1966–7: Experiments on density and turbidity currents. Parts I–III. *Can. J. Earth Sci.* **3,** 523–46, 627–37; **4,** 475–505.

MILLIMAN, J. D. and EMERY, K. O. 1968: Sea levels during the last 35,000 years. *Science* **162,** 1121–3.

MITCHELL, A. H. and READING, H. G. 1969: Continental margins, geosynclines and ocean-floor spreading. *J. Geol.* **77,** 629–46.

1971: Evolution of island arcs. *J. Geol.* **79** (3), 253–84.

MOORE, D. G. 1960: Acoustic-reflection studies of the continental shelf and slope off southern California. *Bull. Geol. Soc. Amer.* **71,** 1121–36.

MOORE, D. G. and CURRAY, J. R. 1963: Structural framework of the continental terrace northwest Gulf of Mexico. *J. Geophys. Res.* **68,** 1725–47.

MOORE, J. R. 1971: Introduction to a series of *Scientific American* reprints. *Oceanography.* San Francisco: Freeman.

MOORE, T. C. and HEATH, G. R. 1966: Manganese nodules, topography and thickness of Quaternary sediment in the central Pacific. *Nature* **221,** 983–5.

MOORE T. C. and HEATH G. R. 1967: Abyssal hills in the central equatorial Pacific: detailed structure of the sea floor and sub-bottom reflectors. *Mar. Geol.* **5,** 161–80.

MOORE, T. C. and KULM, L. D. 1970: A high resolution sub-bottom profiling system for use in ocean basins. *J. Mar. Res.* **28** (2), 271–80.

MORAN, J. M. and BRYSON, R. A. 1969: The contribution of Laurentide ice wastage to the eustatic rise of sea level, 10,000–6,000 B.P. *Arct. and Alp. Res.* **1,** 97–104.

MORGAN, J. P., COLEMAN, J. M. and GAGLIANO, S. M. 1968: Mudlumps: diapiric structures in Mississippi Delta sediments. In Diapirism and diapirs, *Amer. Assn. Petrol. Geol., Memoir* **8,** 145–61.

MORGAN, W. J. 1968: Rises, trenches, great faults and crustal blocks. *J. Geophys. Res.* **73,** 1959–82.

MÖRNER, N. A. 1969a: The late Quaternary history of the Kattegat Sea and the Swedish west coast deglaciation, shoreline displacement, chronology, isostasy and eustasy. *Sver. Geol. Unders.* **63** (3).

1969b: Eustatic and climatic changes during the last 15,000 years. *Geol. Mijnbouw.* **48,** 389–99.

1971a: Eustatic changes during the last 20,000 years and a method of separating the isostatic and eustatic factors in an uplifted area. *Palaeogeog. Palaeoclim. and Palaeoecol.* **9** (3), 153–81.

1971b: Eustatic and climatic oscillations. *Arct. and Alp. Res.* **3** (2), 167–71.

1971c: The positions of the ocean level during the interstadial at about 30,000 B.P. A discussion from climatic and glaciologic point of view. *Can. J. Earth Sci.* **8** (1), 132–43.

MURRAY, J. 1880: On the structure of coral reefs and islands. *Proc. Roy. Soc. Edin.* **10,** 505–18.

MURRAY, J. and HJORT, J. 1912: *Depths of the ocean.* London: John Murray.

MURRAY, J. and RENARD, A. F. 1891: *Report on the scientific results of the voyage of H.M.S. 'Challenger'*, 1873–76. Edinburgh: Neill.

NAFE, J. E. and DRAKE, C. L. 1969: Floor of the north Atlantic—summary of geophysical data. In Marshall Kay (editor), *North Atlantic—geology and structure—a symposium,* Gander, Newfoundland, 1967. Tulsa, Oklahoma: Amer. Assn Petrol. Geol.

NATIONAL SCIENCE FOUNDATION: Initial reports of the deep sea drilling project. Several volumes covering the cruises of the drilling vessel *Glomar Challenger.* Nat. Ocean. Sed. Coring Prog. University of California, Scripps Institute.

NEWMAN, W. S. and RUSNACK, G. A. 1965: Holocene submergence of the eastern shore of Virginia. *Science* **148** (3676), 1464–6.

NIGRINI, C. 1970: Radiolarian assemblages in the north Pacific and their application to a study of Quaternary sediments in Core V 20–130. *Geol. Soc. Amer. Mem.* **126,** 139–83.

NORMARK, W. R. 1970: Growth pattern of deep-sea fans. *Bull. Amer. Assn Petrol. Geol.* **54,** 2170–95.

OPDYKE, N. D. and FOSTER, J. H. 1970: Paleomagnetism of cores from the north Pacific. *Geol. Soc. Amer. Mem.* **126,** 83–119.

OPDYKE, N. D. and GLASS, B. P. 1969: The paleomagnetism of sediment cores from the Indian ocean. *Deep Sea Res.* **16,** 249–57.

OPDYKE, N. D., GLASS, B. P., HAY, J. D., and FOSTER, J. 1966: Paleomagnetic study of Antarctic deep sea cores. *Science* **154,** 349–57.

OROWAN, E. 1965: Convection in a non-Newtonian mantle, continental drift and mountain building. *Phil. Trans. Roy. Soc.* **A 258,** 284–313.

OTVOS, E. G. 1970: Development and migration of barrier islands, north Gulf of Mexico. *Bull. Geol. Soc. Amer.* **81,** 241–6.

PATTULLO, J. G. 1963: Seasonal changes in sea level. In M.N. Hill (editor), *The sea.* Vol. **II.** New York: Wiley.

PENCK, A. 1933: Eustatische Bewegungen der Meeresspiegels während der Eiszeit. *Geogr. Zeitschr.* Leipzig **39,** 329–39.

PHINNEY, R. A. (editor) 1968: *History of the earth's crust.* Princeton, N J.: Princeton University Press.

PIGGOT, C. S. and URRY, W. D. 1942: Time relations in ocean sediments. *Bull. Geol. Soc. Amer.* **53,** 1187–1210.

PIPER, D. J. W. 1970: Transport and deposition of Holocene sediments on La Jolla deep-sea fan, California. *Mar. Geol.* **8,** 211–28.

PITMANN III, W. C. and HEIRTZLER, J. R. 1966: Magnetic anomalies over the Pacific–Antarctic ridge. *Science* **154,** 1164–71.

PITMAN III, W. C., HERRON, E. H., and HEIRTZLER, J. R. 1968: Magnetic anomalies in the south Pacific and sea-floor spreading. *J. Geophys. Res.* **73,** 2069–85.

PRATT, R. M. 1967: The seaward extension of submarine canyons off the northeast coast of the U.S. *Deep Sea Res.* **14,** 409–20.

RAFF, A. D. 1968: Sea-floor spreading—another rift. *J. Geophys. Res.* **73,** 3699–3705.

RAITT, R. W. 1956: Seismic refraction studies of the Pacific ocean basin. Part I: Crustal thickness of the central equatorial Pacific. *Bull. Geol. Soc. Amer.* **67,** 1623–40.

RAITT, R. W., FISHER, R. L., and MASON, R. G. 1955: Tonga trench. *Geol. Soc. Amer. Sp. Pap.* **62,** 237–54.

REDFIELD, A. C. and RUBIN, M. 1962: The age of salt-marsh peat and its relation to recent changes in sea level at Barnstable, Mass. *Proc. Nat. Acad. Sci.* **48** (2, 7), 1728–35.

REVELLE, R. 1944: Marine bottom samples collected in the Pacific Ocean by the *Carnegie* on its 7th cruise. Washington, D.C.: Carnegie Institute, *Pub.* **556.**

1955: On the history of the oceans. *J. Mar. Res.* 14, 446–61.

REVELLE, R., BRAMLETTE, M., ARRHENIUS, G., and GOLDBERG, E. D. 1955: Pelagic sediments of the Pacific. *Geol. Soc. Amer. Sp. Pap.* **62,** 221–36.

RHEA, K., NORTHROP, J., and VON HERZEN, R. P. 1964: Heat-flow measurements between North America and the Hawaiian Islands. *Mar. Geol.* **1,** 220–24.

RICHARDS, A. F. 1967: *International research conference on marine techniques,* Allerton House, 1966. Urbana, Illinois: University of Illinois Press.

RICHARDS, F. A. and REDFIELD, A. C. 1954: A correlation between oxygen content of marine sediments. *Deep Sea Res.* **1,** 279–8.

RIEDEL, W. R., BRAMLETTE, M. N., and PARKER, F. L. 1963: Pliocene-Pleistocene boundary in deep-sea sediments. *Science* **140,** 1238–40.

ROBERTSON, M. I. 1964: Continuous seismic profiles survey of Oceanographer, Gilbert and Lydonia submarine canyons, Georges Bank. *J. Geophys. Res.* **69,** 4779–89.

ROBINSON, A. H. W. 1949: Deep clefts in the Inner Sound of Raasay. *Scot. Geog. Mag.* **65,** 20–25.

1952: Floor of the British seas. *Scot. Geog. Mag.* **68,** 64–79.

1960: Ebb-flood systems in sandy bays and estuaries. *Geog.* **45,** 183–99.

ROBINSON, A. H. W. 1964: Inshore waters, sediment supply and coastal changes of part of Lincolnshire. *East Mid. Geog.* **22,** 307–21.

1966: Residual currents in relation to shoreline evolution of the East Anglian coast. *Mar. Geol.* **4,** 57–84.

RODOLFO, K. S. 1969: Sediments of the Andaman Basin, northeast Indian ocean. *Mar. Geol.* **7,** 371–402.

RONA, P. A. 1970: Submarine canyon origin on the upper continental slope off Cape Hatteras. *J. Geol.* **78,** 141–52.

RONA, P. A., SCHNEIDER, E. D., and HEEZEN, B. C. 1967: Bathymetry of the continental rise off Cape Hatteras. *Deep Sea Res.* **14,** 625–33.

ROSS, D. A. and SHOR, G. G. 1965: Reflection profiles across the middle America trench. *J. Geophys. Res.* **70,** 5551.

ROSSITER, J. R. 1962: Long-term variations in sea level. In M. N. Hill (editor), *The sea.* Vol. **I.** New York: Wiley, chap. 16.

ROTHE, J. P. 1954: La zone séismique médiane Indo-Atlantique. *Proc. Roy. Soc.* **A 222,** 387–97.

RUSSELL, R. J. 1967: *River plains and the sea coast.* Berkeley, California: University of California Press.

RUTFORD, R. H., CRADDOCK, C., and BASTIEN, T. W. 1968: Late Tertiary glaciation and sea-level changes in Antarctica. *Palaeogeog. Palaeoclim. and Palaeoecol.* **5,** 15–39.

SAITO, T., EWING, M., and BURCKLE, L. H. 1966: Tertiary sediment from the mid-Atlantic ridge. *Science* **151,** 1075–9.

SALSMAN, G. G., TOLBERT, W. H., and VILLARS, R. G. 1966: Sand-ridge migration in St Andrews Bay, Florida. *Mar. Geol.* **4,** 11–20.

SCHNEIDER, E. D. and JOHNSON, G. L. 1970: Deep-ocean diapiric structures. In F. M. Delany (editor), *Inst. Geol. Sci. Report* **70/13,** The geology of the east Atlantic continental margin. 1: General and economic papers. London: H.M.S.O. for I.C.S.U./S.C.O.R.

SCHOFIELD, J. C. 1964: Postglacial sea levels and isostatic uplift. *N.Z. J. Geol. Geophys.* **7,** 359–70.

SCHOLL, D. W., BUFFINGTON, E. C., and HOPKINS, D. M. 1968: Geological history of the continental margin of North America in the Bering Sea. *Mar. Geol.* **6,** 297–330.

SCHOLL, D. W., BUFFINGTON, E. C., HOPKINS, D. M., and ALPHA, T. R. 1970: The structure and origin of the large canyons of the Bering Sea. *Mar. Geol.* **8,** 187–210.

SCHOLL, D. W., CHRISTENSEN, M. N., VON HUENE, R., and MARLOW, M. S. 1970: Peru–Chile trench sediments and sea-floor spreading. *Bull. Geol. Soc. Amer.* **81** (5), 1339–60.

SCHOLL, D. W., CRAIGHEAD, F. C., and STUIVER, M. 1969: Florida submergence curve revised: its relation to coastal sedimentation rates. *Science* **163,** 562–4.

SCHOLL, D. W. and STUIVER, M. 1967: Recent submergence of south Florida; a comparison with adjacent coasts and other eustatic data. *Bull. Geol. Soc. Amer.* **78,** 437–54.

SCHWARTZ, M. L. 1971: The multiple causality of barrier islands. *J. Geol.* **79,** 91–4.

SEMPER, C. 1881: *The natural conditions of existence as they affect animal life.* London: Kegan Paul, Trench and Trubner.

SHEPARD, F. P. 1948: *Submarine geology.* 2nd edition, 1963. New York: Harper and Row.

1952: Composite origin of submarine canyons. *J. Geol.* **60,** 84–96.

1959: *The earth beneath the sea.* 2nd edition, 1968. Baltimore: Johns Hopkins Press.

SHEPARD, F. P. 1963: Thirty-five thousand years of sea level. In T. Clements (editor), *Essays in marine geology, in honor of K. O. Emery.* Los Angeles: University of California Press.

1964: Criterion in modern sediment useful in recognizing ancient sedimentary environments. In L. M. J. U. van Straaten (editor), *Developments in sedimentology.* Vol. **I.** 3rd edition, 1973. Amsterdam: Elsevier.

1973: *Submarine morphology.* 3rd edition. New York: Harper and Row.

SHEPARD, F. P. and BUFFINGTON, E. C. 1968: La Jolla submarine fan-valley. *Mar. Geol.* **6,** 107–44.

SHEPARD, F. P. and CURRAY, J. R. 1967: Carbon 14 determination of sea-level change in stable areas. In M. Sears (editor), *Progress in oceanography.* Vol. **4:** Quaternary history of the ocean basins. Oxford: Pergamon.

SHEPARD, F. P. and DILL, R. F. 1966: *Submarine canyons and other sea valleys.* Chicago: Rand McNally.

SHIPEK, C. J. 1960: Photographic study of some deep-sea environments in the east Pacific. *Bull. Geol. Soc. Amer.* **71,** 1067–74.

SHOR, G. G. 1964: Structure of the Bering Sea and the Aleutian ridge. *Mar. Geol.* **1,** 213–19.

SHOR, G. G. and FISHER, R. L. 1961: Middle America trench: seismic refraction studies. *Bull. Geol. Soc. Amer.* **72,** 721–30.

SMITH, A. G. and HALAM, A. 1970: The fit of the southern oceans. *Nature* **225,** 139–44.

SMITH, A. J. and HAMILTON, D. 1970: The origin of the Hurd Deep, English Channel. *Nature* **227,** 828.

STAUDER, W. S. J. 1968: Mechanism of the Rat Island earthquake sequence of Feb. 4, 1963 with relation to island arcs and sea-floor spreading. *J. Geophys. Res.* **73,** 3847–58.

STILLE, H. 1955: Recent deformations of the earth's crust in the light of earlier epochs. *Geol. Soc. Amer. Sp. Pap.* **62,** 171–92.

STODDART, D. R. 1965: The shape of atolls. *Mar. Geol.* **3,** 369–84.

1969: Ecology and morphology of recent coral reefs. *Biol. Rev.* **44,** 433–98.

STRIDE, A. H. 1959a: On the origin of the Dogger Bank. *Geol. Mag.* **96,** 33–44.

1959b: A linear pattern on the sea floor and its interpretation. *J. Mar. Biol. Assn U.K.* **38,** 313–18.

1963: Current-swept sea floors near the southern half of Great Britain. *Quart. J. Geol. Soc.* **119,** 175–99.

1970: Mapping the ocean floor. *Sci. J.* **6** (12), 56–8.

STUIVER, M. and DADDARIO, J. J. 1963: Submergence of the New Jersey coast. *Science* **142,** 951.

SVERDRUP, H. U., JOHNSON, M. W. and FLEMING, R. H. 1946: *The oceans: their physics, chemistry and general biology*. New Jersey: Prentice-Hall.

SWIFT, D. J. P. 1970: Quaternary shelves and the return to grade. *Mar. Geol.* **8,** 5–30.

SWIFT, D. J. P., STANLEY, D. J., and CURRAY, J. R. 1971: Relict sediments on continental shelves: a reconsideration. *J. Geol.* **79** (3), 322–46.

SWINNERTON, H. H. 1931: The postglacial deposits of the Lincolnshire coast. *Quart. J. Geol. Soc.* **87,** 360–75.

SYKES, L. R. 1966: The seismicity and deep structure of island arcs. *J. Geophys. Res.* **71,** 2981–3006.

SYKES, L. R. 1967: Mechanism of earthquakes and nature of faulting in the mid-ocean ridges. *J. Geophys. Res.* **72,** 2131–53.

TALWANI, M. 1964: A review of marine geophysics. *Mar. Geol.* **2,** 29–80.

TANNER, W. F. 1968: Tertiary sea-level symposium—introduction and multiple influences on sea-level changes in the Tertiary. *Palaeogeog. Palaeoclim. and Palaeoecol.* **5,** 7–14, 165–71.

TAYLOR, J. A. 1973: Chronometers and chronicles: a study of palaeo-environments in west central Wales. *Progr. Geogr.* **5,** 247–334.

TAZIEFE, H. 1970: The Afar triangle. In *Oceanography,* introduced by R. J. Moore, *Sci. Amer.* readings. San Francisco: Freeman, chap. 19.

THOM, B. G. 1973: The dilemma of high interstadial sea levels during the last glaciation *Progr. Geogr.* **5,** 167–246.

TOBIN, D. G. and SYKES, L. R. 1968: Seismicity and tectonics of the northeast Pacific Ocean. *J. Geophys. Res.* **73** (12), 3821–45.

TRASK, P. D. (editor) 1955: *Recent marine sediments.* Tulsa, Oklahoma: Amer. Soc. Petrol. Geol.

TUREKIAN, K. K. 1968: *Oceans.* Englewood Cliffs, New Jersey: Prentice-Hall, chaps. 2–4.
(editor) 1971: *Late Cenozoic glacial ages.* New Haven: Yale University Press.

TWENHOFEL, W. H. 1932: *Principles of sedimentation.* New York: McGraw-Hill.

UCHUPI, E. 1967: Slumping on the continental margin southeast of Long Island, New York. *Deep Sea Res.* **14,** 635–9.

UCHUPI, E. and EMERY, K. O. 1967: Structure of the continental margin of the Atlantic coast of the US. *Bull. Assn Amer. Petrol. Geol.* **51,** 223–34.

UMBGROVE, J. H. F. 1946: On the origin of continents and ocean floors. *J. Geol.* **54,** 169–78.
1947: *The pulse of the earth.* 2nd edition. The Hague: Nijhoff.
1950: *The symphony of the earth.* The Hague: Nijhoff.

UNION GEODETIC GEOPHYSIC INTERNATIONALE 1954: Secular variations of sea level. *Assn d'Océanogr. Sci. Pub.* **13.**

UPSON, J. E., LEOPOLD, E. B. and RUBIN, M. 1964: Postglacial change of sea level in New Haven Harbor, Connecticut. *Amer. J. Sci.* **262** (1), 121–32.

URRY, W. D. 1949: Radioactivity of ocean sediments. VI: Concentration of radio-elements in marine sediments of the southern hemisphere. *Amer. J. Sci.* **247,** 257–75.

VALENTIN, H. 1952: *Die Küsten der Erde.* 2nd edition, 1954. Gotha: Petermanns Geog. Mitt., Ergänzungsheft **246.**
1953: Present vertical movements of the British Isles. *Geog. J.* **119,** 299–305.
1954: Der Landverlust in Holderness, Ost-england von 1852 bis 1952. *Die Erde* **6** (3–4), 296–315.

VAN DER LINGEN, G. J. 1969: The turbidite problem. *N.Z. J. Geol. Geophys.* **12,** 7–50.

VAN STRAATEN, L. M. J. U. 1965: Coastal barrier deposits in south and north Holland, in particular in the area around Scheveningen and Ijmuiden. *Med. Geol. Sticht.* NS **17,** 41–75.

VEEH, H. H. 1966: Th^{230}/U^{238} and U^{234}/U^{238} ages of Pleistocene high sea-level stands. *J. Geophys. Res.* **71** (14), 3379–86.

VEENSTRA, H. J. 1969: Gravels of the southern North Sea. *Mar. Geol.* **7,** 449–64.

VENING-MEINESZ, F. A. 1934: Gravity and the hypothesis of convection currents in the earth. *Proc. K. Akad. Wetensch.* Amsterdam, **37.**

VINE, F. J. 1966: Spreading of the ocean floor: new evidence. *Science* **154,** 1405–15.

VON DER BORCH, C. C. 1968: Southern Australia submarine canyons: their distribution and ages. *Mar. Geol.* **6,** 267–80.

1969: Submarine canyons of southeast New Guinea: seismic and bathymetric evidence for their modes of origin. *Deep Sea Res.* **16,** 323–8.

VON HERZEN, R. P. 1964: Ocean floor heat-flow measurements west of the U.S. and Baja California. *Mar. Geol.* **1,** 225–39.

VON HUENE, R. 1969: Geologic structure between the Murray fracture zone and the transverse ranges. *Mar. Geol.* **7,** 475–99.

WALTHER, J. 1926: *Die Methoden der Geologie als historischer und biologischer wissenschaft.* Hanb. Biol. Arbeitsmeth. Abt. **10.** Berlin: Urban und Schwarzenburg.

WARME, J. E., SCANLAND, T. B., AND MARSHALL, N. F. 1971: Submarine canyon erosion: contribution of marine rock burrowers. *Science* **173** (4002), 1127–9.

WEGENER, A. 1922: *The origin of continents and oceans.* Reprinted 1966, translated from the 4th revised German edition by J. Biram. New York: Dover.

WEXLER, H. 1961: Ice budgets for Antarctic and changes of sea level. *J. Glaciol.* **3,** 867–72.

WHITTARD, W. F. and BRADSHAW, R. (editors) 1965: *Submarine geology and geophysics.* Proc. 17th symposium Colston Res. Soc., Bristol. London: Butterworth.

WILSON, TUZO J. 1965a: A new class of faults and their bearing on continental drift. *Nature* **207,** 343–7.

1965b: Evidence from ocean islands suggesting movement in the earth. *Phil. Trans. Roy. Soc.* **A 258,** 145–67.

WINTERER, E. L. 1970: Submarine valley systems around the Coral Sea basin (Australia). *Mar. Geol.* **8,** 229–44.

WINTERER, E. L., CURRAY, J. R., and PETERSEN, M. N. A. 1968: Geological history of the intersection of the Pioneer fracture zone with the Delgado deep-sea fan, northeast Pacific. *Deep Sea Res.* **15,** 509–20.

WISEMAN, J. D. H. 1952: Past temperatures of the upper equatorial Atlantic. In a discussion on the floor of the Atlantic Ocean. Part I: Sediments. *Proc. Roy. Soc.* **A 222,** 287–407.

WISEMAN, J. D. H. and OVEY, C. D. 1950: Recent investigations on the deep-sea floor. *Proc. Geol. Assn* **61,** 28–84.

WOLF, S. C. 1970: Coastal currents and mass transport of surface sediments over the shelf regions of Monterey Bay, California. *Mar. Geol.* **8,** 321–36.

WOLLIN, G., ERICSON, D. B., and EWING, M. 1971: Late Pleistocene climates recorded in Atlantic and Pacific deep-sea sediments. In K. K. Turekian (editor), *Late Cenozoic glacial ages.* New Haven: Yale University Press, chap. 7.

WOODS, J. D. and LYTHGOE, J. N. 1971: *Underwater science—an introduction to experiments by divers*. London: Oxford University Press.

WOOLDRIDGE, S. W. and HENDERSON, H. C. K. 1955: Some aspects of the physiography of the eastern part of the London basin. *Trans. Inst. Brit. Geog.* **21,** 19–31.

WORZEL, J. L. 1965: Deep structure of coastal margins and mid-oceanic ridges. In W. F. Whittard and R. Bradshaw (editors), *Submarine geology and geophysics*. Proc. 17th Colston Res. Soc., Bristol. London: Butterworth.

WORZEL, J. L. and SHURBET, G. L. 1955: Gravity interpretations from standard oceanic and continental crust sections. *Geol. Soc. Amer. Sp. Pap.* **62,** 87–100.

WYVILLE-THOMSON, C. 1873: *Depths of the sea*.

YALKOVSKY, R. 1967: Time-series analysis of Caribbean deep-sea core A 172–6. *J. Geol.* **75,** 224–31.

ZENKOVICH, V. P. 1967: *Processes of coastal development*. Edinburgh: Oliver and Boyd.

Index